Collaborative Bioethics

Volume 5

Series Editor
Insoo Hyun, Harvard Medical School
Boston, USA

The aim of **Collaborative Bioethics** , is to draw attention to an underexplored but increasingly important area of scholarly thought and action: bioethics as a co-creative activity of ethicists working with scientists rather than as always a reaction to biomedical developments after the fact. The scope of this series is determined by each major subfield of science and medicine that raises ethical uncertainties for researchers, regulators, and the public.

Collaborative Bioethics is a series that will provide a central hub for timely publications addressing ethical issues that are emerging right alongside the science. As such, this series will be of interest to a wide swath of readers: bioengineers and scientists at all professional levels; bioethicists intrigued by bioengineering and medical advances; research regulators and funders; and the general public.

Ubaka Ogbogu • Donrich Thaldar
Nishakanthi Gopalan
Editors

Regenerative Medicine and the Bioethics of Globalization, Justice and Hegemony

Springer

Editors
Ubaka Ogbogu
Faculty of Law
University of Alberta
Edmonton, AB, Canada

Donrich Thaldar
Faculty of Law
University of KwaZulu-Natal
Durban, South Africa

Nishakanthi Gopalan
Medical Humanities and Ethics Unit
(MedHEU), Faculty of Medicine
Universiti Malaya
Kuala Lumpur, Malaysia

ISSN 2730-5619 ISSN 2730-5627 (electronic)
Collaborative Bioethics
ISBN 978-3-032-07288-7 ISBN 978-3-032-07289-4 (eBook)
https://doi.org/10.1007/978-3-032-07289-4

This Springer imprint is published by the registered company Springer Nature Switzerland AG
The registered company address is: Gewerbestrasse 11, 6330 Cham, Switzerland

Acknowledgements

This book project was supported by funding from the Government of Canada's New Frontiers in Research Fund—Transformation Stream (NFRFT-2022-00327) and the Stem Cell Network (Canada) Translation and Society Team Award, both awarded to Ubaka Ogbogu. The editors gratefully acknowledge Cala Taylor for providing editorial assistance.

Contents

Part I Foundational Critiques: Epistemic Injustice and Global Asymmetries

1 **Reframing Regenerative Medicine: Global Asymmetries and Ethical Futures** 3
Ubaka Ogbogu, Donrich Thaldar, and Nishakanthi Gopalan

2 **Toward Epistemic Justice: Regenerative Medicine Activity and Governance in Low- and Middle-Income Countries** 17
Ubaka Ogbogu and Lauren Albrecht

3 **A Case for Inclusive Global Governance for Regenerative Medicine** 43
Jonathan O. Chimakonam and Adeolu Oluwaseyi Oyekan

4 **Integrating Regenerative Medicine into Medical Curricula in the Global South: A Tanzanian Perspective** 61
Afadhali D. Russa and Majigo M. Mtebe

Part II National Case Studies: Governance, Context, Constraint, and Endogenous Innovation

5 **Translation and Commercialisation of Regenerative Medicine in South Africa: Navigating the Regulatory Landscape** 81
Marietjie Botes

6 **Regenerative Medicine at the Tip of Africa: Far-Reaching Consequences of Slow Progress** 103
Usri H. Ibrahim, Mari van de Vyver, Yvonne Holt, Carl T. Montague, and Carola U. Niesler

7 **The Kenyan Legal and Policy Framework for Stem Cell Research** ... 123
Paul Ogendi

8 **Stem Cell and Genetic Engineering for Regenerative Medicine: An Indian Perspective** 143
Shambhavi Naik

9 Ethical Crossroads and Regulatory Pathways: Navigating Regenerative Medicine in Malaysia 161
Nishakanthi Gopalan

10 Regenerative Medicine in China: Current Regulatory Challenges and Global Advancements 183
Meng Wang and Li Du

11 Regenerative Medicine and Advanced Therapy Medicinal Products in Ukraine: Research, Regulation, and Challenges in a Time of War 257
Oksana Piven, Sergii Martynenko, and Volodymyr Shablii

Notes on Contributors

Lauren Albrecht is a social scientist and Research Associate in the Faculty of Law at the University of Alberta. She holds a doctorate in Educational Psychology and her research focuses on knowledge mobilisation, health education, research methods, and meta-research. She has worked across a number of fields including education, higher education, nursing, medicine, and law. She is particularly interested in transdisciplinary research approaches to address real-world problems and create social impact with a focus on supporting equity-deserving groups.

Marietjie Botes is a legal scholar and science communicator with a background bridging law, biotechnology, and ethics. After a 20-year career as a practising attorney specialising in health law and biotechnology, she transitioned to academia, focusing on interdisciplinary research at the intersection of science, technology, law, and ethics. She holds the degrees of BProc, LLB, LLM (Intellectual Property Law), and LLD (Biotechnology Law). Her research covers the regulation of regenerative medicines such as gene therapies and stem cell research, health data governance, pandemic preparedness, and digital health technologies. She pays particular attention to emerging neurotechnologies such as brain-computer interfaces and their implications for rights including mental privacy, protection from online behavioural manipulation, and digital constitutionalism. She is based at the Pestilli Neuroscience Lab at the University of Texas at Austin, conducting research under the BRIDGE (Brain Research International Data Governance and Exchange) Project. She has published extensively in national and international journals and has co-edited special issues on digital ethics and legal and ethical reforms for neurotechnology in the Global South. She is an elected board member of the Association for Responsible Research and Innovation in Genome Editing (ARRIGE) and the Academy of Science of South Africa (ASSAf), contributes to the consensus study on gene therapies, and has served as an invited expert and co-author of the South African COVID-19 Country Reports. She is the Country Representative for South Africa in the African Bioethics Network. Outside academia, she is active in science communication, using comics and participatory storytelling to engage diverse audiences on biomedical and digital ethics, and is a performing member of the Western Cape Flute Choir.

Jonathan O. Chimakonam is a Professor of Philosophy at the University of Pretoria, South Africa. He previously taught at the University of Calabar in Nigeria and has been a research fellow at various institutions in Europe and Africa. His teaching and research interests include African philosophy, logic, ethics, philosophy of mind, race studies, and decolonial thinking. He is the author, co-author, editor, or co-editor of more than 140 research works in books, articles, and chapters. He has delivered invited lectures and keynote talks at universities worldwide and has been interviewed by academic blogs and media outlets. His theories and concepts have been the focus of international conferences and panels, and his work has inspired rejoinders, reviews, and theses. His ideas feature in university curricula and philosophy modules. He propounded the theory of Ezumezu logic, developed the method of conversational thinking, and is a leading proponent of conversational philosophy along with several other concepts and principles in the field. He has ongoing and past research collaborations with colleagues at other universities and has received funding from national and international organisations such as the National Institute for the Humanities and Social Sciences, the National Research Foundation, and the John Templeton Foundation. He is a logician, public intellectual, social thinker, and multiple award-winning philosopher.

Li Du is an Associate Professor at the Faculty of Law, University of Macau. He holds dual bachelor's degrees in clinical medicine and law from Wuhan University in China and a PhD in law from the University of Alberta in Canada. His teaching and research interests include international law, food law, biotechnology law and policy, and data protection law. He frequently presents his work at international conferences and publishes regularly in leading academic journals.

Nishakanthi Gopalan is a Senior Lecturer at the Medical Humanities and Ethics Unit (MedHEU), Faculty of Medicine, Universiti Malaya in Kuala Lumpur, Malaysia, and Programme Coordinator for the Master of Health Research Ethics (MOHRE), the first postgraduate programme in health research ethics in Southeast Asia. Developed in collaboration with the Berman Institute of Bioethics at Johns Hopkins University through a Fogarty International Center bioethics training grant, MOHRE focuses on building regional capacity in culturally attuned health research ethics. She holds a PhD in research ethics, and her work centres on bioethics, research ethics, and governance of emerging technologies, with an emphasis on low- and middle-income country contexts. She serves on the Universiti Malaya Research Ethics Committee (UMREC) and the Universiti Malaya Medical Centre–Medical Research Ethics Committee (UMMC-MREC) and is an editorial board member of the *Asian Bioethics Review*. Her research and publications explore the ethical, legal, and social implications of emerging technologies including stem cell research, genetic testing, artificial intelligence in healthcare, and synthetic biology.

Yvonne Holt is the Chief Medical Officer at Next Biosciences, a South African biotech company involved in cryopreservation of stem cells and human tissues, production of human biological products, and genetic testing. She graduated from Wits University as a medical doctor and holds diplomas in Paediatrics and Transfusion

Medicine. She joined Next Biosciences at its founding 19 years ago and has worked with teams combining medicine, science, and technology to create innovative products and services that help people take ownership of their future health. She has been involved in developing cellular therapies from mesenchymal stem cells to exosomes, producing biological dressings from amniotic membrane, formulating specialised eyedrops from cord blood serum, and creating in-house genetic tests in the reproductive field.

Usri Ibrahim is a Postdoctoral Research Fellow specialising in drug delivery, biomaterial functionalisation, and molecular biology. He holds a PhD in Pharmaceutics from the University of KwaZulu-Natal, where he developed expertise in designing advanced stimuli-responsive drug carriers and engineering functionalised biomaterials for therapeutic and diagnostic applications. He has developed a novel multifunctional pharmaceutical excipient and innovative stimuli-responsive drug delivery systems. He collaborates on interdisciplinary projects bridging materials science and biomedicine. His work focuses on developing next-generation therapeutics and diagnostics for diseases such as sepsis and cancer, combining nanotechnology, biofunctionalisation, and molecular tools to enhance precision medicine.

Mtebe Majigo is a Senior Lecturer and consultant clinical microbiologist in the Department of Microbiology and Immunology at the Muhimbili University of Health and Allied Sciences in Dar es Salaam, Tanzania. He has extensive experience in research, disease surveillance, capacity building, consultancy, clinical programme implementation, and medical laboratory management. He has authored and co-authored over 90 articles in peer-reviewed journals. He is currently leading national efforts to combat antimicrobial resistance through surveillance and research.

Sergii Martynenko works in the field of biobanking legislation of human tissues and cells, cellular therapy, and regenerative medicine. He is the founder and General Director of the Institute of Cell Therapy in Ukraine, a private company specialising in cord blood banking, manufacturing, and clinical applications of Advanced Therapy Medicinal Products (ATMPs). He contributed to developing Ukrainian legislation regulating the cryobanking of human tissues and cells and other aspects of the ATMP pipeline. As of 2025, he has organised and sponsored two clinical trials in Ukraine: one for treating acute respiratory distress syndrome (ARDS) caused by COVID-19 and another for knee osteoarthritis, both using perinatal MSC-based medicinal products. He also established the Ukrainian Association of Cord Blood Banks.

Carl Montague is the Managing Director of Shook Services, a consultancy company focused on the biotech and pharmaceutical industries. He holds a PhD in Molecular Medicine from Cambridge University and an MBA from the University of Stellenbosch. Dr Montague has over 30 years of experience in the pharmaceutical and biotech industries in Europe and Africa, including drug discovery and development projects as well as the development and launch of medical devices. As

Managing Director of the biologics division of Next Biosciences, he led the team that developed the AmnioRederm product, an amniotic membrane-based device that promotes accelerated wound healing of chronic non-healing wounds. He is currently working to establish a drug development platform for antiviral drugs targeting RNA viruses prevalent in Africa.

Shambhavi Naik is Chair of the Advanced Biology Programme at the Takshashila Institution, an independent think tank in Bengaluru. She is an external affiliate of the Australian National University and the Centre for Global Science and Epistemic Justice. She holds a PhD in Cancer Biology from the University of Leicester and has undertaken postdoctoral research at the National Centre for Biological Sciences and the Institute for Stem Cell and Regenerative Medicine in Bengaluru. With over seven years of experience at the interface of life sciences and policy, she has worked with Indian government departments, parliamentary committees, and international organisations such as the WHO and the International Science Council. Her research focuses on enabling the adoption of emerging technologies such as genetic editing and regenerative medicine while addressing their ethical and security challenges.

Carola Niesler is an Associate Professor of Biochemistry in the School of Life Sciences, University of KwaZulu-Natal, South Africa, and also serves as an advisor for NEXT Biosciences, a South African biotech company working in the areas of stem cells, biologics, and genetic and pathology testing. She holds a PhD from the Department of Clinical Biochemistry at the University of Cambridge, UK, and has more than 25 years of experience in stem cell research and the application of their cellular products in regenerative medicine. She has supervised more than 60 postgraduate students and published over 40 internationally peer-reviewed articles on regeneration in health and disease.

Ubaka Ogbogu is a Professor and the Katz Chair in Health Law and Science Policy in the Faculty of Law at the University of Alberta. He chairs the University of Alberta's Research Ethics Board 2 and is a member of the Royal Society of Canada's College of New Scholars, Artists, and Scientists. He is a recipient of the Confederation of Alberta Faculty Associations Distinguished Academic Early Career Award. He holds a doctorate in law from the University of Toronto, a Master of Laws from the University of Alberta, and undergraduate degrees in law from the University of Benin in Nigeria and the Nigerian Law School. His research focuses on the ethical, legal, and societal implications of novel and emerging biotechnologies. His recent work addresses the global governance of regenerative medicine, highlighting epistemic injustice, structural inequities, and the underrepresentation of low- and middle-income country perspectives in bioethics and innovation policy. He has served on several boards and councils focused on regenerative medicine governance, including the Council of Canadian Academies Expert Panel on Somatic Gene and Engineered Cell Therapies, the Canadian Institutes of Health Research Stem Cell Oversight Committee, and the International Society for Stem Cell Research Task Force on Guidelines for Stem Cell Research and Clinical Translation.

Paul Ogendi is an advocate of the High Court of Kenya and a Senior Lecturer at the University of Nairobi Faculty of Law. He holds an LLD from the University of Pretoria and completed postdoctoral research at the Chinese University of Hong Kong. He is an Honorary Research Fellow at the University of KwaZulu-Natal. His research interests include trade, intellectual property and health (access to medicines and vaccines), biotechnology law and policy, data governance, AI governance, and blue economy governance.

Adeolu Oyekan holds a PhD in Philosophy from the University of Lagos, Akoka, Nigeria. He is a Research Associate with the Decolonial Research Group at the University of Pretoria in South Africa. He began his academic career in 2008 in the Department of Philosophy at Lagos State University and was a Postdoctoral Research Fellow from 2019 to 2021 in the Identities and Social Cohesion in Africa (ISCIA) programme at Nelson Mandela University in Gqeberha, South Africa. His research interests include African philosophy, philosophy of culture, postcolonial studies, practical ethics, identity politics, restorative justice, and social cohesion.

Oksana Piven works in molecular cardiology at the Institute of Molecular Biology and Genetics of the National Academy of Sciences in Kyiv, Ukraine, focusing on canonical Wnt signalling in cardiac development, maturation, and homeostasis. She led a project to develop new-generation implants from synthetic bone analogues and mesenchymal stem cells for reconstructing damaged bone tissue. Recently, she has been a postdoctoral fellow at the Nencki Institute of Experimental Biology of the Polish Academy of Sciences in Warsaw. She is a member of the Ukrainian Biochemical Society and the Ukrainian Genetic Society, a fellow of the Polish Academy of Sciences (2024 and 2025), recipient of FEBS fellowships for Ukrainian scientists (2023 and 2025), holder of short-term postdoctoral fellowships from the European Molecular Biology Organization (2016 and 2012), and a former visiting research associate at the University of Pennsylvania School of Medicine in the USA (2007 and 2005). She lectures at Kyiv Academic University and at the Faculty of Biology of the Institute of Molecular Biology and Genetics.

Afadhali Denis Russa is an Associate Professor of Human Anatomy, Histology and Embryology at the Muhimbili University of Health and Allied Sciences (MUHAS), a leading medical university in Tanzania. He holds a Bachelor of Veterinary Medicine from Sokoine University in Morogoro, Tanzania, and a Master of Science in Epidemiology and Statistics from the University of Utrecht in the Netherlands. He also earned a PhD in Molecular and Cell Biology from Iwate Medical University in Japan. From 2009 to 2024, he served as Head of the Department of Human Anatomy, Histology and Embryology. During his doctoral research, he discovered that store-operated calcium entry (SOCE) into the cell becomes inactivated during the M-phase of the cell cycle, a mechanism implicated in cancer cell growth and metastasis. His research interests include gross anatomy, stem cells, regenerative medicine, and medical education. He has extensive teaching experience in neuroanatomy, gross anatomy, cell biology, histology, and clinical embryology.

Volodymyr Shablii works in cellular therapy and regenerative medicine, focusing on the immunomodulatory properties of perinatal MSCs in their therapeutic effects in humans. He is Deputy Director of Cryobank, Head of the Placenta Stem Cell Laboratory at the Institute of Cell Therapy, and Senior Researcher at the Institute of Molecular Biology and Genetics of the National Academy of Sciences of Ukraine. His work covers the development, manufacturing, long-term low-temperature storage, and quality control of human investigational products from perinatal tissues in line with Good Manufacturing Practice and ATMP guidelines. He leads translational research aimed at bringing placental and umbilical cord-derived MSCs into clinical application. He is a member of the Ukrainian Biochemical Society, the Ukrainian Cell Biology Society, the International Society for Stem Cell Research, and the International Placenta Stem Cell Society.

Donrich Thaldar is a Professor of Law at the University of KwaZulu-Natal, Durban, where he chairs the Health Law and Ethics Research Interest Group. He is the principal investigator of an NIH-funded project investigating the legal aspects of data science in health innovation in Africa. He also runs a private law practice, specialising in strategic litigation in reproductive law. Before entering academia in 2017, he practised as an advocate at the Pretoria Bar. He has served as legal counsel or amicus curiae in several landmark South African cases in reproductive law, including the first case on 'designer children' (2016), the first case of posthumous conception (2018), the first case of gamete withdrawal from a comatose person (2020), the first case on the enforceability of a sperm donor agreement (2021), and a successful constitutional challenge to the prohibition on non-medical preimplantation sex selection (2022).

Mari van de Vyver is an Associate Professor in the Department of Medicine at Stellenbosch University and a rated researcher with the South African National Research Foundation. She has published more than 50 scientific papers, supervised over 30 postgraduate students, and holds a strong national profile with international recognition in her field. Her expertise spans endocrinology (diabetes), inflammation, stem cells, and regenerative medicine. She actively builds collaborations with partners across academia, industry, and clinical practice in South Africa and abroad, ensuring her team has the expertise needed to achieve impactful outcomes. Through these efforts, her group bridges the gap between laboratory research and clinical application, advancing translational science.

Meng Wang is a PhD candidate in the Faculty of Law at the University of Macau. Her research interests are in international law, health law, and the legal and ethical regulation of biotechnology. She focuses on the protection of patients' rights in the context of emerging biomedical technologies and the intersection between human rights and science and technology governance. She has published several academic articles in leading journals.

Part I

Foundational Critiques: Epistemic Injustice and Global Asymmetries

1 Reframing Regenerative Medicine: Global Asymmetries and Ethical Futures

Ubaka Ogbogu, Donrich Thaldar, and Nishakanthi Gopalan

Abstract

This chapter introduces the volume *Collaborative Bioethics: Regenerative Medicine and the Bioethics of Globalization, Justice and Hegemony*. It outlines the central themes of the collection, including global asymmetries in regenerative medicine research, epistemic injustice, and the need for inclusive and context-sensitive ethical governance. The chapter also summarizes the contributions in the volume, which collectively examine the ethical, legal, and social dimensions of regenerative medicine across diverse national and regional contexts.

Keywords

Regenerative medicine · Global bioethics · Epistemic injustice · Global South · Collaborative governance · Collaborative bioethics · Biomedical research · Global health ethics · Reconnaissance mode of inquiry

U. Ogbogu (✉)
Faculty of Law, University of Alberta, Edmonton, AB, Canada
e-mail: uogbogu@ualberta.ca

D. Thaldar
Faculty of Law, University of KwaZulu-Natal, Durban, South Africa
e-mail: ThaldarD@ukzn.ac.za

N. Gopalan
Medical Humanities and Ethics Unit (MedHEU), Faculty of Medicine, Universiti Malaya, Kuala Lumpur, Malaysia
e-mail: nisha.gopalan@um.edu.my

U. Ogbogu et al. (eds.), *Regenerative Medicine and the Bioethics of Globalization, Justice and Hegemony*, Collaborative Bioethics 5,
https://doi.org/10.1007/978-3-032-07289-4_1

Introduction: Reframing Regenerative Medicine Through a Truly Global and Collaborative Bioethics

Regenerative medicine (RM) refers to a range of innovative therapies, clinical interventions, and research and development (R&D) activities aimed at repairing or replacing damaged or diseased human cells, tissues, and organs, with a primary focus on conditions that are currently considered incurable [1]. As RM progresses through critical stages of clinical study, regulatory review, product development, and clinical uptake, it is increasingly evident that globally scaled social, ethical, and scientific considerations, or what might be termed geosocial, geoethical, and geoscientific dimensions, are central, not peripheral, to its advancement [2–4]. Yet these dimensions remain underexamined in many mainstream accounts of RM, especially those that prioritize technical and market readiness considerations [5, 6]. Central among these underexamined areas is the global distribution of RM R&D activities and the uneven ethical and governance landscapes in which such activities take place.

Much of the existing literature on RM, including on its bioethical implications and governance frameworks designed to guide it, draws from contexts rooted in high-income countries [7–12]. These include jurisdictions with advanced biomedical infrastructures, robust public and private funding mechanisms, established regulatory pathways, and deep integration into international scientific networks. Such countries, often grouped under the umbrella term "Global North," continue to function as both the geographic and epistemic hubs of biomedical innovation. Accordingly, research agendas, ethical frameworks, and regulatory instruments that originate in or cater to these environments often become the default global references, even when their applicability elsewhere is uncertain or inappropriate.

In contrast, significantly less is known about the state of RM research and development in low- and middle-income countries (LMICs), especially those located in the "Global South," including the ethical, legal, and social issues specific to these settings. Although efforts have been made to document RM activities in low-resource settings, they have largely promoted decontextualized understandings of crisis and scarcity, or imposed Western-centric issues and frameworks onto these contexts that may not reflect local realities [2, 13]. Aggregated context-appropriate accounts that move beyond crisis- or scarcity-driven narratives are rare and are either dated or relegated to less visible channels of knowledge dissemination [14–16].

This gap is not incidental. It reflects a set of systemic conditions that shape visibility, participation, and recognition within the global biomedical enterprise. These conditions include, among others, the disproportionate attention paid by scholars and analysts to high-income settings, the marginalization or outright exclusion of LMIC-driven initiatives from key global platforms and dialogues, and the clustering of R&D investment and infrastructure in countries perceived to be more capable of moving biomedical innovations through what is often called the "valley of death," the critical phase in the innovation lifecycle where a substantial infusion of capital and technical support is necessary to avoid failure. As a result, promising initiatives in LMICs may remain underresourced, underdocumented, or misunderstood, which in turn reinforces the impression that little of relevance is taking place in these regions.

The consequences of these dynamics are neither abstract nor confined to questions of equity in resource allocation. They influence the shape and direction of global health innovation, the legitimacy of ethical and governance frameworks, and the development of sustainable international collaborations. One consequence is the missed opportunity to expand and diversify the global RM R&D ecosystem, including the knowledge systems and value orientations that inform it [3, 17, 18]. Another is the reproduction of epistemic injustice, a term that refers to forms of injustice related to knowledge production, recognition, and credibility. In this context, epistemic injustice manifests through the systemic neglect of LMIC perspectives, priorities, and expertise, leading to narrow, exclusionary understandings of what counts as legitimate inquiry or responsible innovation [19–21]. These issues intersect with broader concerns about global justice, particularly the uneven distribution of the benefits and burdens associated with emergent health technologies.

This volume, *Collaborative Bioethics: Regenerative Medicine and the Bioethics of Globalization, Justice and Hegemony*, responds to these concerns by critically examining the conditions under which RM is developed, governed, and evaluated across diverse geopolitical contexts. It brings together interdisciplinary perspectives from contributors working in different regions and institutional environments to explore how RM intersects with longstanding and emerging challenges in global health ethics. In particular, the volume foregrounds the importance of collaborative approaches to bioethics. These are approaches that do not assume consensus but instead make space for complexity, asymmetry, and contested values. Rather than treating ethics as an afterthought to technological progress, the contributions collected here position ethical reflection as intrinsic to how innovation is imagined, operationalized, and distributed.

This chapter sets the stage for the discussions that follow by framing the global inequities and asymmetries that currently characterize the field of RM. It emphasizes that ethical governance of RM cannot be adequately addressed through frameworks developed exclusively within and for high-income countries. Nor can it rely on assumptions of universal applicability that ignore the sociopolitical and economic conditions shaping research and clinical practice in other parts of the world. Collaborative bioethics requires careful attention to power, positionality, and voice. It demands a willingness to reexamine dominant narratives and to recognize the epistemic and normative contributions of actors and institutions beyond the traditional centers of influence.

The task, then, is not simply to expand participation or redistribute resources, although both remain essential. It is also to question the structures through which legitimacy, credibility, and value are assigned in global bioethics and biomedical innovation. As RM continues to evolve, and as global challenges such as inequality, migration, environmental degradation, and political instability shape its uptake and development, the need for a more inclusive, reflexive, and dialogical bioethics becomes ever more pressing. This volume aims to contribute to that project by opening up space for critical inquiry, cross-regional engagement, and sustained attention to the ethical and political dimensions of RM in a globalized world.

Regenerative medicine, and scientific innovation more broadly, cannot be ethically or sustainably governed within frameworks that are siloed or exclusionary. A truly collaborative bioethics, one that reflects the complexities of globalization and remains attuned to historical and structural inequities, provides an essential foundation for advancing innovation in ways that are just, inclusive, and responsive to the realities of our interconnected world.

To achieve its objectives, the collection adopts and adapts the reconnaissance mode of inquiry, a methodological stance explicitly oriented toward revealing underexplored or marginalized realities [22, 23]. Conceived as a politically and strategically engaged research endeavour, this mode of inquiry systematically challenges dominant and hegemonic knowledge frames. In the context of RM, reconnaissance involves illuminating diverse practices, situations, ethical considerations, and governance dynamics within LMICs without assuming their alignment or coherence with more dominant frameworks, and without attempting to tame or discipline either the substantive integrity of what is described or the expressive style and analytical approaches of the contributors. Put simply, the contributions in this volume were not solicited or expected to conform to any predetermined style or perspective, but rather to reflect the authentic voices and approaches of their authors.

By enabling scholars intimately familiar with the complexities and specificities of RM in the overlooked regions to share their insights, the volume will significantly enhance the visibility of underrecognized knowledge and lived realities. Furthermore, these contributions aim to critically interrogate, diversify, and expand the dominant bioethical discourse, identifying both commonalities and divergences that can inform more equitable, inclusive, and globally responsive bioethical frameworks in the rapidly evolving field of RM. Rather than treating bioethics as a fixed set of rules or universally applicable standards, this volume approaches it as an ongoing, situated conversation shaped by power, history, and context. It reflects a commitment to listening closely to voices and experiences that have too often been left out of global discussions on RM.

This volume emerged from a workshop titled *Regenerative Medicine Innovation and Governance in the Global South*, held in Ballito, South Africa, from April 16–18, 2024 and supported by funding from two major research grants [24, 25]. The workshop convened a diverse group of experts based in the Global South and diaspora, including scientists, legal scholars, bioethicists, social scientists, philosophers, health equity advocates, and industry representatives. The workshop examined the current state of RM innovation and governance in LMICs in the Global South, reflecting on opportunities and challenges, contributions of the Global South to international governance, and questions of epistemic justice, accessibility, and affordability. This volume captures the insights and perspectives shared at the workshop and extends the conversation through contributions that interrogate and expand the frameworks governing RM in a global context. As editors, we acknowledge that our own positionalities shape the framing of this volume: two of us are scholars based in the Global South, and one is a member of the academic diaspora. This reflexive standpoint underpins our commitment to centering knowledge, practices, and priorities that are often marginalized within dominant global bioethical and biomedical discourses.

In this volume, the terms *Global North* and *Global South* are used as heuristic categories to describe broad, albeit contested, groupings of countries that share certain economic, political, and historical characteristics [26–29]. Originating in part from the Brandt Report's conceptualization of a global divide, geographically represented by the so-called "Brandt Line," these terms have come to denote, respectively, wealthier, industrialized nations often located in the Northern Hemisphere, and less industrialized LMICs predominantly situated in the Southern Hemisphere. While we acknowledge the limitations and critiques of these terms, including their potential to oversimplify complex geopolitical realities and reinforce reductive binaries, we use them here and in the volume with critical awareness and specificity. In particular, *Global South* primarily refers to countries that are members of the Group of 77 and China, a coalition of LMICs at the United Nations that seeks to advance their shared economic interests and strengthen their collective voice in global negotiations [30]. In contrast, *Global North* refers to countries that are either not part of this group or are classified as high-income nations. The terms function in this volume not as rigid designations but as lenses for examining the structural asymmetries, epistemic exclusions, and governance disparities that shape the global landscape of RM. They help foreground the differing conditions under which innovation unfolds, while emphasizing the importance of inclusive, pluralistic, and context-sensitive approaches to bioethics and biomedical governance.

The chapters in this volume are organized into two parts, each comprising interrelated thematic arcs. Part I presents foundational critiques that examine epistemic injustice and global asymmetries. It includes three chapters that collectively define the political and epistemological terrain on which the rest of the volume builds. These chapters provide a theoretical grounding for the national case studies that follow. Part II features country-specific analyses that explore the governance and ethical challenges of RM in diverse contexts. These chapters demonstrate how global dynamics manifest within distinct regulatory, social, and institutional environments, translating the broader critiques of Part I into situated practice. Across both parts, the chapters also address the cultural, religious, and educational forces that shape RM and its governance, while proposing strategies for developing more inclusive, pluralistic, and responsive approaches to global bioethics.

Part I: Foundational Critiques: Epistemic Injustice and Global Asymmetries

In the first substantive chapter, Ubaka Ogbogu and Lauren Albrecht offer a critical and empirically grounded exploration of epistemic injustice in the global governance and documentation of RM. Drawing on systematic mapping of clinical trials, regulatory environments, and R&D activity across LMICs in Africa and Asia, the chapter challenges the dominant narrative that positions these countries as marginal or inactive in the field of RM.

The authors argue that the invisibility of LMIC contributions is not the result of empirical absence but of structural frameworks that systematically exclude these regions from global systems and narratives of credibility and recognition. This exclusion is both testimonial and hermeneutical. LMIC actors are not only denied a voice in global policy forums, but the frameworks themselves are ill-equipped to interpret or value knowledge generated outside dominant domains of expertise.

What distinguishes this chapter is its insistence that the underdocumentation of LMIC activity is not a benign gap but an active form of epistemic violence. This violence is perpetuated through research practices that privilege Global North agendas, publication norms that marginalize non-Western authorship, and governance models that fail to accommodate alternative epistemologies.

The chapter concludes by advocating for bioethics that is epistemically plural, structurally inclusive, and grounded in reciprocal partnerships. By surfacing both the contributions and the exclusions of LMIC actors, this chapter does more than critique; it provides a roadmap for restorative and epistemic justice in the global RM landscape.

In their jointly authored chapter, Jonathan Chimakonam and Adeolu Oyekan interrogate the dominant ethical paradigms shaping global discussions of RM, arguing that these are often grounded in Euro-American worldviews that fail to account for the lived realities and normative frameworks of the Global South. They propose an alternative model of collaborative bioethics that foregrounds epistemic pluralism and prioritizes African intellectual traditions. The chapter critiques the prevailing tendency to universalize ethical standards that originate in the Global North and calls instead for a multipolar approach that respects diverse moral foundations.

At the core of their argument is the concept of trivalent ethics, a framework derived from African philosophy that seeks to integrate individual, communal, and spiritual dimensions of moral reasoning. This approach challenges the binary logic that often underpins dominant bioethical thinking and offers a more flexible and inclusive foundation for engaging with the complexities of RM. By emphasizing complementarity rather than consensus, the authors advocate for an ethical model that can accommodate deep value differences without erasing them.

The chapter's significance lies in its invitation to reimagine global bioethics as a genuinely intercultural and dialogical project. Rather than assimilating LMIC perspectives into preexisting Northern frameworks, Chimakonam and Oyekan call for a restructuring of the ethical field itself, one that begins from a position of mutual respect, epistemic humility, and recognition of historically excluded knowledge systems. In doing so, their chapter provides a theoretical anchor for the volume's broader commitment to collaborative, situated, and justice-oriented approaches to RM governance.

In the last chapter in Part I, Afadhali Russa and Majigo Mtebe bring a pedagogical lens to the volume, interrogating the glaring absence of RM content in medical curricula, with a focus on Tanzania. Their chapter underscores a critical but often overlooked site of epistemic exclusion: professional education. In a healthcare future where regenerative therapies are poised to become standard interventions for noncommunicable diseases and antimicrobial-resistant infections, the continued omission of these technologies from training programs represents a profound ethical and strategic oversight.

The authors detail how Tanzanian medical students receive little to no exposure to RM training, a gap that not only limits their clinical competence but also constrains possibilities for local innovation and advocacy. This educational deficit is not the result of deliberate resistance but rather a consequence of underinvestment in curriculum development, a lack of trained faculty, and the absence of institutional commitment to biomedical modernization.

Yet the chapter also situates this lacuna within a broader political economy of knowledge production, wherein the Global North dictates educational standards and content through licensing requirements, funding structures, and research collaborations. As a result, curricula in many LMICs often remain tethered to outdated paradigms, unable to keep pace with scientific developments or respond to local disease burdens.

Russa and Mtebe advocate for the systematic integration of RM into undergraduate and postgraduate medical training, positioning education as both a site of resistance and transformation. Their argument is not limited to Tanzania but resonates across the Global South, where curricular reform represents a foundational step toward epistemic inclusion and health system resilience. This chapter reorients our understanding of bioethics as not only a matter of policy and practice but also of pedagogy and professional formation.

Part II: National Case Studies: Governance, Context, Constraint, and Endogenous Innovation

Part II opens with Marietjie Botes' chapter, *Translation and Commercialisation of Regenerative Medicine in South Africa: Navigating the Regulatory Landscape*. The chapter offers a detailed analysis of South Africa's outdated and fragmented regulatory framework for RM translation and commercialization, focusing in particular on the limitations of the National Health Act and the Medicines and Related Substances Act. These statutes, originally designed for more conventional biomedical products and services, have not kept pace with the scientific and clinical realities of RM. As a result, regulatory gaps have emerged that hinder responsible innovation, deter investment, and allow unproven interventions to circulate unchecked.

Botes emphasizes that the core issue is not simply legislative delay, but the absence of regulatory systems capable of adapting to the unique demands of cell- and tissue-based therapies. Drawing attention to global examples, she argues for the development of agile regulatory frameworks that can support timely, ethical, and contextually appropriate oversight. The chapter underscores how static or ill-fitting regulatory models not only stall clinical translation but may also undermine patient safety by failing to distinguish legitimate innovation from opportunistic exploitation.

In response, Botes proposes a series of reforms tailored to South Africa's regulatory environment. These include the creation of a dedicated Regenerative Medicine Office, the introduction of conditional licensing mechanisms, and the institutionalization of structured stakeholder engagement. Taken together, these proposals aim to build a governance framework that is responsive, ethically grounded, and innovation-enabling. The chapter offers a compelling case for regulatory agility as a necessary condition for advancing RM in low- and middle-income settings.

Following Botes' examination of South Africa's regulatory landscape, the chapter by Usri Ibrahim, Mari van de Vyver, Yvonne Holt, Carl Montague, and Carola Niesler provides a complementary perspective on the country's clinical and translational challenges with RM and their broader implications for global biomedical equity. While Botes focuses on regulatory inertia, this chapter shifts attention to the systemic barriers that prevent the translation of research into therapeutic applications. The authors argue that despite significant scientific expertise within academic institutions, South Africa's RM ecosystem is constrained by infrastructural deficiencies, fragmented oversight, and inadequate investment pathways.

These challenges, the authors contend, are compounded by stark disparities in access to healthcare and the persistence of a dual health economy. Even when regenerative therapies are technically available, their benefits are typically limited to a small, privileged segment of the population. The chapter situates this exclusion not simply as a matter of affordability, but as a consequence of chronic underinvestment in translational infrastructure and the marginalization of African populations within global research agendas.

A key contribution of the chapter is its emphasis on the scientific costs of exclusion. Ibrahim and colleagues highlight how the systematic omission of genetically diverse populations from clinical trials compromises both the efficacy and ethical legitimacy of RM. They call for a structural reorientation of research priorities, funding models, and policy frameworks to ensure that inclusion is embedded in the foundations of global innovation. In doing so, the chapter aligns with the broader concerns of the volume, demonstrating how epistemic and infrastructural inequities intersect to shape the future of RM in the Global South.

In his informative chapter, Paul Ogendi offers a critical assessment of Kenya's nascent but evolving legal landscape for stem cell research, situating it within broader debates about the governance of RM in Africa and beyond. While the establishment of a national stem cell laboratory at the Kenya Medical Research Institute marks a significant institutional step forward, the chapter reveals that Kenya's legal architecture remains ill-equipped to regulate the ethical and scientific complexities of RM.

Ogendi identifies a fragmented legal terrain shaped by overlapping, and at times contradictory, legislative instruments, from health and science policy to criminal law and data protection statutes. Rather than facilitating innovation, this legal patchwork creates uncertainty for researchers and clinicians, stifles legitimate experimentation, and opens space for regulatory avoidance. Particularly acute are the ethical and moral controversies surrounding embryonic stem cell research, which remain unresolved in public discourse and largely unaddressed in statutory form.

The chapter suggests that these governance gaps are not merely technical deficiencies, but instead reflect deeper structural dynamics that marginalize African regulatory perspectives. Kenya's default deference to foreign models and its failure to articulate a locally grounded governance vision point to the persistence of colonial logics in biomedical policymaking. In this sense, the chapter contributes to the volume's broader argument that the exclusion of LMIC perspectives from global bioethics is not accidental but structurally produced.

Ogendi calls for a coherent and ethically robust national framework that draws on local norms, jurisprudence, and public engagement. He suggests that Kenya need not start from scratch, pointing to the South African experience as a proximate model of context-sensitive legal development. This chapter serves as a reminder that legal frameworks are not neutral scaffolds but normative instruments that can either reproduce exclusion or facilitate epistemic and scientific self-determination.

Shambhavi Naik's chapter presents a timely and layered examination of India's engagement with RM through the twin lenses of stem cell science and genetic engineering. India, positioned as a rising biomedical power with a high burden of non-communicable diseases, becomes a crucial site for exploring the ethical and infrastructural tensions that define the Global South's encounter with advanced biotechnologies. Naik does not romanticize India's ascent but instead unpacks the contradictions within its research ecosystem, where scientific ambition often outpaces regulatory maturity.

Central to the chapter is the argument that India's biomedical promise remains constrained by affordability gaps, uneven public awareness, and infrastructural bottlenecks. These limitations are not incidental; they are symptomatic of a broader political economy in which emergent markets are simultaneously courted as sites of innovation and neglected as spaces of regulatory precarity. The result is a biomedical landscape where high-end therapies coexist with deep disparities in access and capacity.

Naik also interrogates the cultural and religious dimensions of biomedical ethics in India, highlighting how attitudes toward genome editing and stem cell therapy are shaped by plural, sometimes conflicting, belief systems. Rather than treating these as barriers to progress, the chapter suggests that these perspectives offer valuable ethical orientations that ought to inform policy and practice. The call here is not for alignment with Global North standards but for the cultivation of a governance architecture that is both scientifically robust and culturally attuned.

In contributing a grounded, context-specific analysis of India's regulatory and ethical terrain, this chapter enriches the volume's commitment to a globally responsive bioethics. It illustrates how the governance of RM in the Global South must reckon with entangled legacies of inequality, pluralism, and technological aspiration.

Nishakanthi Gopalan's chapter offers a nuanced examination of Malaysia's complex biomedical landscape, revealing both the possibilities and perils of RM in a rapidly modernizing LMIC. Against the backdrop of Malaysia's strategic investment in biotech and its growing appeal as a destination for medical tourism, Gopalan uncovers the ethical crosscurrents shaping the country's engagement with advanced therapies.

What sets this chapter apart is its deft tracing of how commercialization, religious pluralism, and transnational policy adoption intersect to produce a regulatory environment that is at once ambitious and ambivalent. Malaysia emerges not as a passive recipient of Northern models, but as a site of hybrid governance, negotiating between global standards and local ethical norms, particularly around issues of consent, equity, and religious legitimacy.

Yet, the chapter also warns of the dangers of superficial harmonization: the adoption of international best practices without adequate adaptation to sociocultural realities can deepen existing exclusions. Gopalan calls for a governance framework that is not only compliant but critically reflexive and capable of mediating between innovation imperatives and moral complexity.

In situating Malaysia as both an exemplar and cautionary tale, this chapter contributes to the volume's larger argument about the need for ethically grounded, context-sensitive governance of RM. It shows that the ethical stakes of RM are not limited to research ethics or clinical risk but include the political economy of biomedical modernization itself.

Meng Wang and Li Du provide a detailed and critical analysis of China's engagement with RM, focusing on the interplay between national ambition, scientific infrastructure, and ethical governance. As a country that now ranks second globally in clinical trials involving cell and gene therapies, China represents a complex case of a Global South actor rapidly ascending the RM hierarchy.

The chapter charts China's substantial investments in RM, including its approval of CAR-T therapies and the integration of artificial intelligence into tissue engineering workflows. These developments mark a shift from follower to leader status in global biomedical innovation. Yet, the authors argue that China's rise is shadowed by persistent ethical and legal challenges, particularly in relation to unproven therapies, data governance, and informed consent.

What emerges is a dual narrative. On one hand, China is proactively developing its own regulatory standards, aligning them with international best practices while also asserting its own strategic priorities. On the other hand, it remains entangled in the broader dynamics of biomedical globalization, including the exploitation of regulatory loopholes by private clinics, the rise of "stem cell tourism," and the difficulty of enforcing oversight in a fast-moving and decentralized sector.

The chapter's most significant contribution lies in its framing of China not as an outlier but as a node of global ethical complexity. It challenges simplistic binaries between the Global North and South, showing how power, expertise, and risk circulate in nonlinear and contested ways. By analyzing both national reforms and international responsibilities, the authors call for a form of bioethical governance that is capable of addressing asymmetries not only between countries but also within them.

In the final chapter in the volume, Oksana Piven and her coauthors offer a deeply contextualized analysis of the regulatory and biotechnological landscape of RM in Ukraine. Situated at the intersection of scientific ambition, war and geopolitical instability, and comparative legal inquiry, this chapter presents Ukraine as both a case study and a mirror for the broader regulatory challenges facing LMICs with transitional economies.

The authors trace the country's engagement with advanced therapy medicinal products (ATMPs), stem cell derivatives, and tissue-engineered products, while critically examining the gaps in governance that threaten to undermine public trust and scientific integrity. Unlike many LMICs, Ukraine has developed a relatively advanced infrastructure for cell-based therapy, but this progress is tempered by a lack of enforceable guidelines, fragmented oversight, and the pressures of an ongoing military conflict that diverts resources and destabilizes institutional continuity.

Importantly, the chapter does not treat Ukraine in isolation. It offers a comparative analysis of regulatory frameworks across Europe, highlighting divergences in how quality, safety, and ethical compliance are defined and enforced. In doing so, the authors make a broader argument about the unevenness of global governance regimes and the vulnerabilities of LMICs that must navigate both external dependencies and internal limitations.

The chapter concludes with a call for regulatory coherence grounded in local scientific capacity and ethical norms, rather than the passive importation of foreign models. In its attention to both technical detail and normative stakes, the chapter aligns with the volume's broader argument: that effective governance of RM cannot be disentangled from questions of sovereignty, stability, and epistemic self-determination.

Conclusion: Toward Ethical Futures and a Global Reconnaissance Bioethics for Regenerative Medicine

This volume seeks to reframe the ethical governance of RM through a global lens that is alert to asymmetries in power, recognition, infrastructure, and knowledge production. Across diverse contributions, it has become clear that the ethical futures of RM cannot be adequately shaped through frameworks conceived within, or exclusively for, high-income settings. The persistent marginalization of voices, priorities, and practices from low-resource settings, particularly in the Global South, not only undermines global equity but also constrains the epistemic and normative foundations of the field itself.

A central thread running through the chapters is the concept of epistemic injustice, not merely as an abstract philosophical concern, but as a lived reality that manifests in the underrepresentation of LMIC contexts in scientific publications, clinical trials, regulatory models, and education. The included contributions aim to show that the invisibility of LMIC contributions is not due to a lack of activity or expertise but is the result of systemic exclusions embedded in global biomedical infrastructures and governance regimes. From South Africa's regulatory inertia and Kenya's fragmented legal frameworks to India's ethical pluralism and Malaysia's hybrid governance, the chapters lay bare how RM development is entangled in complex, and often contradictory, national and global dynamics.

This book also foregrounds the limitations of universalist ethics that presume neutrality while reproducing Northern epistemologies and governance norms. Instead, contributors advocate for ethical pluralism and trivalent bioethics that resists homogenization and emphasizes dialogue between diverse normative traditions. Whether it is African communal ethics, religious pluralism in Southeast Asia, or postcolonial critiques of knowledge hierarchies, these perspectives push back against the hegemony of technocratic rationalism and offer alternative moral architectures that are better attuned to local needs and values.

What emerges from these insights is the importance of a reconnaissance approach to bioethics as a critical, politically engaged, and exploratory mode of inquiry that not only exposes excluded realities but also amplifies the voices and practices shaping RM beyond the strongholds of power. This approach to bioethics does not seek tidy consensus; rather, it embraces the tensions, asymmetries, and complexities inherent in efforts to democratize science and innovation. It demands attentiveness to history, to structural conditions, and to the recursive relationships between science, society, and ethics.

This collection also brings into view the institutional and pedagogical dimensions of bioethics. From the exclusion of RM in medical curricula to the inadequacy of regulatory systems designed for older biomedical paradigms, multiple chapters reveal how systemic neglect in education and infrastructure limits the capacity of LMICs to shape and benefit from emerging biotechnologies. Reform here is not simply about content but about reclaiming institutional sovereignty, realigning global research priorities, and building capacities for context-sensitive innovation.

As RM continues to evolve, so too must the frameworks through which it is governed. The challenges posed by regenerative technologies, ranging from equitable access and consent to legitimacy and transnational regulation, cannot be resolved within exclusionary or static models. A just and globally responsive bioethics must be dialogical, situated, and dynamically engaged with both the material conditions and the moral landscapes of the communities it seeks to serve.

In closing, this volume calls for a shift from viewing LMICs as sites of ethical deficiency or regulatory lag, to recognizing them as critical nodes of innovation, knowledge, and moral insight. By embracing epistemic plurality, fostering reciprocal partnerships, and advocating for structural transformation, this collection contributes to the ongoing project of decolonizing global bioethics. Only through such an inclusive and reflexive approach can RM realize its promise, not just as a frontier of science, but as a catalyst for global justice and shared human flourishing.

References

1. Mason C, Dunnill P (2007) A brief definition of regenerative medicine. Regen Med 3(1):1–5. https://doi.org/10.2217/17460751.3.1.1
2. Chan S (2017) Current and emerging global themes in the bioethics of regenerative medicine: the tangled web of stem cell translation. Regen Med 12(7):839–851. https://doi.org/10.2217/rme-2017-0065
3. Ogbogu U, Hardcastle L (2021) Bioethics and practical justice in the post-COVID-19 era. Dev World Bioeth 21(1):31–35. https://doi.org/10.1111/dewb.12279
4. Hayath M, Bowers O (2024) Cultural relativity and acceptance of embryonic stem cell research. Voices Bioeth 10. https://doi.org/10.52214/vib.v10i.12685

5. Beheshtizadeh N, Gharibshahian M, Pazhouhnia Z et al (2022) Commercialization and regulation of regenerative medicine products: promises, advances and challenges. Biomed Pharmacother 153:113431. https://doi.org/10.1016/j.biopha.2022.113431
6. Doxzen KW, Adair JE, Bazzo YMF et al (2024) The translational gap for gene therapies in low- and middle-income countries. Sci Transl Med 16(746). https://doi.org/10.1126/scitranslmed.adn1902
7. Barfoot J, Rosemann A, Blackburn CC (2017) Special focus issue on regenerative medicine in society: interdisciplinary perspectives (part II)—foreword. Regen Med 12(7):733–736. https://doi.org/10.2217/rme-2017-0136
8. Rennie S, Mupenda B (2008) Living apart together: reflections on bioethics, global inequality and social justice. Philos Ethics Humanit Med 3:25. https://doi.org/10.1186/1747-5341-3-25
9. Rosemann A, Barfoot J, Blackburn C (2017) Special focus issue on regenerative medicine in society: interdisciplinary perspectives (part I)—foreword. Regen Med 12(6):577–580. https://doi.org/10.2217/rme-2017-0117
10. Salter B (2009) State strategies and the geopolitics of the global knowledge economy: China, India and the case of regenerative medicine. Geopolitics 14(1):47–78. https://doi.org/10.1080/14650040802578666
11. Salter B, Faulkner A (2011) State strategies of governance in biomedical innovation: aligning conceptual approaches for understanding 'rising powers' in the global context. Glob Health 7:3. https://doi.org/10.1186/1744-8603-7-3
12. Thomson Reuters (2011) A bibliometric analysis of regenerative medicine. Thomson Reuters, Leeds. https://assets.publishing.service.gov.uk/media/5a79a92c40f0b642860d9c7e/11-1059-bibliometric-analysis-of-regenerative-medicine.pdf. Accessed 5 May 2025
13. Isasi RM, Knoppers BM (2006) Beyond the permissibility of embryonic and stem cell research: substantive requirements and procedural safeguards. Hum Reprod 21(10):2474–2481. https://doi.org/10.1093/humrep/del235
14. Greenwood HL, Singer PA, Downey GP, Martin DK, Thorsteinsdóttir H, Daar AS (2006) Regenerative medicine and the developing world. PLoS Med 3(9):e381. https://doi.org/10.1371/journal.pmed.0030381
15. Ogbogu U, Botes M, Chimakonam J, Du L, Edgcumbe A, Naik S, Ogendi P, Piven O, Russa D, Simão-Silva DP, Thaldar D, Gopalan N, Holt Y, Niesler C, Nair P, Otim G (2024) Regenerative medicine innovation and governance: insights from the global south. SSRN. https://doi.org/10.2139/ssrn.5254428
16. Millett P, Naik S, Otim G, Piven O, Roca MM, Shozi B, Simão-Silva DP, Vavitsas K, Wang Z (2023) Somatic genome editing governance approaches and regulatory capacity in different countries. SSRN. https://doi.org/10.2139/ssrn.4375726
17. Popejoy AB, Fullerton SM (2016) Genomics is failing on diversity. Nature 538(7624):161–164. https://doi.org/10.1038/538161a
18. Sirugo G, Williams SM, Tishkoff SA (2019) The missing diversity in human genetic studies. Cell 177(1):26–31. https://doi.org/10.1016/j.cell.2019.02.048
19. Abimbola S (2023) Knowledge from the global south is in the global south. J Med Ethics 49(5):337–338. https://doi.org/10.1136/jme-2023-109089
20. Bhakuni H, Abimbola S (2021) Epistemic injustice in academic global health. Lancet Glob Health 9(10):e1465–e1470. https://doi.org/10.1016/S2214-109X(21)00301-6
21. de Vries J, Pratt B (2023) Epistemic justice in bioethics: interculturality and the possibility of reparations. J Med Ethics 49(5):347–347. https://doi.org/10.1136/jme-2023-109149
22. Armstrong T, Nuttall A (2022) Reconnaissance work in educational research: an exploration under pandemic conditions. Practice 4(3):226–238. https://doi.org/10.1080/25783858.2022.2147016
23. McKay I (2008) Reasoning otherwise: leftists and the people's enlightenment in Canada, 1890–1920. Between the Lines, Toronto
24. Schuurmans C, Aubert I, del Sol A, Kurrasch D, Mattar P, McLaurin J, Morshead C, Near J, Slavcev R, Stefanovic B, Badea I, Bethell J, Blakney A, Castro D, Drouin-Ouellet J, Gibbings D, Götz M, Hynynen K, Jacobs-LeVan J, Leong H, McAiney C, Ogbogu U, Stricker S,

Yang G (2023–2029) A pipeline approach for the rational design of neuronal lineage conversion approaches to treat neurological disorders. Government of Canada's New Frontiers in Research Fund (NFRF) – Transformation Stream. https://sshrc-crsh.canada.ca/funding-financement/nfrf-fnfr/transformation/2022/award_recipients-titulaires_subvention-eng.aspx. Accessed 8 June 2025
25. Zarzeczny A, Ogbogu U, Caulfield T (2022–2025) Law, public policy and social license for next-generation regenerative medicine. Stem Cell Network Translation and Society Team Award. https://stemcellnetwork.ca/wp-content/uploads/2022/05/2022-2025-Funding-Results.pdf. Accessed 8 June 2025
26. Dados N, Connell R (2012) The global south. Contexts 11(1):12–13. https://doi.org/10.1177/1536504212436479
27. Hollington A, Salverda T, Schwarz T, Tappe O (2015) Concepts of the global south. Global South Studies Center, Cologne. https://gssc.uni-koeln.de/en/wissenstransfer/voices-from-around-the-world/concepts-of-the-global-south. Accessed 8 June 2025
28. Patrick S, Huggins A (2023) The term "global south" is surging. It should be retired. https://carnegieendowment.org/posts/2023/08/the-term-global-south-is-surging-it-should-be-retired?lang=en. Accessed 8 June 2025
29. Prashad V (2013) The poorer nations: a possible history of the global south. Verso, London
30. Group of 77 (nd) About the Group of 77. https://www.g77.org/doc/index.html#aim. Accessed 8 June 2025

Toward Epistemic Justice: Regenerative Medicine Activity and Governance in Low- and Middle-Income Countries

2

Ubaka Ogbogu and Lauren Albrecht

Abstract

As regenerative medicine (RM) transitions into a critical translational phase, there remains limited understanding of its global research and development landscape, particularly in low- and middle-income countries (LMICs). This chapter examines how the under-documentation of RM activity in LMICs perpetuates epistemic injustice, i.e., systematic exclusion from knowledge production and recognition. Through a detailed analysis of clinical trial activity and regulatory frameworks across Africa and Asia, the chapter reveals that LMICs are not passive recipients but active, though often unrecognized, contributors to global RM innovation. It highlights how challenges with funding, infrastructure, and regulatory capacity obscure LMIC achievements and hinder equitable participation in the knowledge economy. Drawing on mapping studies, the chapter identifies both the existence of robust, locally adapted RM innovation in several LMICs and the regulatory and infrastructural gaps that challenge commercialization and global recognition. The chapter aims to show that these disparities are not only technical but deeply epistemic, reinforcing broader inequalities in health and scientific development. It calls for an expanded commitment to epistemic justice and global health equity, urging reforms in governance, funding, and institutional partnerships that better include LMIC perspectives. Ultimately, it proposes that recognizing and integrating LMIC contributions is essential for a more inclusive and effective global RM ecosystem, and for ensuring that the benefits of RM are accessible to diverse populations worldwide.

U. Ogbogu (✉) · L. Albrecht
Faculty of Law, University of Alberta, Edmonton, AB, Canada
e-mail: uogbogu@ualberta.ca; lauren.albrecht@ualberta.ca

U. Ogbogu et al. (eds.), *Regenerative Medicine and the Bioethics of Globalization, Justice and Hegemony*, Collaborative Bioethics 5,
https://doi.org/10.1007/978-3-032-07289-4_2

Keywords

Regenerative medicine · Epistemic injustice · Global health equity · Low- and middle-income countries · Regulatory frameworks · Biomedical innovation

Introduction

Regenerative medicine (RM) is an emerging field focused on developing therapies and products that restore or replace damaged or diseased human cells, genes, tissues, organs, or metabolic functions [1]. It encompasses a range of approaches, including cell therapies, gene therapies, and tissue-engineered products [1].

As RM enters the critical stages of clinical studies, product development, and marketing authorization, limited information exists about the global distribution of RM research and development (R&D) activities and associated governance frameworks [2]. Most of the available data pertain to high-income countries (HICs) [3–7]. The reasons for this gap are understudied but likely include an outsized emphasis on HICs, the concentration of financial and infrastructural opportunities in HICs, a hegemonic biomedical paradigm of "expertise, governance and values" driven by HICs [8], and the relative invisibility of R&D and related governance efforts in LMICs.

Whether due to these or other factors, poor knowledge of the scope of RM activities in LMICs or unfounded assumptions about its paucity can carry serious consequences for the global advancement of RM. Two main concerns stand out. First, these gaps may obscure opportunities to scale RM R&D in LMICs or to integrate LMIC perspectives and approaches into the field and its related governance activities. Second, they may constitute a form of epistemic injustice, a term used here to refer to the ways in which individuals or groups can be unfairly excluded from participation in knowledge production and recognition [9]. This second concern highlights broader issues of inequity, particularly in how burdens and benefits of cutting-edge health technologies are produced and distributed across the globe. This chapter focuses on the second concern: how the limited representation of LMICs in RM perpetuates epistemic injustice.

Biomedical and pharmaceutical R&D activities in LMICs are often under-documented, resulting in an incomplete picture of global innovation [4, 10, 11]. Multiple factors contribute to this under-documentation, including the disproportionate focus on HIC settings by major research institutions and funding agencies, the relatively limited local infrastructure available for tracking and publishing detailed data, and the challenges that LMIC-based researchers face in securing funding or publishing their work in high-impact venues [12–14]. Collectively, these issues perpetuate a skewed global narrative, making it difficult to recognize, represent, and fully integrate LMIC contributions in broader research and policy agendas.

This chapter seeks to address this under-documentation and the associated invisibility as they relate to RM R&D in LMICs. The primary aim of this chapter is to challenge the exclusion of LMICs from discussions on RM, a pattern often justified, in our experience as scholars in the field based in a HIC, by the assumption that little or no RM R&D occurs in these settings. This chapter's objectives are twofold: (1) to frame this exclusionary assumption as an instance of epistemic injustice, and (2) to challenge it empirically through a systematic review of RM R&D activity and governance frameworks in LMICs.

We focus on RM because it is at a pivotal translational stage, poised to deliver transformative health technologies in areas ranging from tissue engineering to advanced cell and gene therapies. Demonstrating that LMICs either have low levels of RM activity or, conversely, exhibit robust but under-recognized innovation has significant implications for epistemic justice and global health equity. If LMICs are in fact excluded from RM R&D and the resulting knowledge economies, it constitutes an inequity in global health innovation. Worse still, their exclusion on the basis of erroneous assumptions regarding knowledge deficits amplifies the injustice.

Epistemic Justice in Regenerative Medicine

The term "epistemic injustice" was popularized by Miranda Fricker in her work *Epistemic Injustice: Power and the Ethics of Knowing*. Epistemic injustice assumes that epistemic contribution, meaning the need to be heard and understood as a knower, is a central aspect of the human experience. Applying a "failure-first methodology," Fricker looks to understand justice by first examining injustice [15]. She distinguishes between "testimonial injustice," which occurs when prejudice causes a hearer to give a deflated level of credibility to a speaker's word, and "hermeneutical injustice," which arises from structural gaps in collective interpretive resources. In other words, certain groups can be systematically excluded from important epistemic practices, or their knowledge can be devalued because of biases embedded in social institutions.

Scholars such as José Medina [16] and Gaile Pohlhaus Jr [17] have further expanded these concepts to show how epistemic injustices reinforce broader patterns of oppression, marginalizing entire communities from knowledge-making processes, which can result in hermeneutical death—an extreme form of hermeneutical injustice that erases a group's capacity to be recognized as a knower. Boaventura de Sousa Santos coined the term "epistemicide" to articulate the destruction and erasure of traditional and indigenous sources of knowledge and the imposition of Western, Eurocentric epistemology [18]. He argues that global social justice is impossible without global cognitive justice, specifically the reclamation of epistemological diversity.

The field of global health assumes that health issues transcend territorial boundaries and the capacity of states to effectively address them in isolation [19]. However, Bhakuni and Abimbola [12] highlight common global health practices that can lead to or exacerbate epistemic injustice, resulting in the marginalization of valuable perspectives, particularly those from LMICs, indigenous populations, and other underrepresented groups. These unfair knowledge practices include relegating LMIC academics to providing samples or conducting fieldwork, not addressing power imbalances between researchers and participants in data collection methods, working through translators and across languages, not incorporating local knowledge or interpretive devices into evaluation techniques, misapplication/universal application of WEIRD (e.g., Western Educated Industrialized, Rich, Democratic) theories, not listing LMIC authors in privileged authorship positions, not prioritizing local audiences for the purpose of local learning, funding decisions, and more [12, 14, 20]. These forms of exclusion can create an unjust research cycle spawning policies and interventions that do not fully address the needs of diverse communities.

One framework for epistemic justice in global health ethics proposed by Pratt and de Vries [21], asks us to consider: (1) who produces ethics knowledge? (2) what theories and concepts are applied? and (3) whose voices are sought and used? Ensuring perspectives and experiences of people from underrepresented groups are actively sought and incorporated into each of these dimensions is pivotal to fostering a more inclusive and equitable approach to global health ethics.

Addressing structural epistemic injustice requires institutional reform that promotes integration, encourages participation, and prevents and corrects systemic biases, thereby fostering epistemic democracy and enabling equal participation in knowledge production [22].

In the context of RM, epistemic injustice arises when researchers, clinicians, and policymakers from LMICs are systematically overlooked or deemed irrelevant to global R&D and related policy discussions. Biases about perceived deficits in infrastructure or expertise often lead international agencies and funders to prioritize HIC-based initiatives, sidelining valuable insights from LMICs. Such exclusion not only distorts the collective understanding of RM's development and potential but also reinforces broader inequities in how scientific knowledge is valued and applied worldwide.

Addressing epistemic injustice in RM entails recognizing and documenting the full scope of scientific, technological, and governance contributions from LMICs. As Abimbola [23] wisely noted, "knowledge from the global South is in the global South." Rather than being absent or incomplete, knowledge from the Global South is rich, present, and well understood in the Global South and in peripheral spaces. Creating and strengthening platforms to propagate this knowledge broadly can help inform more equitable resource allocation, foster collaborative partnerships, and, ultimately, ensure that breakthroughs in RM benefit a truly global patient population.

The following sections underscore the need for a more inclusive account of RM activities by first briefly examining the global distribution of R&D, and then focusing more closely on developments in LMICs across Africa and Asia, including advancements in R&D activity and legal and ethical governance frameworks. Building on emerging scholarship that systematically documents RM R&D activities and governance frameworks in LMICs [4], our goal is to contribute to this limited but growing body of work and help bridge critical knowledge gaps that impede inclusive global policy development and equitable access to RM innovations [24].

The Global Distribution of Regenerative Medicine R&D

Available information indicates that the global RM landscape is dominated by HICs, particularly Australia and those in North America and Europe, as well as Japan and South Korea in Asia [3, 5]. In 2022, investment in RM clinical development was estimated at over USD 10.8 billion in North America, USD 3.2 billion in Europe, and USD 2.2 billion in the Asia-Pacific region [25]. By contrast, LMICs often face barriers such as limited financing, misaligned regulatory frameworks, and the dominance of HIC-based governance models [4, 24], all of which contribute to the under-documentation of both RM R&D activities and related governance frameworks in many parts of Africa, South America, and Southeast Asia.

Despite these obstacles, RM innovation in LMICs does not simply replicate the linear trajectory followed by HICs. Instead, many LMICs practice a type of "home-grown adaptation," innovating in ways that circumvent conventional developmental stages. This approach is evident when research targets diseases that disproportionately affect LMIC populations, conditions such as cerebral malaria, hepatitis B, and HIV/AIDS, where RM-based interventions have been comparatively neglected in global R&D pipelines [26]. Although non-communicable diseases dominate global RM investment, the unique epidemiological profiles of many LMICs present opportunities to advance regenerative interventions for communicable diseases, thereby addressing crucial global health needs [27].

Even in HICs, RM R&D faces significant hurdles, including fragmented intellectual property regimes, high manufacturing costs, and insufficient planning for affordability and patient access [28–30]. Although these challenges are well documented in HIC contexts, their impact on LMICs has received far less attention. Given that many LMICs lack the financial and regulatory support structures often taken for granted in HICs, it is safe to assume that these obstacles also affect the progress and sustainability of RM R&D in low-resource settings.

Still, understanding these dynamics outside of HIC contexts is indispensable for developing inclusive policies and ensuring that next-generation RM therapies, such as advanced cell and gene therapies, are accessible to diverse populations.

Against this background, the next section explores the landscape of RM R&D in selected LMICs in Africa and Asia, with particular attention to mid- to late-stage clinical research (Phases II–IV) and related downstream activities. The analysis is based primarily on a report completed by our team in March 2024 [31]. This report draws on data gathered through a mixed-methods approach that combined desk-based research with targeted expert consultation. Sources included publicly accessible online clinical trial registries (such as ClinicalTrials.gov, the Pan African Clinical Trials Registry, and others), regulatory and company websites, and a range of digital and paper-based primary and secondary literature. Supplementary insights were obtained through personal communications with subject matter experts and regulators via phone and email, which served to confirm existing findings and direct us to additional relevant materials. Data collection was conducted between January and February 2023 and focused exclusively on RM clinical trials and products in Phases II to IV, excluding those that had been withdrawn or terminated.

Regenerative Medicine Activity in Select LMICs in Africa and Asia

Africa

The need for RM therapies is acute in Africa, given high burdens of both communicable and non-communicable diseases, as well as low overall ratings on healthcare quality indices. African LMICs are emerging as contributors to RM research and clinical trials. Egypt, South Africa, and Tunisia stand out for their growing engagements in stem cell therapies, tissue engineering, and related areas.

Pharmaceutical manufacturing on the African continent traces back to the late nineteenth century, driven initially by multinational subsidiaries [32, 33]. Today, most LMICs in Africa have laws to certify drugs for public safety [34], but regulatory capacities and institutional strengths vary considerably [33]. This unevenness also affects more complex therapies, including RM, where infrastructure deficits and limited human capital remain significant bottlenecks [35, 36].

Some nations have achieved relatively high regulatory maturity. Egypt and Nigeria, for example, have gained World Health Organization (WHO) accreditation levels that signify alignment with international standards in areas like vaccine or medicines regulation [34, 37]. Yet such progress often does not extend to RM-specific protocols, leaving many advanced therapies regulated under the same framework as conventional drugs. The African Medicines Regulatory Harmonization (AMRH) initiative seeks to encourage national regulatory authorities to share technical requirements, conduct joint inspections, and streamline market authorization processes, potentially benefiting future RM product oversight [38].

African LMICs often face constraints in infrastructure, specialized skill sets, and preclinical capacity, resulting in an environment where RM research can quickly stall unless bolstered by foreign investment or cross-border collaborations [35, 39]. Despite incremental reforms, unproven stem cell clinics have appeared in countries like South Africa, Nigeria, Uganda, and Egypt, exploiting regulatory gaps [40, 41]. These developments emphasize the need for clearer frameworks that explicitly address cell-based interventions and gene therapies.

Addressing these constraints will require not just regulatory reforms, but also capacity-building initiatives, international collaborations, training, and infrastructure support that ensure African researchers are integrated into global RM networks. Commercialization pathways and comprehensive regulatory frameworks remain unevenly developed, which hinders broader adoption. Local manufacturing facilities, ideally GMP-certified, can help anchor production, reduce costs, and make future RM products more widely available across the continent.

Egypt

Egypt has become a regional leader in RM. Our mapping efforts identified over 20 RM clinical trials in Phases II through IV, by far the largest cluster in Africa [31]. Public universities are the primary sponsors, with a smaller subset led by private clinics, hospitals, and governmental organizations. Many of Egypt's trials focus on platelet-rich fibrin (PRF), an inexpensive, minimally manipulated therapy often used to address dental and maxillofacial conditions such as periodontal disease, pulp necrosis, and gingival pigmentation [42]. Other investigations target conditions ranging from premature ovarian failure to azoospermia and retinoblastoma.

Notably, Egypt stands out in having several Phase IV trials, suggesting that some RM therapies might be poised for market entry in the near future [43]. Five of these late-stage trials use PRF in dental or craniofacial applications, an area in which local academic expertise appears especially strong. One Phase IV trial explores adipose-derived stem cells, which require large-scale manipulation and care during their preparation, banking, and manufacturing, for the treatment of androgenetic alopecia [44]. None has yet conclusively reported results or obtained clear commercialization pathways, so it remains to be seen whether these trials will translate into marketed RM products [43].

At present, Egypt does not have an established GMP system specific to RM or advanced therapies. It remains unclear how new products, if proven safe and effective, will move from trial to market [45, 46].

The Egypt Center for Research and Regenerative Medicine (ECRRM) in Cairo could help address these downstream challenges. Officially tasked with fostering RM R&D, it may offer a much-needed nexus between university-led clinical studies and the supportive infrastructure needed for commercialization [47]. However, ECRRM's actual role in ongoing trials has not been publicized.

More broadly, the absence of formal regulatory guidelines for cell and gene therapies means that even successful trial results risk stalling before reaching full market adoption. Egypt exemplifies the gap between high-level regulatory capacity and incomplete RM governance. While the Egyptian Medical Syndicate permits research only on adult stem cells [40, 48], there have been recent efforts to expand domestic infrastructure and oversight. In 2022, the Egyptian Armed Forces signed an agreement with the US-based company CEMEX to establish the International Center for Research and Treatment of Stem Cells. The center, developed in partnership with the Armed Forces' Medical Research and Regenerative Medicine Center, aims to support scientific research, clinical translation, and advanced medical services in line with Egypt's broader health system goals. It will include a Research and Development Center, a Clinical Trials Unit, and a Biomaterials Production Facility, and is positioned as part of Egypt's vision for a more integrated and modern healthcare system [46, 49]. Nonetheless, the broader regulatory framework for advanced therapies remains limited.

The passage in 2020 of Law No. 214 Regulating Clinical Medical Research introduced prohibitions on commercializing human samples, an obstacle for developing and marketing stem cell and gene therapies [46]. In practice, RM products exist in a gray area, as they are neither fully integrated into conventional drug pathways nor subject to distinct regulations. Despite Egypt's relatively mature drug regulation system in other domains, the slow formulation of specific RM guidelines means domestic researchers and international partners face persistent uncertainties about how or when novel therapies can move from clinical studies to market. If Egypt can develop a coherent regulatory framework, including GMP-certified facilities, its strong academic base and growing pool of trained personnel could translate into meaningful clinical uptake of RM therapies in the near future.

South Africa

South Africa's participation in RM clinical trials, often as part of multinational or pharmaceutical-sponsored initiatives, has benefited from a relatively strong clinical research infrastructure. While South Africa hosts the continent's largest number of registered trials overall, fewer than five mid- to late-stage RM trials have been documented, indicating that RM remains a relatively small part of the country's broader research portfolio [31].

Most RM trials identified in South Africa focus on hemophilia (A and B) or spinal muscular atrophy and are led by major multinational pharmaceutical companies. Sanofi has sponsored multiple studies on fitusiran (an RNA-targeting therapy) for hemophilia, and Novartis has tested its onasemnogene abeparvovec (a gene therapy) for pediatric spinal muscular atrophy. These studies reflect South Africa's attractiveness as a trial site, thanks to a strong clinical research infrastructure. However, local universities, biotech ventures, or public research hospitals appear less active in sponsoring advanced RM trials compared with Egypt.

Although South Africa's regulatory framework for conventional drugs is relatively robust, it lacks specific regulations tailored to complex cell- and gene-based therapies. South Africa's national regulator, the South African Health Products Regulatory Authority (SAHPRA), has a demonstrated commitment to ensuring drug safety across various therapeutic categories [50]. However, current legislation, anchored by the Medicines and Related Substances Control Act of 1965, does not clearly categorize RM therapies, creating legal ambiguity for RM interventions [51].

SAHPRA provides guidance on biological medicines, encompassing products such as vaccines, blood components, biotherapeutics, biosimilars, and recombinant proteins [52]. However, the extent to which SAHPRA's guidance comprehensively covers the full spectrum of RM products, including cell therapies, gene therapies, and tissue-engineered products, remains somewhat ambiguous [51]. While SAHPRA has mechanisms, such as recognition pathways, to evaluate and register industrially manufactured gene- and cell-based therapies, there is a recognized need for a more defined and comprehensive regulatory framework tailored specifically to the unique characteristics of RM products [51, 53]. This includes considerations for their classification, manufacturing processes, and clinical applications [51].

Given local demand for cutting-edge treatments, particularly for serious or terminal conditions, scholars have recommended establishing a dedicated cell and gene therapy society, enacting standards for cell-processing facilities, and designing an authorization pathway akin to those for pharmaceuticals [54]. Such measures could help South Africa more effectively oversee RM research and curb unauthorized clinical use or marketing of unapproved RM products or therapies. At present, many stakeholders rely on broad pharmaceutical guidelines that often do not capture the technical complexities of RM products.

As advanced RM products progress to later phases, new rules may be required to address unique considerations such as long-term follow-up, high manufacturing costs, and specialized distribution logistics. Given that multinational sponsors are already active in the country, South Africa may be well-positioned to adapt existing processes and expand them to RM, provided the government and industry stakeholders see sufficient impetus to do so.

Tunisia

Tunisia has a modest overall volume of clinical research with only one RM trial identified [31]. Public registries list a now-completed Phase III study investigating autologous peripheral blood stem cells for multiple myeloma. Sponsored by the National Bone Marrow Transplant Center, this trial concluded years ago without evident follow-up or progression toward a Phase IV investigation [31].

Some RM applications, such as hematopoietic stem cell transplantation, have been integrated into clinical care, with public support for associated treatments [55]. However, comprehensive clinical documentation and reporting on the use and outcomes of advanced therapies remain limited. Observed complications (e.g., bronchiolitis obliterans after certain stem cell transplants) raise concerns about oversight and highlight a need for clearer regulation [56, 57].

The lack of dedicated RM regulatory pathways in Tunisia and inadequate financing for clinical research impede broader engagement with RM. In Tunisia, the regulation and registration of pharmaceuticals fall under the authority of the Directorate of Pharmacy and Medicines (Direction du Médicament et de la Pharmacie), operating within the Ministry of Health [58]. This body oversees marketing authorizations for both imported and domestically manufactured products. While existing regulatory guidelines provide a general framework for biologics, they do not yet establish a distinct category for RM therapies and interventions, such as cell- and gene-based therapies [59, 60]. As a result, novel biologics, including RM interventions already approved elsewhere, are reviewed under conventional procedures that may not adequately address their unique scientific and clinical characteristics. This regulatory gap presents challenges for the evaluation, authorization, and integration of emerging RM therapies into the health system.

Given that Tunisia also aims to stimulate broader research and manufacturing, bridging this regulatory gap may prove crucial to avoid ad hoc or ambiguous approvals.

Uganda

Uganda also has a modest volume of clinical research with a single Phase II RM trial involving autologous adipose-derived stem cells for treating keloids, associated with Makerere University [31]. This institution has participated in numerous other clinical studies, suggesting potential local capacity for advanced research. However, the specific status of the keloid therapy trial is unknown, and no additional RM projects have been publicly announced.

Uganda's institutional framework for drug regulation involves multiple entities, such as the Uganda National Bureau of Standards and the National Drug Authority, but does not feature explicit RM guidelines [61, 62]. Clinical trials of any kind require NDA authorization, ethics review, and a permit from the Uganda National

Council for Science and Technology. While these layers can ensure fundamental safety checks, they do not directly address the unique manufacturing, traceability, or long-term follow-up demands typical of regenerative therapies [63].

Most Ugandan drug regulations focus on conventional pharmaceuticals, leaving advanced cell or gene therapies in a regulatory void. Any potential RM product, whether locally developed or imported, faces uncertainty regarding classification, data requirements, and marketing pathways. Without further regulations, Uganda may lack the tools to evaluate novel therapies effectively or prevent the proliferation of unproven interventions.

Asia

Shifting attention to Asia, a different picture emerges. Four LMICs, China, India, Malaysia, and Turkey, collectively host over 100 mid- to late-stage RM clinical trials, a number that outpaces the African count [31]. In our survey, China and India each account for 44 trials, reflecting governmental prioritization of biotech innovation and growing rosters of homegrown biotech enterprises [31]. Notably, a majority are in Phase II (over 70% by some estimates), suggesting that while interest is high, the field is still working to confirm early-stage safety and efficacy before moving on to widespread commercialization.

Government-led programs, evolving regulatory oversight, and local manufacturing have helped these countries advance, though concerns about affordability, ethics, and unequal access persist. Regulators in Asian LMICs increasingly recognize the need for specialized guidelines for cell and gene therapies, aligning with the WHO or International Council for Harmonisation of Technical Requirements for Pharmaceuticals for Human Use (ICH) standards, to ensure that the expansion of RM proceeds in an ethically and scientifically robust manner. Biotech companies, many of which are local or regionally based, appear well-placed to accelerate clinical translation, but must also commit to responsible pricing and access strategies for local populations. Even so, these countries are far from uniform in their capacity, regulatory frameworks, or commercial readiness.

China

China's "Healthy China 2030" initiative is part of a push to modernize healthcare and broaden access. It has also heightened the emphasis on rapid approval and commercialization of innovative therapies, including RM [64, 65]. China leads the Asian LMICs in both the scale and scope of RM activities, propelled by substantial government funding, reportedly around 1.4 billion USD, and a growing network of GMP-certified facilities. Over forty mid- to late-stage trials have been reported, most of which target conditions such as hematological disorders (hemophilia, beta thalassemia), cancers of various types, and rare genetic diseases [31, 66, 67]. China's progress is further evidenced by the two Phase IV trials for rAd-p53 gene therapy, sponsored by Shenzhen SiBiono GeneTech [68, 69]. Although the status of both trials has not been updated in some time, Shenzhen SiBiono was responsible for Gendicine, the world's first approved gene therapy for cancer.

Regulatory frameworks continue to evolve to align with international standards and encourage domestic innovation, driving a marked increase in local approvals over the last decade. The National Medical Products Administration plays a central role in assessing trial applications and awarding market approvals. In addition to conditional approvals and expedited reviews for priority products, China now recognizes the potential of "real world evidence" to accelerate licensing decisions [67, 70].

Amid these expansions, concerns persist about the proliferation of unauthorized stem cell interventions, an issue magnified during early stages of the COVID-19 pandemic, when unproven RM approaches were publicly promoted as quick fixes [71]. China's recent regulations restricting the online sale of high-risk medical items underscore a growing awareness of potential abuse [72].

Taken together, these developments signal that China is balancing a drive for rapid innovation with the recognition that advanced cell and gene therapies pose unique safety and ethical complexities.

India

India matches China in the number of mid- to late-phase RM clinical trials, with many of India's trials focused on therapies for diabetes, liver cirrhosis, multiple neurodegenerative diseases, and other conditions [31, 73]. A variety of sponsors, including private hospitals, biotechnology ventures, contract research organizations, and a handful of large pharmaceutical companies, reflect the country's diverse R&D ecosystem [73]. While India has one identified Phase IV trial (an autologous mesenchymal stem cell infusion for liver cirrhosis), no results have been published at the time of writing this chapter, and the manufacturing requirements remain unclear [74]. However, Indian medical and research organizations are forming strategic partnerships to gain access to manufacturing and clinical capabilities not available locally, potentially paving the way for a smoother path to commercialization. For example, AIG Hospitals in India has entered a strategic collaboration with the Mayo Clinic Care Network in the US to access advanced clinical expertise, decision-support tools, and professional development opportunities not readily available locally—an initiative that may strengthen clinical capabilities and facilitate a more streamlined path to advanced care delivery and potential commercialization of biomedical innovations [75].

While India does not offer a specific legal definition for "regenerative medicine," cell- and gene-based therapies typically fall under the pharmaceutical category [76]. Oversight is primarily handled by the Central Drugs Standard Control Organization, with specialized committees like the Gene Therapy Advisory and Evaluation Committee contributing expert guidance [73, 77].

Several initiatives illustrate India's evolving approach. The 2019 National Guidelines for Gene Therapy Product Development and Clinical Trials clarify requirements for gene therapies, while an accelerated approval process allows conditional or fast-track approvals for products addressing urgent medical needs

[78–80]. These frameworks aim to spur innovation while mitigating the risks of unauthorized stem cell clinics, several of which reportedly market secretome-based treatments directly to patients [81].

India is also in the midst of updating its GMP regulations to align more closely with global expectations. Coupled with incoming private investments (e.g., from StemCures) to build large-scale stem cell manufacturing labs, these measures could lower production barriers and accelerate the arrival of new RM products on the Indian market. Observers caution that while Indian regulators are evolving their GMP standards and accelerated-approval pathways, the actual deployment of these therapies may hinge on how quickly local manufacturing facilities can meet more stringent guidelines [78, 80].

Malaysia

Malaysia hosts a smaller but still significant cluster of mid-stage RM clinical trials [31]. Local biotech companies drive most of these efforts, focusing on treatments for osteoarthritis, hematological conditions, immune-related disorders, and certain genetic diseases.

In 2016, Malaysia's National Pharmaceutical Regulatory Agency (NPRA) introduced the Guidance Document and Guidelines for Registration of Cell and Gene Therapy Products [82]. Despite being framed as a comprehensive roadmap, these guidelines treat RM similarly to other biologics, utilizing standard Investigational New Drug procedures. Lower-risk products (minimal manipulation, homologous use) may circumvent extensive pre-market review, creating gray zones when classifying more complex cell-based therapies [83].

Malaysia is also a well-known destination for stem cell tourism, where private hospitals and clinics offer a range of stem cell "treatments," some of which do not meet robust international scientific or regulatory standards, to international patients seeking therapies not readily available in their home countries [84, 85]. Companies like Cytopeutics Sdn. Bhd. have begun to integrate Shariah compliance into their manufacturing and clinical processes. This positions them to serve Muslim-majority populations domestically and from jurisdictions where religious considerations are central to healthcare decisions. The influx of foreign patients from wealthier countries exerts upward pressure on prices, often rendering RM interventions inaccessible to local populations. Government support for Malaysia's stem cell industry has improved the quality of private clinical facilities, but the unequal distribution of resources remains a concern.

Stakeholder engagement in developing RM regulations has been described as partial. Private clinics offering direct-to-consumer therapies and their patient clientele were not fully included [86]. Meanwhile, calls for a more enforceable legal structure persist, as many RM guidelines lack binding force [87]. Given Malaysia's burgeoning medical tourism sector, clearer and more robust regulatory authority could help distinguish evidence-based RM from the unregulated services that risk tarnishing the country's reputation and endangering patients.

Turkey

Turkey has a notable pipeline of mid- to late-stage RM clinical trials, most focusing on mesenchymal stem cell-based therapies for conditions like graft-versus-host disease, inherited retinal dystrophies, liver cirrhosis, and hemophilia [31]. Public universities remain highly active, and partnerships with global pharmaceutical companies help finance and guide trials through advanced phases.

Turkey's broader pharmaceutical regulation resides under the Turkish Medicines and Medical Devices Agency, which has introduced measures for advanced therapy medicinal products but remains somewhat inconsistent regarding stem cell and embryonic research [88]. Historically, embryonic stem cell work was restricted [89, 90], while adult stem cell research was permitted within specific parameters. In 2021, new rules tightened oversight, especially for gene therapy products that rely on human autologous, allogeneic, or xenogeneic systems [91]. Recent updates also mandate more rigorous risk management and data quality for medical devices [92], though it remains unclear how thoroughly these changes apply to RM therapies. Practical enforcement of these regulations remains a work in progress [91, 92].

Turkey also sees a fair amount of stem cell tourism, similar to Malaysia. Research capacity is bolstered by institutions like the Turkish Academy of Sciences (which previously convened a stem cell Working Group) and by GMP-certified facilities such as the Regenerative Therapy Centre. With stronger clinical infrastructure than many LMICs and an active domestic biotech sector, Turkey may be well-prepared to translate mid-stage trials into commercial products, assuming regulatory and ethical standards keep pace. Clarity on long-term affordability and equitable access remains limited, especially in light of continued interest from international "medical tourists" seeking quicker or less-regulated treatments.

Observations on RM R&D Activity in African and Asian LMICs

In LMIC contexts, including those discussed above, RM therapies continue to evolve, revealing a landscape that, though still fragmented, shows notable pockets of innovative clinical trials. In Africa, mid- to late-stage RM trials are spread across a handful of countries. Many of these trials focus on either autologous stem cell therapies or platelet-rich products, and they are typically sponsored by public universities and hospitals rather than big pharmaceutical companies. This reliance on academic and public funding underscores a key driver of RM in Africa: local researchers tackling diseases and conditions of immediate concern, including certain dental disorders, metabolic diseases, and specific malignancies [42, 43].

What stands out most in the African context is how unevenly these clinical trials are distributed. A large majority originate in Egypt, but despite this relative concentration of efforts, many of Egypt's trials must still navigate a vague regulatory environment. Elsewhere in Africa, the overall count of RM studies is modest. Countries like South Africa do host large volumes of clinical research, yet only a few mid- to late-stage RM trials have been identified.

Despite the limited number of trials and uneven spread, the overall direction is one of growth and diversification. University-led or hospital-led trials in Africa are increasingly joined by biotech-sponsored ventures, suggesting that private-sector engagement could become a bigger driving force.

In Asia, collaborative efforts between public institutes and multinational companies have the potential to accelerate mid- to late-stage trials into mainstream care, provided ongoing regulatory reforms continue to encourage robust quality checks [27].

The relatively small number of Phase IV trials that exist in LMICs underscores how rarely RM research progresses beyond early-stage clinical findings into actual commercial availability. While some projects, such as those in Egypt or India, suggest that novel therapies might soon reach the market, these trials confront practical hurdles in scaling up manufacturing and distribution. A few countries, including Egypt, China, and India, show more visible near-market activities, indicating potential for new locally developed therapies. Others, such as South Africa, Malaysia, or Turkey, have well-established clinical research ecosystems but grapple with gaps in advanced therapy legislation, cost barriers, or the predominance of foreign sponsors. Nations like Tunisia and Uganda, with very limited RM pipelines, could benefit from targeted capacity building and policy support to move promising early-phase work into more advanced phases. Viewed together, these examples illustrate the growing involvement of LMICs in RM research, but also highlight substantial variability in regulatory, infrastructural, and economic contexts.

Many of the challenges to translating RM therapies into clinical practice and commercial success (high costs, regulatory complexity, lack of manufacturing capacity, and scarcity of trained personnel) are not unique to LMICs. HICs face similar barriers, though they often have more resources to address them. LMICs can learn from these experiences and design more efficient or context-appropriate models for RM development.

Key strategies include:

1. Training and collaboration: Building scientific expertise through local and international training programs, collaborative research networks, and partnerships with established biotech firms.
2. Strengthening regulation: Developing clear, transparent GMP requirements tailored to RM, along with robust ethical and safety oversight.
3. Scaling manufacturing: Investing in GMP-compliant manufacturing sites, possibly through public-private partnerships, to produce and distribute RM products at lower cost.
4. Computational modeling: Where laboratory resources remain limited, computational approaches can help refine manufacturing processes and assess economic feasibility before building large-scale facilities.
5. Equitable access: Ensuring local populations can benefit from emerging RM therapies by controlling price inflation, especially in nations with thriving medical tourism sectors.

RM Governance and Regulatory Frameworks in LMICs

Regulatory systems for RM in LMICs often mirror broader challenges in governing advanced health technologies. Four interrelated factors shape how RM products are regulated across different jurisdictions:

1. *Characteristics and Origin of the RM Product*
   The source (i.e., human, animal, synthetic) and the methods used to modify or create a RM product influence its regulatory path [7]. Because many LMICs have historically focused on conventional pharmaceuticals, policies for more complex biological therapies, including cell, gene, or tissue-engineered products, may be incomplete [66, 73].
2. *Unique or Distinctive Features*
   Each RM product has specific mechanisms of action, safety considerations, and intended uses. Regulatory bodies must determine whether such products fall under existing drug or device frameworks or require dedicated pathways. Moreover, the status of an RM product, whether in clinical trials, pending approval, or already authorized elsewhere, can influence local perceptions of its credibility and the adequacy of regulatory oversight. [93–95].
3. *Presence (or Absence) of a Specific RM Regulatory Framework*
   Some LMICs have enacted explicit RM regulations (e.g., detailed guidelines for stem cell or gene therapy), whereas others continue to regulate these products under more general pharmaceutical or medical device rules [96]. The clarity of the legal framework can critically influence both legitimate development and the emergence of unregulated stem cell clinics.
4. *Broader Social and Societal Context*
   Cultural, religious, and ethical dimensions can influence the acceptance of advanced therapies like RM. In some Middle Eastern countries, for example, religious beliefs have impacted the uptake of assisted reproductive technologies, an area overlapping with RM techniques [97, 98]. In Turkey and Egypt, cultural norms and paternalistic medical traditions further complicate how regulators approach clinical innovations [99, 100].

Generally, regulatory landscapes for RM clinical research and marketing authorization in LMICs show a patchwork of frameworks designed for traditional drug approval, anchored by laws introduced in the twentieth century to manage chemical or small-molecule medicines rather than highly specialized biotherapies [73, 101]. Although the specifics vary across jurisdictions, a common theme is the tension between established regulatory norms for conventional pharmaceuticals and the emerging demands of RM products, such as cell- and gene-based therapies.

Where specialized RM guidelines do exist, they frequently borrow from frameworks established for biologics or medical devices. These guidelines often set out classification tiers for low-risk, minimally manipulated products versus high-risk products requiring more stringent review [83]. In principle, such risk-based approaches can streamline approvals for simpler therapies, such as platelet-rich plasma, while applying heightened scrutiny to gene-edited cells or extensively engineered tissues). In practice, however, many LMIC regulators struggle to define key boundaries, like what constitutes "minimal manipulation" or "homologous use." This ambiguity can lead to inconsistent enforcement [86, 87].

Although many LMICs have instituted robust controls for conventional pharmaceuticals [34], RM products often fall under regulatory gray zones due to at least four distinct factors: the biological attributes of each RM therapy, the therapy's unique mechanisms or risk profiles, the presence or absence of specialized frameworks, and local societal contexts [7, 96, 97]. The reliance on existing rules can limit how effectively regulators address RM particularities.

Compounding this issue is the fact that biopharmaceutical capacities and legal infrastructures vary widely across LMICs [33]. In parts of Africa, for example, countries with relatively strong institutional maturity, such as Egypt or Nigeria, have implemented protocols that align in important respects with international requirements [34, 37]. National or regional initiatives, such as AMRH, encourage reliance on shared technical standards, collective inspections, and joint marketing reviews [38]. These initiatives, alongside the WHO's benchmarking for medical products, could provide templates for standardizing some aspects of RM oversight and help smaller regulatory bodies tap into broader expertise and reduce duplicative review processes for RM products. Even so, the actual monitoring of advanced regenerative therapies may remain under-specified, often because legislation has not caught up to their technical complexities [40, 41].

In Asia, various pathways, such as conditional or accelerated approval pathways, have emerged where regulators see potential for high clinical impact and an acceptable risk–benefit trade-off [78, 80]. This indicates a willingness to foster innovation, although close monitoring is necessary to safeguard patient welfare. Where dedicated RM regulations remain lacking, local researchers and foreign sponsors alike face uncertainty about trial approvals, marketing channels, and compliance requirements.

The cultural and societal environment shapes the pace and acceptance of RM regulation. Religious or ethical considerations surrounding the origin of cells, and broader historical hesitations about embryo manipulation, can influence whether a given jurisdiction permits extensive stem cell research or only narrowly sanctioned protocols [98, 100]. In some locales, paternalistic norms in physician–patient interactions complicate efforts to strengthen patient rights and informed consent for novel therapies [99]. Layered onto these cultural dimensions is the fact that many LMICs lack public funding to build and staff dedicated oversight bodies, leaving their general drug authorities to juggle RM products alongside vaccines, antibiotics, and other urgent priorities [35, 102].

Ultimately, establishing effective governance for RM in LMICs is not merely a technical exercise. Policymakers must accommodate new scientific paradigms, address entrenched infrastructural gaps, and incorporate cultural and ethical sensitivities. As LMICs modernize their frameworks to accommodate advanced therapies and expand their capacity to regulate RM, they have an opportunity to craft inclusive frameworks that support equitable access, while also guarding against exploitation, misinformation, and lax oversight that too often accompany rapidly developing fields.

Another prominent shift is the increasing emphasis on local manufacturing, especially for cell-based products, which often cannot be easily imported due to viability constraints [103]. Some nations, aware of the need for GMP-compliant facilities, have begun incentivizing public–private partnerships to construct specialized labs and production centers. Others rely on external partnerships or multinational sponsors, who already possess the necessary know-how and infrastructure. Where manufacturing does take place locally, regulators must also adapt to new modes of oversight, ranging from real-time supervision of cell-processing labs to clarifying how intellectual property rights intersect with local research activity [104].

Across all LMIC settings, a recurring challenge is the lack of comprehensive frameworks for manufacturing (GMP), long-term safety monitoring, and intellectual property, which can hamper commercialization even for trials that successfully reach later phases [44]. In parallel, the business models required for distributing advanced therapies, especially autologous or gene-based interventions, are capital-intensive and hinge on specialized manufacturing pipelines that are not yet widespread in many LMICs [39]. At the same time, the relative novelty of RM allows these countries to draw lessons from HIC settings, where developers have struggled with financial viability, scalability, and equitable access, and to adapt infrastructure or policies from the outset. If LMICs can align supportive regulatory reforms with growing scientific expertise, the RM therapies on the horizon in each locale may well expand global access to advanced cell and gene therapies that benefit both local and international patient populations.

One common challenge across both Africa and Asia is the coexistence of legitimate RM research with the rapid proliferation of direct-to-consumer clinics marketing unverified stem cell-based "therapies" [93, 94]. In locales where regulatory clarity is patchy, unscrupulous operators can market high-cost, experimental interventions to local or international patients without providing rigorous evidence of safety or efficacy [93, 94]. As China, India, Malaysia, and Turkey expand their RM markets, authorities will have to grapple with how to deter misleading claims while still promoting legitimate R&D [71, 72, 105]. For instance, sponsors might claim that a treatment in "Phase II or III" overseas is already proven, obfuscating differences in the local regulatory approval process, undermining public trust, and complicating efforts of legitimate researchers [95]. Resolving this tension often involves a combination of better public outreach, stricter online advertising rules, and explicit sanctions for clinics that flout ethical and scientific norms [95, 106].

Reflections on Implications of RM Contexts in LMICs for Epistemic Justice

The documented patterns of RM development across LMICs in Africa and Asia reveal more than technical disparities in innovation capacity. They reflect entrenched hierarchies in global knowledge production that routinely sideline LMIC contributions, even when those contributions are substantive and impactful. Although RM and biomedical R&D face significant challenges in LMICs, these difficulties should not be mistaken for an absence of innovation, expertise, or normative engagement with RM. The under-recognition of LMIC research in global RM discourse not only limits access to collaboration and funding but also perpetuates hermeneutical injustice. LMICs are often framed as lacking innovation, not due to actual deficiencies, but because global systems of recognition are oriented toward the standards and infrastructure in HICs.

This epistemic marginalization appears in how RM research is documented, interpreted, and circulated. Global mechanisms of trial registration, academic publication, and regulatory recognition are largely shaped by institutions based in the Global North [107–110]. As a result, RM activities in LMICs often remain invisible or are devalued unless channeled through HIC-affiliated networks. For instance, the scale of Egypt's Phase IV RM trials and India's growing network of GMP-certified facilities have not fundamentally altered international perceptions of where expertise resides. This invisibility is not neutral. It limits access to resources, shapes research agendas, and contributes to a distorted understanding of where innovation is occurring and what problems are prioritized.

The problem is not only one of visibility, but also of structural control over the terms of participation. LMIC regulatory frameworks, even when robust or aligned with international standards, are often viewed as inadequate unless they fully replicate HIC models [109, 111–113]. This amounts to a form of "governance epistemicide," whereby local regulatory knowledge, priorities, and institutional practices in LMICs are systematically denied normative value and legitimacy through the imposition of HIC models as the default standard. Even when LMIC approaches are contextually effective or tailored to local needs, they are often disregarded unless they mirror dominant global frameworks, reinforcing structural inequities in how regulatory authority and expertise are recognized. Along these lines, LMIC researchers are frequently restricted to implementing roles in international collaborations, rather than being recognized as epistemic leaders [114]. These constraints undermine the development of context-specific RM ecosystems and discourage sustained investment in locally defined research priorities.

Achieving epistemic justice in RM requires more than symbolic inclusion. It demands reform of the institutions, funding structures, and evaluative norms that currently disadvantage LMIC actors. Governance and ethics review frameworks should support co-leadership models that center LMIC expertise. International agencies and research funders must prioritize long-term investment in LMIC institutions, rather than short-term partnerships that leave little lasting infrastructure. Likewise, scientific journals, editorial boards, and conferences must adopt explicit policies to elevate voices from underrepresented settings and recognize diverse forms of scientific contribution.

More inclusive global participation in RM is not only an ethical imperative, but a practical necessity. Many of the conditions that RM seeks to address, from sickle cell anemia to liver fibrosis, disproportionately affect LMIC populations. Research solutions that are affordable, logistically feasible, and culturally appropriate are most likely to emerge from within these settings. Elevating LMIC contributions is essential to ensuring that RM develops as a field grounded in global equity and scientific pluralism, rather than repeating patterns of extractive innovation. Only through intentional epistemic inclusion can RM fulfill its promise to deliver meaningful and just health interventions worldwide.

As LMICs continue to contribute to RM innovation, it is imperative to acknowledge their efforts and integrate them into the global scientific discourse. Doing so is key to mitigating epistemic injustice in RM.

Conclusions and Looking Forward

In sum, while African and Asian LMICs each have unique barriers and opportunities, they share a common need to integrate scientific, regulatory, and commercial infrastructures in ways that mitigate epistemic injustice and ensure the equitable development of RM. Moving forward, multiple stakeholders, governments, research institutions, patient groups, religious leaders, and industry will likely continue negotiating the contours of RM oversight. Collaborative initiatives that engage local health researchers and communities may help address ethical complexities and cultural attitudes toward embryo use, genetically modified tissues, or transplanted biomaterials [97]. Pragmatic measures, such as robust data-sharing platforms, joint review processes, and internationally harmonized definitions, can reduce the administrative burden on LMIC regulators. Meanwhile, concerted efforts to align with high-level global frameworks (for instance, the WHO's maturity-level benchmarking or the ICH guidelines) can ensure that LMIC-produced RM therapies meet quality and safety benchmarks recognized worldwide [34, 70].

Ultimately, it is clear that LMICs have begun to stake out distinctive niches in RM, targeting diseases prevalent in their populations while also aligning with global research trends in cancer, orthopedics, ophthalmology, reproductive medicine, and beyond. Yet the governance of RM in LMICs is not simply a technical undertaking. It reflects an interplay between cutting-edge science, healthcare inequities, cultural values, and economic considerations. The pace of progress hinges on addressing infrastructural deficits, bridging regulatory gaps, and, crucially, mitigating epistemic injustice by ensuring accurate recognition of these activities in international policy and discourse. If these hurdles can be addressed, the growing body of RM in LMICs may well enrich the worldwide pipeline of next-generation therapies, potentially making them more accessible, more affordable, and more responsive to the varied healthcare challenges faced by patients outside the usual high-income research corridors, enabling LMICs to be central, rather than peripheral, participants in a rapidly evolving global field.

To fully realize this promise, global actors must go beyond mere technical assistance and engage in genuine co-creation of knowledge, infrastructure, and standards. Institutional funding models should be recalibrated to prioritize LMIC-led initiatives, with mechanisms that explicitly value context-sensitive innovation, local health priorities, and capacity-building. Academic and policy discourses on RM must broaden their evaluative lenses to include indicators of success that resonate with LMIC realities: public-sector leadership in clinical trials, locally relevant disease focus, incremental regulatory development, and ethical responsiveness to cultural norms. In doing so, the field can move toward a model of distributed innovation that is both scientifically rigorous and socially just.

Furthermore, LMICs themselves must be empowered to define what success in RM looks like. This includes the authority to shape ethical frameworks, set research priorities, and determine acceptable trade-offs between innovation speed, access, and safety. Such autonomy is critical not only for epistemic justice but also for the sustainability of RM ecosystems that serve local populations. Countries such as Egypt, India, and Malaysia illustrate that with deliberate investment in local expertise, targeted regulatory development, and protection from exploitative market forces, LMICs can chart their own course in this domain. Their experiences offer valuable lessons for other jurisdictions seeking to balance innovation with inclusion.

Looking ahead, the global RM community should embrace a pluriversal model, one that recognizes multiple centers of innovation and governance rather than a single dominant paradigm. This shift requires a reimagination of scientific collaboration and regulatory harmonization, grounded in mutual respect and epistemic humility. It also entails the proactive dismantling of barriers that have historically excluded LMICs from full participation in RM. By supporting frameworks that value diverse knowledge systems and promoting meaningful representation in global RM governance, the field can fulfill its potential as a truly transformative and equitable domain of medical science.

References

1. Altyar AE, El-Sayed A, Abdeen A, Piscopo M, Mousa SA, Najda A, Abdel-Daim MM (2023) Future regenerative medicine developments and their therapeutic applications. Biomed Pharmacother 158:114131. https://doi.org/10.1016/j.biopha.2022.114131
2. Greenwood HL, Singer PA, Downey GP, Martin DK, Thorsteinsdottir H, Daar AS (2006) Regenerative medicine and the developing world. PLoS Med 3(9):e381. https://doi.org/10.1371/journal.pmed.0030381
3. Beheshtizadeh N, Gharibshahian M, Pazhouhnia Z et al (2022) Commercialization and regulation of regenerative medicine products: promises, advances and challenges. Biomed Pharmacother 153:113431. https://doi.org/10.1016/j.biopha.2022.113431
4. Doxzen KW, Adair JE, Bazzo YMF et al (2024) The translational gap for gene therapies in low- and middle-income countries. Sci Transl Med 16(746). https://doi.org/10.1126/scitranslmed.adn1902

5. GlobalData (2021) Australia's regenerative medicine global pipeline tracker. https://www.ausbiotech.org/documents/item/673. Accessed 5 May 2025
6. Li MD, Atkins H, Bubela T (2013) The global landscape of stem cell clinical trials. Regen Med 9:27–39. https://doi.org/10.2217/rme.13.80
7. Qiu T, Hanna E, Dabbous M et al (2020) Regenerative medicine regulatory policies: a systematic review and international comparison. Health Policy 124(7):701–713. https://doi.org/10.1016/j.healthpol.2020.05.004
8. Salter B, Zhou Y, Datta S (2015) Hegemony in the marketplace of biomedical innovation: consumer demand and stem cell science. Soc Sci Med 131:156–163. https://doi.org/10.1016/j.socscimed.2015.03.015
9. Fricker M (2007) Epistemic injustice: power and the ethics of knowing. Oxford University Press, New York. https://doi.org/10.1093/analys/anp028
10. Vieira M, Andia T, Karim O et al (2023) Rising pharmaceutical innovation in the global south: a landscape study. J Pharm Policy Pract 16:155. https://doi.org/10.1186/s40545-023-00669-3
11. Wenham C, Wouters O, Jones C et al (2021) Measuring health science research and development in Africa: mapping the available data. Health Res Policy Syst 19:142. https://doi.org/10.1186/s12961-021-00778-y
12. Bhakuni H, Abimbola S (2021) Epistemic injustice in academic global health. Lancet Glob Health 9(10):e1465–e1470. https://doi.org/10.1016/S2214-109X(21)00301-6
13. Charani E, Abimbola S, Pai M et al (2022) Funders: the missing link in equitable global health research? PLOS Glob Public Health 2(6):e0000583. https://doi.org/10.1371/journal/pgph.0000583
14. Koum Besson ES (2022) How to identify epistemic injustice in global health research funding practices: a decolonial guide. BMJ Glob Health 7(4):e008950. https://doi.org/10.1136/bmjgh-2022-008950
15. Fricker M (2015) Epistemic contribution as a central human capability. In: Hull G (ed) The equal society: essays on equality in theory and practice. Lexington Books, Maryland, pp 73–90
16. Medina J (2017) Varieties of hermeneutical injustice. In: Kidd IJ, Medina J, Pohlhaus G Jr (eds) The Routledge handbook of epistemic injustice. Routledge, New York, pp 53–60. eBook ISBN: 9781315212043
17. Pohlhaus G Jr (2017) Varieties of epistemic injustice. In: Kidd IJ, Medina J, Pohlhaus G Jr (eds) The Routledge handbook of epistemic injustice. Routledge, New York, pp 13–26. eBook ISBN: 9781315212043
18. de Sousa SB (2014) Epistemologies of the south: justice against epistemicide. Routledge, New York. https://doi.org/10.4324/9781315634876
19. Hill PS (2011) Understanding global health as a complex adaptive system. Glob Public Health 6(6):593–605. https://doi.org/10.1080/17441691003762108
20. Gilmore B (2019) Realist evaluations in low- and middle-income countries: reflections and recommendations from the experiences of a foreign researcher. BMJ Glob Health 4:e001638. https://doi.org/10.1136/bmjgh-2019-001638
21. Pratt B, de Vries J (2023) Where is knowledge from the global south? An account of epistemic justice for a global bioethics. J Med Ethics 49(5):325–334. https://doi.org/10.1136/jme-2022-108291
22. Anderson E (2012) Epistemic justice as a virtue of social institutions. Soc Epistemol 26(2):163–173. https://doi.org/10.1080/02691728.2011.652211
23. Abimbola S (2023) Knowledge from the global south is in the global south. J Med Ethics 49(5):337–338. https://doi.org/10.1136/jme-2023-109089
24. Rouce RH, Grilley BJ (2025) How to democratize cell and gene therapy: a global approach. Mol Ther 33(5):2082–2090. https://doi.org/10.1016/j.ymthe.2025.03.061
25. Alliance for Regenerative Medicine (2023) Sector snapshot: April 2023: A new vision (Issue April). https://alliancerm.org/sector-snapshot-december-2023/. Accessed 5 May 2025
26. Ellison B (2016) Stem cell research and social justice: aligning scientific progress with social need. Curr Stem Cell Rep 2(4):328–335. https://doi.org/10.1007/s40778-016-0063-3

27. Deinsberger J, Reisinger D, Weber B (2020) Global trends in clinical trials involving pluripotent stem cells: a systematic multi-database analysis. NPJ Regen Med 5(1):1–13. https://doi.org/10.1038/s41536-020-00100-4
28. Council of Canadian Academies (2020) From research to reality: the expert panel on the approval and use of somatic gene therapies in Canada. https://cca-reports.ca/wp-content/uploads/2019/08/Report-From-Research-to-Reality-EN.pdf. Accessed 5 May 2025
29. Gold ER (2021) The fall of the innovation empire and its possible rise through open science. Res Policy 50:104226. https://doi.org/10.1016/j.respol.2021.104226
30. Ogbogu U, Albrecht L (2025) Making advanced therapies affordable and accessible: Two stragetic approaches. Dev World Bioeth. https://doi.org/10.1111/dewb.70004
31. Ogbogu U, Case N (2025) Report on regenerative medicine clinical research activities and product pipelines in low- and middle-income countries in Africa and Asia. Zenodo. https://doi.org/10.5281/zenodo.15362145
32. Banda G (2023) The political economy of the African pharmaceutical sector's "industrial underdevelopment" lock-in: the importance of understanding the impact of persistent colonial extractive institutions. Front Res Metr 8. https://doi.org/10.3389/frma.2023.1020588
33. Banda G, Mugwagwa J, Mackintosh M et al (2022) The localisation of medical manufacturing in Africa. https://www.iej.org.za/wp-content/uploads/2022/11/IEJ-LoMMiA-report_Nov2022.pdf. Accessed 5 May 2025
34. World Health Organization (2022) Egypt and Nigeria medicines regulators achieve high maturity level in WHO classification and WHO launches list of regulatory authorities that meet international standards. https://www.who.int/news/item/30-03-2022-egypt-and-nigeria-medicines-regulators-achieve-high-maturity-level-in-who-classification-and-who-launches-list-of-regulatory-authorities-that-meet-international-standards. Accessed 5 May 2025
35. Arbuthnot P, Maepa MB, Ely A et al (2017) The state of gene therapy research in Africa, its significance and implications for the future. Gene Ther 24(9):581–589. https://doi.org/10.1038/gt.2017.57
36. Gaobotse G (2018) Stem cell research in Africa: legislation and challenges. J Regen Med 7(1):1–93. https://doi.org/10.4172/2325-9620.1000142
37. World Health Organization (2024) WHO Global Benchmarking Tool (GBT) for evaluation of national regulatory systems. https://www.who.int/tools/global-benchmarking-tools. Accessed 5 May 2025
38. Ndomondo-Sigonda M, Azatyan S, Doerr P et al (2023) Best practices in the African Medicines Regulatory Harmonization initiative: perspectives of regulators and medicines manufacturers. PLOS Glob Public Health 3(4):e0001651. https://doi.org/10.1371/journal.pgph.0001651
39. Azeez IA, Awogbindin IO, Olayinka JN et al (2022) Neural stem cell research in Africa: current realities and future prospects. Biol Open 11(11). https://doi.org/10.1242/bio.059574
40. Mahmoud A (2022) Islamic bioethics: National regulations and guidelines of human stem cell research in the Muslim world. https://digitalcommons.chapman.edu/international_studies_theses/14/. Accessed 5 May 2025
41. Slabbert MN, Pepper MS (2015) A global comparative overview of the legal regulation of stem cell research and therapy: lessons for South Africa. S Afr J Bioeth Law 8(2):12–22. https://www.ajol.info/index.php/sajbl/article/view/125135
42. Karimi K, Rockwell H (2019) The benefits of platelet-rich fibrin. Facial Plast Surg Clin North Am 27(3):331–340. https://doi.org/10.1016/j.fsc.2019.03.005
43. Nejad AR, Hamidieh AA, Amirkhani MA et al (2021) Update review on five top clinical applications of human amniotic membrane in regenerative medicine. Placenta 103:104–119. https://doi.org/10.1016/j.placenta.2020.10.026
44. Mazini L, Ezzoubi M, Malka G (2021) Overview of current adipose-derived stem cell (ADSCs) processing involved in therapeutic advancements: flow chart and regulation updates before and after COVID-19. Stem Cell Res Ther 12(1):1–17. https://doi.org/10.1186/s13287-020-02006-w

45. African Union Development Agency (2022a) African Medicines Regulatory Harmonization Programme—Egypt. https://amrh.nepad.org/amrh-countries/egypt. Accessed 5 May 2025
46. Matar A, Silverman HJ (2022) Ethical analysis of Egypt's law regulating clinical research. J Empir Res Hum Res Ethics 17(4):494–503. https://doi.org/10.1177/15562646221096188
47. Egypt Center for Research and Regenerative Medicine (2020) About us: introduction. http://www.ecrrm.ac.eg/Introduction.aspx. Accessed 5 May 2025
48. Sleem H (2012) Ethical and practice issues in stem cell research and therapy—"Egyptian situation". https://www.aub.edu.lb/fm/shbpp/ar/Documents/Ethical-and-practice-Issues-in-SC-research-Egyptian-Situation_Sleem.pdf. Accessed 5 May 2025
49. Egyptian Armed Forces Ministry of Defense (2022) The armed forces signed a contract to establish the International Center for Research and Treatment of Stem Cells. https://www.mod.gov.eg/ModWebSite/NewsDetails.aspx?id=42277. Accessed 5 May 2025
50. SAHPRA (South African Health Products Regulatory Authority) (2024) SAHPRA encourages the safe use of medicines and reporting of suspected side effects this #MedSafetyWeek. https://www.sahpra.org.za/news-and-updates/sahpra-encourages-the-safe-use-of-medicines-and-reporting-of-suspected-side-effects-this-medsafetyweek/. Accessed 5 May 2025
51. Viljoen IM, Hendricks CL, Malherbe HL et al (2022) Regenerative medicines: a new regulatory paradigm for South Africa. Biochimie 196:123–130. https://doi.org/10.1016/j.biochi.2022.02.010
52. SAHPRA (South African Health Products Regulatory Authority) (n.d.) Biological medicines. https://www.sahpra.org.za/biological-products/. Accessed 5 May 2025
53. Skeen M, Eksteen C, Pepper M (2019) Stem cell tourism and spinal cord injury in South Africa. S Afr Med J 109(8b):17–23. https://doi.org/10.7196/SAMJ.2019.v109i8b.13824
54. Hendricks CL, Alessandrini M, Pepper MS (2023) Equitable access to cell and gene therapies in South Africa: opportunities and hurdles. Gene Ther 30:180–186. https://doi.org/10.1038/s41434-021-00309-y
55. Khrouf MR, Achour L, Thabti A et al (2017) Direct cost analysis of the second year post-allogeneic hematopoietic stem cell transplantation in the Bone Marrow Transplant Centre of Tunisia. JMAHP 5(1):1335161. https://doi.org/10.1080/20016689.2017.1335161
56. Mechergui A, Achour W, Mathlouthi S et al (2019) Prevalence of infectious multi-drug resistant bacteria isolated from immunocompromised patients in Tunisia. Afr Health Sci 19(2):2021–2025. https://doi.org/10.4314/ahs.v19i2.25
57. Touihri M, Torjeman L, Kaabi H et al (2019) Bronchiolitis obliterans after allogeneic hematopoietic stem cell transplantation: the effect of NOD2/CARD15 mutations in a Tunisian population. Hum Immunol 80(3):163–168. https://doi.org/10.1016/j.humimm.2018.12.005
58. African Union Development Agency (2022b) African Medicines Regulatory Harmonization Programme—Tunisia. https://amrh.nepad.org/amrh-countries/tunisia. Accessed 5 May 2025
59. Product Life Group (2023) An overview of the pharmaceutical regulations in Tunisia. https://www.productlifegroup.com/an-overview-of-the-pharmaceutical-regulations-in-tunisia/. Accessed 5 May 2025
60. Republic of Tunisia Ministry of Health – Pharmacy and Drug Directorate (2016) Medicinal products registration guide in Tunisia. https://perma.cc/NK2V-XU39. Accessed 5 May 2025
61. African Union Development Agency (2022c) African Medicines Regulatory Harmonization Programme—Uganda. https://amrh.nepad.org/amrh-countries/uganda. Accessed 5 May 2025
62. Nakandi BT, Muhimbise O, Djuhadi A et al (2023) Experiences of medical device innovators as they navigate the regulatory system in Uganda. Front Med 5:1162174. https://doi.org/10.3389/fmed.2023.1162174
63. Uganda National Drug Authority (2023) Guidelines on submission of documentation for registration of a pharmaceutical product for human use. https://perma.cc/VG4N-47CM. Accessed 5 May 2025
64. Bayer Global (2023) Healthy China 2030: China's heealthcare journey. https://www.bayer.com/en/pharma/healthy-china-2030-chinas-healthcare-journey. Accessed 5 May 2025
65. Li B (ed) (2020) Tutorial for outline of the healthy China 2030 plan. Springer, Beijing. https://doi.org/10.1007/978-981-32-9603-9

66. National Medical Products Administration (2019) Main responsibilities of the National Medical Products Administration. https://english.nmpa.gov.cn/2019-07/18/c_377587.htm. Accessed 5 May 2025
67. Zou L, Qi Y, Jiang Y et al (2023) Criteria and regulatory considerations for the conditional approval of innovative antitumor drugs in China: from the perspective of clinical reviewers. Cancer Commun 43(2):171–176. https://doi.org/10.1002/cac2
68. National Library of Medicine (2012a) rAd-p53 gene therapy for advanced malignant thyroid tumors. ClinicalTrials.gov. https://clinicaltrials.gov/study/NCT00902122?locStr=China&country=China&cond=Thyroid%20Tumor&aggFilters=phase:4,studyType:int&rank=1. Accessed 5 May 2025
69. National Library of Medicine (2012b) rAd-p53 gene therapy for advanced oral and maxillofacial malignant tumors (rAd-p53). ClinicalTrials.gov. https://clinicaltrials.gov/study/NCT00902083?locStr=China&country=China&cond=Maxillofacial%20Tumors&aggFilters=phase:4,studyType:int&rank=1. Accessed 5 May 2025
70. Zhang JY (2017) Lost in translation? Accountability and governance of clinical stem cell research in China. Regen Med 12(6):647–656. https://doi.org/10.2217/rme-2017-0035
71. Turner L (2020) Preying on public fears and anxieties in a pandemic: businesses selling unproven and unlicensed "stem cell treatments" for COVID-19. Cell Stem Cell 26(6):806–810. https://doi.org/10.1016/j.stem.2020.05.003
72. Gao J, Gao C (2022) Development and regulation of stem cell-based therapies in China. Cell Prolif 55(8):e13217. https://doi.org/10.1111/cpr.13217
73. Sivagourounadin K, Ravichandran M, Rajendran P (2021) National guidelines for gene therapy product (2019): a road-map to gene therapy products development and clinical trials. Perspect Clin Res 12(3):118–125. https://doi.org/10.4103/picr.PICR_189_20
74. National Library of Medicine (2020) Combination of autologous MSC and HSC infusion in patients with decompensated cirrhosis. ClinicalTrials.gov. https://clinicaltrials.gov/study/NCT04243681?locStr=India&country=India&cond=Liver%20Cirrhosis&aggFilters=phase:4,studyType:int&rank=1#study-record-dates. Accessed 5 May 2025
75. Madson R (2019) AIG Hospitals in India joins Mayo Clinic Care Network. Mayo Clinic News Network. https://newsnetwork.mayoclinic.org/discussion/aig-hospitals-in-india-joins-mayo-clinic-care-network/. Accessed 5 May 2025
76. Tiwari SS, Raman S, Martin P (2017) Regenerative medicine in India: trends and challenges in innovation and regulation. Regen Med 12(7). https://doi.org/10.2217/rme-2017-0094
77. Basu M (2019) Genetics labs come under watch, govt to monitor gene therapy research across India. https://theprint.in/health/genetics-labs-come-under-watch-govt-to-monitor-gene-therapy-research-across-india/276840/. Accessed 5 May 2025
78. Cherukuri VP, Balakrishna CS, Golla M et al (2021) Regulatory framework for regenerative medicine in India. IJDRA - Int J Drug Regul Aff 9(3):25–31. https://doi.org/10.22270/ijdra.v9i3.479
79. Indian Council of Medical Research (2019) National guidelines for gene therapy product development & clinical trials. https://www.icmr.gov.in/icmrobject/custom_data/pdf/resource-guidelines/guidelines_GTP.pdf. Accessed 5 May 2025
80. Pradhan AK (2023) Regulatory measures to promote fast patients access in India. https://www.pmda.go.jp/files/000250284.pdf. Accessed 5 May 2025
81. Asadpour A, Yahaya BH, Bicknell K et al (2023) Uncovering the gray zone: mapping the global landscape of direct-to-consumer businesses offering interventions based on secretomes, extracellular vesicles, and exosomes. Stem Cell Res Ther 14(1):1–7. https://doi.org/10.1186/s13287-023-03335-2
82. National Pharmaceutical Regulatory Agency (2016) Guidance Document and guidelines for registration of Cell and Gene Therapy Products (CGTPs) in Malaysia. https://www.npra.gov.my/index.php/en/guideline-bio/1527314-guidance-document-and-guidelines-for-registration-of-cell-and-gene-therapy-products-cgtps-in-malaysia-2.html. Accessed 5 May 2025

83. Gopalan N, Nor SNM, Mohamed MS (2020) Regulation of stem cell technology in Malaysia: current status and recommendations. Sci Eng Ethics 26:1–25. https://doi.org/10.1007/s11948-019-00111-5
84. Gopalan N, Noor SNM, Mohamed MS (2021) The pro-medical tourism stance of Malaysia and how it affects stem cell tourism industry. SAGE Open 11(2):1–15. https://doi.org/10.1177/21582440211016837
85. Imran SAM, Hamizul MHAM, Khairul Bariah AAN et al (2022) Regenerative medicine therapy in Malaysia: an update. Front Bioeng Biotechnol 10(April):1–15. https://doi.org/10.3389/fbioe.2022.789644
86. Loh EYX, Goh PS, Mannan AMM et al (2021) Cell and gene therapy products in Malaysia: a snapshot of the industry's current regulation preparedness. Cytotherapy 23(12):1108–1113. https://doi.org/10.1016/j.jcyt.2021.06.006
87. Radzi MAZA, Zawawi M, Sha'ban M et al (2020) Legal aspects of articular cartilage tissue engineering experimentation: a review on Malaysian acts, regulations and guidelines. Malays J Med Health Sci 16(3):272–284. eISSN 2636-9346
88. Arda B, Aciduman A (2009) An evaluation regarding the current situation of stem cell studies in Turkey. Stem Cell Rev 5(2):130–134. https://doi.org/10.1007/s12015-009-9072-2
89. Enserink M (2006) Biomedicine: selling the stem cell dream. Science 313(5784):160–163. https://doi.org/10.1126/science.313.5784.160
90. European Spinal Cord Injury Federation (2020) Unproven therapies. https://www.escif.org/unproven-therapies/. Accessed 5 May 2025
91. Turkish Medicines and Medical Devices Agency (2021) Regulation on the registration of medicinal products for human use. https://titck.gov.tr/storage/legislation/QEv7VEBH.pdf. Accessed 5 May 2025
92. Herdem Attorneys at Law (2023) Turkish Medicine and Medical Devices Authority (TITCK) updates regulations in 2023. https://www.mondaq.com/turkey/healthcare/1332354/turkish-medicine-and-medical-devices-authority-titck-updates-regulations-in-2023. Accessed 5 May 2025
93. Bin Abdul Aziz MF, Morrison M, Kaye J (2018) Regulating human stem cell research and therapy in low-and middle-income countries: Malaysian perspectives. New Genet Soc 37(1):28–45. https://doi.org/10.1080/14636778.2017.1415137
94. Bloom G, Henson S, Peters DH (2014) Innovation in regulation of rapidly changing health markets. Glob Health 10:53. https://doi.org/10.1186/1744-8603-10-53
95. Zhong A, Darren B, Loiseau B et al (2021) Ethical, social, and cultural issues related to clinical genetic testing and counseling in low- and middle-income countries: a systematic review. Genet Med 23(12):2270–2280. https://pubmed.ncbi.nlm.nih.gov/30072741/
96. George B (2011) Regulations and guidelines governing stem cell based products: clinical considerations. Perspect Clin Res 2(3):94–99. https://doi.org/10.4103/2229-3485.83228
97. Inhorn MC, Birenbaum-Carmeli D, Tremayne S et al (2017) Assisted reproduction and Middle East kinship: a regional and religious comparison. Reprod Biomed Online 4:41–51. https://doi.org/10.1016/j.rbms.2017.06.003
98. Serour GI, Serour AG (2021) The impact of religion and culture on medically assisted reproduction in the Middle East and Europe. Reprod Biomed Online 43(3):421–433. https://pubmed.ncbi.nlm.nih.gov/34344602/
99. Guven T, Sert G (2010) Advance directives in Turkey's cultural context: examining the potential benefits for the implementation of patient rights. Bioethics 24(3):127–133. https://doi.org/10.1111/j.1467-8519.2009.01789.x
100. Turkmen HO, Arda B (2008) Ethical and legal aspects of stem cell practices in Turkey: where are we? J Med Ethics 34(12):833–837. https://doi.org/10.1136/jme.2007.022806
101. Ogbogu U, Noohi F (2025) Report on regulatory frameworks for clinical research on and marketing authorization of regenerative medicine products in low- and middle-income countries. Zenodo. https://doi.org/10.5281/zenodo.15362273
102. Development Reimagined (2022) Report: every African country depends on pharmaceutical imports—here is a new way out. https://developmentreimagined.com/

new-report-every-african-country-depends-on-pharmaceutical-imports-here-is-a-new-way-out/. Accessed 5 May 2025
103. Horner R (2022) Global value chains, import orientation, and the state: South Africa's pharmaceutical industry. J Int Bus Policy 5(1):68–87. https://doi.org/10.1057/s42214-021-00103-y
104. Mostafa DR (2021) The appropriateness of patents legislations for attaining sustainable development in Egypt. JIPIM 4(4):47–62. https://doi.org/10.21608/JIPIM.2021.231034
105. Kepekci AH (2019) The level of awareness among ENT clinicians about the studies of stem cells in Turkey. EJMI 3(4):310–314. https://doi.org/10.14744/ejmi.2019.22027
106. Ogbogu U (2015) Combatting unlicensed stem cell interventions through truthful advertising law: a survey of regulatory trends. McGill J Law Health 9:311. https://canlii.ca/t/6zr
107. Baker S (2023) North–south publishing data show stark inequities in global research. Nature 624(7991):S1. https://doi.org/10.1038/d41586-023-03901-x
108. Kariuki T, Marincola E (2023) How to level the global publishing playing field. Nature 624(7991):S13. https://doi.org/10.1038/d41586-023-03905-7
109. Nakamura G, Soares B, Pillar V (2023) Three pathways to better recognize the expertise of global south researchers. npj Biodivers 17(2). https://doi.org/10.1038/s44185-023-00021-7
110. Sumathipala A, Siribaddana S, Patel V (2004) Under-representation of developing countries in the research literature: ethical issues arising from a survey of five leading medical journals. BMC Med Ethics 5(5). https://doi.org/10.1186/1472-6939-5-5
111. Chan S (2017) Current and emerging global themes in the bioethics of regenerative medicine: the tangled web of stem cell translation. Regen Med 12(7):839–851. https://doi.org/10.2217/rme-2017-0065
112. Lyons S, Salgaonkar S, Flaherty G (2022) International stem cell tourism: a critical literature review and evidence-based recommendations. Int Health 14(2):132–141. https://doi.org/10.1093/inthealth/ihab050
113. Meissner-Roloff M, Pepper M (2013) Curbing stem cell tourism in South Africa. Appl Transl Genom 2:22–27. https://doi.org/10.1016/j.atg.2013.05.001
114. Polejack A, Goveas J, Robinson S, Flink T, Ferreira G (2022) Where is the global south in the science diplomacy narrative? https://ssrn.com/abstract=4278557 or https://doi.org/10.2139/ssrn.4278557

A Case for Inclusive Global Governance for Regenerative Medicine

3

Jonathan O. Chimakonam
and Adeolu Oluwaseyi Oyekan

Abstract

Scientific and technological breakthroughs are increasingly reshaping human life in profound and far-reaching ways. As new knowledge emerges across diverse fields, its influence on every aspect of human existence continues to grow. To ensure these advancements are applied responsibly and avoid harmful consequences, it is essential to establish robust ethical policies and regulatory safeguards. While some national and regional regulatory frameworks exist, most are either too weak or inadequate for the effective regulation of the field of regenerative medicine. The gap we will attempt to fill in this work is to advance an argument not just in favour of a global regulatory framework for regenerative medicine but for an inclusive one. To achieve the above, we will demonstrate the importance of global governance in regenerative medicine and show that the existing initiatives are not global, pointing to the conspicuous omission of the ethical parameters from worldviews in the Global South. Further, we will outline some of the existing barriers to inclusive governance. Evoking the frameworks of global justice and decoloniality, we will argue for accommodating various ethical parameters, using African ethics as a case in point.

Keywords

Regenerative medicine · Global governance · Inclusive ethics · African philosophy, · Decoloniality

J. O. Chimakonam (✉)
Department of Philosophy, University of Pretoria, Pretoria, South Africa
e-mail: Jonathan.okeke@up.ac.za

A. O. Oyekan
Decoloniality Research Group, University of Pretoria, Pretoria, South Africa
e-mail: adeolu.oyekan@up.ac.za

U. Ogbogu et al. (eds.), *Regenerative Medicine and the Bioethics of Globalization, Justice and Hegemony*, Collaborative Bioethics 5,
https://doi.org/10.1007/978-3-032-07289-4_3

Introduction

The rapid advancement of scientific and technological research continues to impact and reshape human lives in many profound ways. The emergence of new knowledge in diverse fields has deep implications for the totality of human existence, and it is not unreasonable to expect an increased level of influence of new technologies in the years to come. A sturdy ethical policy is required to keep these advancements in check and prevent a possible mishap in the application of these technologies resulting from poor or inordinate regulation. While some national and regional regulatory frameworks exist, most are either too weak or are inadequate for effective regenerative medicine (RM) regulation [1].

The gap we will attempt to fill in this work is to advance an argument not just in favour of a global regulatory framework but for an inclusive one. To achieve the above, we will demonstrate the importance of global governance in RM and show that the existing initiatives are not global, pointing to the conspicuous omission of the ethical parameters from worldviews in the Global South. Further, we will outline some of the existing barriers to inclusive governance. Evoking the frameworks of global justice and decoloniality, we will argue for the accommodation of various ethical parameters, using African ethics as a case in point.

RM has emerged as an area where the prospects of life-altering technological interventions are increasingly becoming possible. A basic description of RM is that it focuses on repairing, replacing, and regenerating damaged human tissues and organs in order to restore them to their normal functions [2]. RM is further concerned with alleviating symptoms from various diseases, including preventing the onset of such diseases and slowing their progression. The therapies developed within RM aim to provide new hope for conditions previously considered untreatable. This expanding field of innovation has also shown great promise in treating common chronic diseases like stroke and renal failure.

However, the prospects of regenerative therapies to alter human biological limitations through processes such as stem cell treatments, tissue engineering, and gene editing extend beyond individual health benefits to broader social, economic, and ethical implications, necessitating a robust framework of governance. History is replete with beneficial innovations that became threats when left unregulated—nuclear technology being a key example. Therefore, it is imperative that ethical and governance issues in RM receive adequate attention. Some of these ethical issues include the conceptualization and design of RM therapies, accessibility, concerns with research and experiments, and so on.

Given these serious ethical implications, we argue that due to the nature of RM, there is a need for effective governance to ensure the field is developed and all resulting innovations are applied in a responsible manner. Specifically, there are many areas of RM that require ethical governance and we will mention a few in this chapter. First, governance is required on how regenerative treatments are conceptualized and designed. In this aspect, focusing on the end goal of specific designs, as well as their unintended and intended implications, is crucial. Second, the manner in which research and experiments are conducted is also of primary concern, such that underprivileged demographics from economic or cultural standpoints do not become the default guinea pigs of those who are privileged. Third, the issue of access is closely connected to the conceptualization and design stages of RM treatments along with a consideration of the implications of these treatments.

The issue of accessibility necessarily poses important questions about the way RM treatments are designed, marketed, and ultimately administered to the greater public. How are regenerative innovations to be dispensed or administered? Should they be exclusive to those who are wealthy and are able to afford it? Should society treat RM as luxury or necessity? What are the implications of unequal distribution of therapies and treatments that are capable of enhancing the quality and duration of life? Who determines who needs regenerative treatments, as well as the specific type or types of treatment needed?

All of these are questions which must be considered—particularly when many nations are still in the early stages of developing RM-focused legal and regulatory frameworks. In a world where people reject blood transfusions to save their own lives, it cannot be merely assumed that RM is 'good' for everyone, nor can it be assumed that everyone will be keen to adopt and utilize RM treatments and therapies. Questions about autonomy, choice, and informed consent, therefore, arise in the administration of RM. From the above, it is obvious that the governance of RM requires the imposition of certain normative protocols throughout the many diverse aspects of the process, from research to manufacturing and application.

RM research has advanced rapidly in developed countries, especially in the West, compared with less developed ones, made up mostly by many African countries, among others [3]. Given existing inequalities between the developed and less developed parts of the world and the capacity of RM to alter life qualities on a mass scale, it is apposite to wonder whether the possibility of wider inequality does not exist. The implication of this inequality then is that the governance system of RM ought to be geographically, culturally, and epistemically inclusive in order to be deemed fair and just. As the world becomes increasingly interconnected, policies in one region can significantly impact other regions, making global considerations both salient and essential for birthing global development and progress. Hence a robust framework of global governance is necessary, though we will outline some of the barriers in creating and implementing such an inclusive framework later in the work.

In this chapter, therefore, our focus is twofold. First, we will make the case that as opposed to the current approach in which developed countries govern RM in isolation, sometimes with minimal collaboration (such as between the United States and Europe), there is therefore a need for a global framework of governance. Second, we will also explore what an African ethical contribution to the global governance of RM might look like.

This chapter is organized into four sections. The first section will explore the origin and meaning of RM, providing a foundation for understanding its development and diverse approaches. With the examination of the evolution of RM, we can better understand the context in which global governance is necessary. The second section will focus on the current state of governance in RM globally. The third section will address why, and defend the need for global governance in RM. The fourth section will outline barriers in the path towards inclusive governance and explore what a global governance framework in RM should look like. This framework will also be considered in the interest of mitigating any harm presented by this emerging field.

Regenerative Medicine: Meaning and History

RM harnesses the body's natural healing processes, often enhanced by advanced technologies, to develop new therapeutic strategies for better health outcomes. According to Edgar et al. [4], these therapeutic strategies aim to achieve a state where immunosuppression or anti-rejection treatment is not necessary, hence, enhancing the quality of life, minimizing complications and toxicities, and eliminating the expenses associated with such lifelong treatments. Cutting-edge approaches to healing include the use of decellularization to create acellular scaffolds for organ manufacturing, three-dimensional printing, and interspecies blastocyst complementation [4].

A key aspect of RM is the utilization of stem cells, biomaterials, and bioengineering techniques to create functional tissues and organs, marking a significant shift in the treatment of injuries and diseases. According to Barr et al. [5], RM can be understood in multiple ways. Literally, the word regenerative means the renewal or replenishing of something that has been depleted, while medicine in this context means a curative measure or treatment of diseases. In a systematic definition, RM, according to Barr et al. [5], can be described as that which focuses on creating living, functional tissues to restore or replace tissue or organ functions lost due to aging, disease, damage, or congenital defects. This field aims to regenerate damaged tissues and organs by stimulating the body's natural healing processes, enabling scientists to grow tissues and organs in laboratories for safe implantation when natural healing is insufficient. Advocates believe that RM could potentially solve the shortage of organs available for donation relative to the number of patients needing transplants, while also addressing the issue of organ transplant rejection, as the cells of the transplanted organ would be a genetic match to the patient [6, 7].

Effective management of chronic wounds has become crucial not only for enhancing the quality of life but also for extending life expectancy. This has spurred a drive towards developing more sophisticated and biomimetic strategies. These advancements are driven by the exploration of various emerging technologies aimed at addressing the complex nature of chronic wounds in particular [8]. Thus, the emergence of RM can be seen as crucial to humankind's perennial health challenges.

Historically, the evolution of RM was marked by several significant events, such as the early identification and characterization of mesenchymal stem cells (MSCs) in the 1960s, which laid the groundwork for conceptualizing and understanding the potential of stem cells to differentiate into various cell types [9]. The modern era was catalyzed by Langer and Vacanti's principles of tissue engineering in the 1980s, which combined scaffolds and growth factors to engineer tissues [10]. Furthermore, a groundbreaking advancement came with the discovery of induced pluripotent stem cells (iPSCs) by Shinya Yamanaka in 2006, demonstrating that adult cells can be reprogrammed to a pluripotent state [11]. This discovery revolutionized RM by providing a source of patient-specific pluripotent stem cells, paving the way for personalized therapies.

In recent years, significant advancements have continued in RM. One notable development is the progress in 3D bioprinting, which allows for the creation of complex human tissue constructs. For example, Murphy and Atala [12] highlight the potential of 3D bioprinting to create organs for transplantation, demonstrating a transformative impact on RM. 3D bioprinting allows for the precise layering of cells to construct complex tissue structures, addressing the critical shortage of donor organs. In addition to organ creation, 3D bioprinting has shown promise in producing customized prosthetics and in developing patient-specific tissue models for drug testing. According to Jovic et al. [13], this advancement reduces reliance on animal models and enhances the accuracy of preclinical trials. The preceding is particularly significant because it eliminates problems associated with animal models, such as infections, rejection, immune reactions, and religious taboos. When combined with stem cell technology, 3D bioprinting has further potential because it can produce living tissues that can grow and integrate with the patient's body. This combination offers a more sustainable and long-term solution for tissue repair and regeneration. The convergence of 3D bioprinting and stem cell technologies in particular highlights the need for robust regulatory frameworks and ethical guidelines to ensure safe and equal access to these advanced medical therapies, without which the cost of malpractice could be high.

Thus, stem cell therapies have progressed significantly, as detailed in a comprehensive review by Trounson and McDonald [14], including advancements in treating conditions such as spinal cord injuries, heart diseases, and degenerative disorders, which account for many deaths and low quality of life in the afflicted persons. These advancements expose the significant strides made in the field, alongside the many regulatory and ethical challenges that must be addressed to ensure safe and effective applications.

Regenerative Medicine and Governance

According to the US Government Accountability Office [15], there must be standardization, regulation, and monitoring policy guiding developments in RM in the country. However, on a global scale, there seems to be an absence of a unified approach for addressing RM policy, resulting in the absence of standardized regulations. According to Sawarkar and Bapat [16], the regulation of RM differs across the globe because each country, region, and continent has its own set of guidelines. These guidelines differ widely, leading to various classifications of nations based on their approach to cell research and RM. These classifications range from very permissive, permissive with restrictions, restricted, very restricted, to unregulated [16].

Human embryonic stem cell (hESC) research in particular illustrates the vastly different approaches to RM by region. This research is permitted in different parts of Europe, but only a few nations allow the production of human embryos specifically for certain purposes. For instance, countries such as Belgium and the United Kingdom are regarded as more open in terms of the rules that allow research and treatment using hESCs under certain conditions (see [2, 17–19]). On the other hand, countries such as Austria and Lithuania have stricter regulations, including outright bans on hESC practices [16]. From the above, it is clear that these differences are due to each country's unique ethical, legal, and political views—all of which make it challenging to create a unified set of rules across Europe, let alone globally, because each nation prioritizes its independence on the right to decide these matters. However, as science advances amidst ongoing ethical and moral discussions, the regulatory frameworks which govern these advances must evolve in a concentrated and thoughtful manner in order to keep pace with these advancements.

In Asia, countries like India, China, and Saudi Arabia have established rules guiding some aspects of RM but these rules vary by country. China, in particular, is considered to be less regulated than India and Saudi Arabia (see [17, 19, 20]. In India, RM and stem cell practices are regulated by the guidelines of the Indian Council of Medical Research (ICMR) and the Department of Biotechnology (See [19, 21]). According to Cherukuri et al. [21], in December 2019, India proactively introduced guidelines aimed at preventing the misuse of gene therapy by scientists. The apex health research body, the ICMR, issued national guidelines outlining the procedures for the development and administration of gene therapies to address inherited genetic or rare diseases in the country. However, this regulatory body (the ICMR together with the Department of Biotechnology) has played more of an advisory role than its supposed, intended legal role (see [21, 22]). Thus, because of the lack of legal force behind these guidelines, some physicians are likely to misuse these therapies—and such misuse includes deceptive advertising practices. In other words, this poor regulatory response means that unproven stem cell treatments are marketed without proper evaluation and oversight, which undeniably reduces the trust in the effectiveness of the regulations.

In China, RM governance is complicated by the unclear classification of stem cell treatments as either medicines or medical technologies [20, 23]. This uncertainty has negative effects on the country's policy implementation processes, which render their regulatory and monitoring processes inconsistent (see [19, 24]). Saudi Arabia has recorded significant progress in regulating stem cell research through Sharia laws and research ethics laws, among other associated regulations. The country incorporates ethical principles together with its Sharia law provisions into its regulatory policies [25], and supports these ethical principles with the policies and principles proposed by the research institutions [16]. Saudi Arabia has sought to prevent ethical breaches and to ensure a more organized and structured regulatory framework with well-monitored ethical conduct in stem cell research and applications.

In Africa, technological innovations are at a comparatively low stage, meaning that the continent significantly lags behind many other jurisdictions, including in clinical stem cell research. Africa contributes only 2.5% of global MSC clinical trials on the global stage [16, 26], thus making the continent appear less relevant as far as technological innovation is concerned. While there are different efforts underway to actually build on stem cell research for treating diseases such as cardiac disease [27], the fact still remains that most stem cell applications in Africa are still at the experimental level rather than the therapeutic level, which according to the UNESCO Education Report [28] is hinged on the fact that in Africa, stem cell research is driven more by cultural and spiritual motivations, whereas in other regions, such research is primarily driven by profit/the private sector/collaboration with other developed nations etc.

Notably, South Africa has succeeded in establishing laws to regulate stem cell services, but these laws are weak, which in turn opens the way for the misuse of regulations by medical professionals involved in medical tourism [28]. Countries like Egypt and Tunisia have different stances on stem cell research, with Egypt having one of the most liberal approaches given that they allow for stem cell generation through cloning [16]. On the other hand, Tunisia has an outright ban on experimental activities concerning embryos [28]. Therefore, it can be said that Africa currently contributes less to the world's technological innovation in stem cell research, and thus the governing structure is at its early stage.

In the United States, guidelines and principles for stem cell products and regenerative therapies have been established and approved by the Food and Drug Administration (FDA) [29, 30]. These guidelines, according to Turner [31], have the potential to greatly improve the treatment of catastrophic injuries and chronic diseases. However, despite the guidelines approved by the FDA, a significant number of treatments offered by US clinics remain unapproved and have caused severe effects on patients, including life-threatening infections and, in some cases, death (see [17, 19]). According to Turner's report [31], over 700 clinics in the US offer these unapproved treatments to their patients, including treatments for Alzheimer's disease, muscular dystrophy, autism, and spinal cord injuries. However, it cannot be denied that the FDA plays a significant role in attempting to regulate these therapies, although the entity lacks the authority to regulate medical practice directly, and due to this, the FDA has been working to refine and enforce existing regulations for products derived from human cells and tissues [24]. In other words, the United States does have a regulatory framework to guide the use of stem cell research and RM; however, the framework's capability in practice is still lagging behind, and the poor efficiency and overall inability to effectively monitor RM treatments remain major challenges.

The Need for Global Governance of Regenerative Medicine

Rapid advancements in RM necessitate robust global governance frameworks to address the ethical, regulatory, and socio-economic challenges in order to ensure equal access and safety across different regions.

The concept of global governance is complex and involves various dimensions. According to a definition provided by Sinclair [32], global governance includes more than just the formal institutions and organizations that manage international affairs; it encompasses systems of rule at all levels of human activity, from families to international organizations, where the pursuit of goals through control has transnational impacts. This view aligns with Lawrence S. Finkelstein's [33] definition of global governance as the management of relationships that go beyond national borders without sovereign authority, performing internationally what governments do within their own countries. Global governance also refers to the intricate system of institutions, norms, and processes responsible for regulating international relations and addressing global challenges [34]. This places ethics and matters of ethical regulations at the center of global governance initiatives.

Global governance in RM is crucial for several compelling reasons, including for its ability to address health risks, safety, equal access, ethical issues, and credible applications of new technologies. RM offers tremendous potential for transforming healthcare by repairing and replacing damaged tissues and organs, but the rapid advancements also bring significant risks and provoke ethical concerns, both considerations that necessitate a robust global governance framework to ensure safety, efficacy, and equal access for RM across different regions.

One of the primary reasons global governance in RM is needed is to ensure the safety and efficacy of various therapies [35]. The complex and potentially high-risk procedures involved in regenerative treatments require rigorous standards and regulations to protect patients. International guidelines are crucial for the harmonization of regulatory policies across countries, thereby facilitating the development and distribution of safe and effective regenerative treatments. According to Qiu et al. [2], harmonizing these policies is essential to mitigating the risks associated with inconsistent regulations, which can lead to differing levels of safety and effectiveness. Without a unified approach, patients could be exposed to unsafe or ineffective treatments, undermining the potential benefits of RM.

In a world where capitalist inclination has permeated every aspect of society, vulnerable and less privileged individuals are at risk of exploitation if the management of RM is not properly regulated. Ensuring that the interests of all people are protected becomes a crucial concern in this context. Without adequate oversight and regulation, the benefits of RM could be unevenly distributed, exacerbating existing inequalities and marginalizing those who are already disadvantaged [2]. Therefore, robust global governance is essential to safeguard the welfare of all individuals and ensure that advancements in RM are equally accessible and ethically managed.

In addition, RM often involves the use of hESCs, gene-editing technologies such as clustered regularly interspaced short palindromic repeats-associated protein 9 (CRISPR-Cas9), and other sensitive materials. These technologies raise numerous ethical issues, including those surrounding consent, privacy, and the potential for genetic modification. In a study by Lovell-Badge et al. [36], they emphasize the importance of global guidelines to address these ethical concerns. For instance, the ethical boundaries of human genetic engineering have been a significant point of debate, particularly with the advent of CRISPR-Cas9 [37]. The revolutionary potential of CRISPR-Cas9 to edit genes with unprecedented precision opens up vast possibilities for treating genetic disorders, improving crop resilience, and even eradicating certain diseases. But these advancements come with profound ethical concerns, including the potential for unintended genetic consequences, the risk of creating genetic inequalities, and the moral implications of altering the human germline. Hence, a robust global governance framework becomes imperative, as it could help in the establishment of ethical standards that promote innovation while also preventing misuse and protecting individual rights. Such a framework would necessitate the creation of international guidelines that ensure CRISPR-Cas9, for example, and other gene-editing technologies, are used responsibly [37]. These proposed guidelines would need to address issues of consent, privacy, and the long-term impacts of genetic modifications on both individuals as well as the broader human gene pool.

Further, the socio-economic implications of RM also call for comprehensive global governance. Regenerative therapies can be expensive, and without proper policies, there is a risk of exacerbating healthcare inequalities between different socio-economic groups and regions. Global governance should aim to create mechanisms that ensure affordable and accessible treatments for all populations. These mechanisms are crucial to prevent a scenario where only wealthy individuals or nations can benefit from advancements in RM. As Hanna et al. [38] note, international cooperation and funding are vital to support research and development in low and middle-income countries, helping to bridge the gap in access to these cutting-edge therapies.

Thus, global governance becomes necessary to successfully navigate the ethical, regulatory, and socio-economic challenges associated with the rapidly advancing field of RM. Establishing international standards, addressing ethical concerns, promoting equal access, and engaging the public for their views and contributions are critical components of a robust governance framework. Measures that ensure the benefits of RM are realized safely and inclusively, in turn, ensure innovation while protecting the interests of all people, irrespective of their geographical origin or class.

However, this type of inclusive governance gives rise to other barriers, which we will outline and discuss in the next section. Admittedly, global governance can sometimes be complex due to the question of who enforces formulated policies and how to ensure compliance from all nations. Also, reaching a normative consensus in a world of disparate moral visions and beliefs can be very herculean. Nevertheless, the challenges and dangers associated with the proliferation of regulations are real, as there is a very strong possibility of each society or region tailoring regulations towards narrow competitive advantages that are detrimental to other regions. The resulting chaos from such an individualized approach would stifle both needed collaboration and further advancement across societies, skewing the benefits of RM in favour of countries that are most advanced.

Toward a Global Governance in Regenerative Medicine: Overcoming Some Barriers to an Inclusive Framework

In the preceding sections, we have shown that some of the existing national and regional regulatory frameworks for RM are either weak or inadequate, leaving considerable concerns for the rapidly progressing research and technological developments in the field. The worrying lack of an adequate global framework inspires two crucial questions. Firstly, how inclusive should the global governance initiative on RM be? Secondly, what type of barriers obstruct the formulation of an inclusive governance framework? These questions will inform our inquiries in this section. Because no ethical framework can be truly, wholly comprehensive, it follows that any regulatory initiative based on one system of ethics will be inadequate. What is considered taboo in one ethical worldview may not be so in another. For example, an Islamic country is unlikely to accept regenerative therapies derived from cells from an impure animal, or from organs grown in swine. In the Qur'an, which is the source of Islamic ethics, pork is designated as impure [39, 40]. However, deriving cells from these sources will not be an issue in Western countries, at least on religious grounds, because their ethics codes do not treat pork or swine as taboo.

The above example indicates that developing an effective and inclusive global governance for RM will be difficult, largely because of the vast differences in ethical codes across countries and cultures. We will outline and discuss four of the factors that can militate against an inclusive governance initiative and propose strategies for overcoming them.

Unique ethical dilemmas There are many points of ethical conflicts among the cultures of the world, which may constitute the ground for irreconcilable differences and thwart any effort for an effective RM global governance. However, it is also possible that we can devise a strategy for navigating these differences to create an adequate and effective global regulatory framework. Here, we formulate and call such a mechanism the Taking Advantage of Our Disadvantage or the TAOD Strategy for short. This strategy stipulates that:

> In any disadvantageous situation, one should always look out for opportunities presented rather than losses incurred.

The TAOD strategy can enable us to develop RM global governance that is inclusive and effective, while also allowing each culture the leeway to override its unique ethical dilemmas. For example, if pigs prove to be the best animals for growing human organs, then an Islamic country does not have to deploy this technology, but a Muslim in need of an organ can obtain this treatment in another country where such a practice is not culturally tabooed. Critics may observe that the receiver of the organ would have contravened the Qur'an and rendered their body impure. This is a credible criticism, but there are two conditions that can override it. The first is that Islam has a ritual for self-purification, and the second is that it holds human life as sacred. For the first condition, a passage in the Qur'an says, "[h]e who purifies himself, who remembers the name of his Lord and prays, shall indeed be successful" [41]. For the second condition, an Islamic scholar Mohammad Ali Shomali [42] argues, "[n]o doubt life in its all [sic] forms enjoys a very high status in Islam. Human life is one of the most sacred creatures of God. Therefore, it must be appreciated, respected and protected." These verses suggest a direct injunction from Allah for individuals to protect their own life. To protect one's life seems to be an overriding duty compared to keeping away from impurities.

The aim of the ritual seems clear too; Allah, being all-knowing, understands that his children will stray from time to time, hence the inclusion of self-purification rituals in the Qur'an. From the foregoing, a Muslim who obtains an organ that was grown in a pig, or therapies that contained materials from an impure animal, has put Allah's injunction to protect their own life first and can remedy the desecration through self-purification. Using this culturally informed logic, it is possible to overcome unique cultural dilemmas to forge a comprehensive, effective, and inclusive global governance framework by considering opportunities rather than losses.

Cultural chauvinism The post-Second World War era is characterized by global institutions regulating various practices, though these institutions' basic norms are derived almost exclusively from the West instead of a broad spectrum that covers multiple cultures. This practice leaves out or even ignores the ethical traditions of cultures outside the West, though this chauvinistic posturing is increasingly being challenged and criticized. For example, the Cameroonian thinker Meinrad Hebga remarks that:

> The dogma of one standard and of one all-embracing prototype for civilization and culture is losing its backers right along. If the fact of having arrived at the atom, of having probed nature in its depths and furthermost reaches gives legitimate pride to the discoverers and consecrates the civilizations which have produced them, all this still leaves an important place to the other cultures, embryonic though they may be called, a place which they occupy humbly, but at the same time to the advantage of all. [43]

The above captures the advocacy of many scholars for global epistemic justice. For instance, Chimakonam [44] contends that the epistemologies of cultures in the Global South are often sidelined in the formulation of global ethical frameworks. He challenges this practice and observes its negative consequences not only for the marginalization of the cultures affected but also for the growth of human knowledge. As Hebga notes above, every culture, no matter how embryonic, has something to contribute to world civilization, and cultural chauvinism seems to be an obstacle to a kind of epistemic egalitarianism—allowing or tolerating multiple perspectives—that can allow for a thousand flowers to bloom.

At the root of cultural chauvinism are some baseless assumptions. For example, Lucien Levy-Bruhl [45] and Robin Horton [46] argue that the African episteme is backward and useless for modern needs. While Levy-Bruhl specifically challenges Africans' intellectual capacities to create rational knowledge, Horton flatly describes their epistemic accumulation as a "closed predicament," by which he means that their accumulations are a pile of superstitions that can neither be questioned, falsified, nor revised. Even though Levy-Bruhl's assumptions have been discredited, many continue to see wisdom in Horton's claim, as evidenced by the continuing disregard for the African knowledge system by the leading presses and scientific societies in the West. The same treatment is extended to other cultures in the Global South, making the chauvinism of Western culture more apparent in our time. We argue that all actors in the RM/scientific sphere must consciously seek to avoid the mistake of cultural chauvinism if they are to establish a credible, adequate, and inclusive global governance initiative for RM.

One way to get around the obstacle of cultural chauvinism is through epistemic humility. This is a cognitive exercise, where one consciously undertakes to overcome their idiosyncrasies and biases against others. It involves a reorientation that lowers one's over-bloated or elevated self-concept that impacts their values and practices. However, because cultural chauvinism creates an opposite effect—cultural timidity—in those who bear the brunt of it, there is also a need to address this effect. This can be done through epistemic empowerment, which is a cognitive exercise that enables individuals to elevate their deflated self-concept, imbuing them with confidence to accept, revise, and assert their epistemic vision for the good of humanity.

The above two strategies can only be truly implemented by those directly experiencing the obstacles. This is why it is a conscious, personal activity. Thus, unlike advocates of global justice debate who blame the hegemonic West and impose on the West the sole responsibility of remedying the situation, we argue that both sides have a role to play in creating an atmosphere for an inclusive and complementary global governance in RM. Stakeholders in RM must therefore undergo orientations in epistemic humility and epistemic empowerment to be able to pursue an inclusive global governance in RM.

The decolonial challenge This challenge is not only about the resistance to decoloniality but also the construal of decoloniality as a displacement narrative. In the resistance to decoloniality, the hegemonic forces from the West are not backing down from the arrogant idea that the Western episteme is superior and supreme. These forces treat the intellectual accumulation of the West as acontextual and absolute, leaving no room for the necessity of other epistemologies or for the incomprehensiveness of Western epistemology. In this connection, Tsenay Serequeberhan [47] argues that:

> In the name of the universality of values, European colonialism violently universalized its own singular particularity and annihilated the historicality of the colonized. In this context, Western philosophy—in the guise of a disinterested, universalistic, transcendental, speculative discourse—served the indispensable function of being the ultimate veracious buttress of European conquest.

From the above excerpt, we can see that much of Western scholarship is fundamentally opposed to the decolonization of knowledge given the West's failure to recognize that the existence of different epistemologies and protocols does not make any such epistemologies inferior and that what it regards as global knowledge is actually the Western contribution to knowledge. The success of colonial conquest erroneously cemented the assumption that those conquered were inferior to the conqueror, not only in terms of arms conflict but also in terms of learned culture and variegated intellectual accumulations. Innocent Asouzu [48] cautions against this type of mindset because it leads to what he calls "unintended ethnocentric commitments," which is a biased and jingoistic mentality that seeks to superiorize one's position while residualizing those of others. No matter how glorious our positionality might be, Asouzu [48] enjoins that we must avoid elevating any world-immanent missing link to an absolute instance. He further argues that each epistemic standpoint is a missing link in a web of other missing links, and that complementarity is a better, more efficient approach for cross-cultural initiatives [48]. The realization that no epistemic standpoint is acontextual or absolute will pave the way for the acceptance of a decolonial strategy. Epistemic humility, as discussed above, can help make this acceptance possible.

Additionally, advocates of decoloniality, largely from the marginalized cultures of the Global South, must also realize that decoloniality is not about displacing the Western episteme. Often, suppressed people carry a heavy burden of exclusivity instigated by their long experience of marginalization. They seek to carve a path for themselves that does not cross with that of the oppressor, even when the oppression has ended. Under this influence, decoloniality could easily be interpreted as a displacement agenda of the oppressed aimed at bringing down the epistemic edifice of the hegemonic culture. This is an improper mindset that must be addressed. Our strategy in this instance hinges on the successful implementation of epistemic empowerment. Both epistemic humility and empowerment have the necessary cognitive resources to promote a mindset that is not opposed to decoloniality and which does not seek to inordinately displace Western episteme.

The preceding imposes a responsibility on African researchers to articulate ideas that are relevant to contemporary challenges as equal partners in a global quest to address current challenges, just like their Western counterparts. Such contributions, even when indigenous or inspired by traditional, normative ideas, must transcend the need to merely debunk Western scholarly chauvinism but should also address modern challenges [49]. This makes epistemic humility and empowerment veritable strategies for creating an inclusive framework in any social sphere, including RM.

The bivalent limitation Of all the obstacles militating against an inclusive regulatory framework in RM, the bivalence implicated in dominant Western ethics is the biggest one. The thesis of bivalence states that a proposition can only be evaluated as true or false, which then eliminates the possibility of intermediate values. In the ethical regulation of a field like RM, this thesis postulates that options will be limited, not only for researchers and technology developers but also for consumers. For example, in Western far-right and far-left political spectra, it may be controversial to permit research with embryos. The same controversy exists within some religious nations with strong teachings on the sanctity of life; this is because of the moral burden implicated in their laws on human rights and religious beliefs. This type of bottleneck can be addressed by appealing to trivalent ethical systems, like the one in some notable African thought systems.

Some eminent African ethicists have demonstrated the trivalent structure of African ethics [50, 51]. In the spectrum of moral valuation, their formulations indicate that besides the polar values of true and false, there is an intermediate value, which is the complementation of the extreme two. We can glean this from their statements of the parameters of right action, which stipulate that a right action is one that protects both the collective and individual interests. The collective is an assemblage of disparate individuals, which, in logical terms, is represented by the intermediate value [52].

Three main principles undergird the African trivalent system: relationality, contextuality, and complementarity. Relationality stipulates that, "variables necessarily interrelate, all things considered because no variable is an ego solus." Contextuality states that the "relationships between variables occur in specific contexts because context upsets facts," while complementarity states that opposites can have a relationship of complementation besides contradiction [53]. In light of global governance for RM, African ethics can come in handy in many ways. Using the example of an embryo above, African ethics can justify such experimentations with the principle of contextuality. If the situation is such that embryos are crucial for breakthrough research, including research that could lead to permanent solutions for cancer or other deadly diseases, thus saving millions, if not billions of lives, it then follows that accommodations can reasonably be made.

Western and religious ethics based on bivalent logical structure often do not permit this flexibility, as evidence of some existing national and regional regulatory frameworks discussed earlier shows. African ethics is largely derived from the African worldview, and it prizes the value of life in whatever form, just like many other worldviews and ethics. But Africans are always able to consider context in their moral evaluations because of the centrality of the concept of relationship. To illustrate, a man who takes his neighbour's yam tuber to feed his wife, who has just delivered a child, is not regarded as a thief. Although the neighbour returns home to learn that he has lost something of value, he is expected to understand. Similarly, embryos have values in themselves, but if experimenting with some of them can possibly lead to solutions to mortal diseases, then the latter context outweighs the former.

Additionally, the principles of relationality and complementarity justify the imperative for cross-cultural cooperation in setting up an inclusive ethical framework for RM. Where the ethical values of a country or region are fundamentally opposed to a crucial research technique or technology, another country or region whose values are not opposed to it should then host the research and deliver therapies to consumers. Additionally, relationality and complementarity demonstrate the necessity for global research collaboration, risk management, ethical decision-making, and disaster prevention in a field that continues to evolve amidst possibilities and dangers.

Conclusion

RM has moved from theory to practice; every day, new technologies and techniques from this field are emerging and are increasingly deployed on an experimental basis. Most of these developments are taking place in the West and in developing economies like Malaysia and Thailand. However, one general concern for many nations is the lack of adequate regulation for RM. This is a problem not just at the national and regional levels but also at the global level, since some clinics offer their services internationally even without having undergone proper clinical trials or adequate testing. The danger of a large-scale application of these therapies is immense, necessitating serious globally coordinated regulation. Creating this global-level framework will be difficult, especially given existing epistemic diversity and disagreement. However, we have charted these epistemic differences as barriers and in turn devised strategies that can ameliorate these barriers and help bring about an inclusive global governance initiative in RM.

Critics may challenge our proposal as too optimistic, citing the complexities of cultural chauvinism and epistemic disagreements, but we will respond by stating that complex situations are not necessarily impossible ones. We wish to add that the African ethical approach we have proposed is open to dialogue and improvement in a way that is sensitive to cultural diversity and other modes of knowing. Given its relational nature, African ethics is undergirded by a reflexive disposition that does not propel it towards supplanting one dominant system with claims to universality with another.

We believe that awareness of the steep challenges posed by the non-regulation or poor regulation of RM is enough to compel stakeholders to set aside petty differences and find common ground. Additionally, if an inclusive framework demonstrates the potential for providing RM services to people irrespective of their economic status, racial backgrounds, or geographic location, then these actors are likely to support it. What we do not know yet is the long-term effects of some of these therapies on the human genetic pool and diversity, how they would impact human genetic structure and diversity, and whether the potential for immortality is something we would come to appreciate or regret should this potential materialize. However, these inquiries constitute issues for future research.

References

1. Yoon J et al (2025) Brief summary of the regulatory frameworks of regenerative medicine therapies. Front Pharmacol 15:1486812. https://doi.org/10.3389/fphar.2024.1486812
2. Qiu T et al (2020) Regenerative medicine regulatory policies: a systematic review and international comparison. Health Policy 124(10):1032–1040. https://doi.org/10.1016/j.healthpol.2020.08.001
3. World Health Organization (2017) Global atlas of medical devices. World Health Organization, Geneva. https://iris.who.int/handle/10665/255181. Accessed 21 May 2025

4. Edgar L, Pu T, Porter B, Aziz JM, La Pointe C, Asthana A, Orlando G (2020) Regenerative medicine, organ bioengineering and transplantation. Br J Surg 107(7):793–800. https://doi.org/10.1002/bjs.11686
5. Barr SW, Rodger MJ, Kelly S (2019) Developmental perspective on regenerative medicine: an update. Sci Insights 2019(2):1–7
6. Atala A (2011) Tissue engineering of human bladder. Br Med Bull 97(1):81–104. https://doi.org/10.1093/bmb/ldr003
7. Mason C, Dunnill P (2008) A brief definition of regenerative medicine. Regen Med 3(1):1–5. https://doi.org/10.2217/17460751.3.1.1
8. Rosemann A, Vasen F, Bortz G (2019) Global diversification in medicine regulation: insights from regenerative stem cell medicine. Sci Cult 28(2):223–249. https://doi.org/10.1080/09505431.2018.1556253
9. Wakitani S et al (2024) Review of Caplan (1991) on cell-based therapeutic technology using mesenchymal stem cells. J ISAKOS 9(3):426–430. https://doi.org/10.1016/j.jisako.2023.08.010
10. Mathew AP et al (2016) Tissue engineering: principles, recent trends and the future. In: Nanomedicine and tissue engineering: state of the art and recent trends, pp 31–82
11. Karagiannis P et al (2019) Induced pluripotent stem cells and their use in human models of disease and development. Physiol Rev 99(1):79–114. https://doi.org/10.1152/physrev.00039.2017
12. Murphy SV, Atala A (2014) 3D bioprinting of tissues and organs. Nat Biotechnol 32(8):773–785. https://doi.org/10.1038/nbt.2958
13. Jovic TH et al (2020) 3D bioprinting and the future of surgery. Front Surg 7:609836. https://doi.org/10.3389/fsurg.2020.609836
14. Trounson A, McDonald C (2015) Stem cell therapies in clinical trials: progress and challenges. Cell Stem Cell 17(1):11–22. https://doi.org/10.1016/j.stem.2015.06.007
15. United States Government Accountability Office (2023) Regenerative medicine: therapeutic applications, challenges, and policy options (Report No. GAO-23-105430). https://www.gao.gov/products/GAO-23-105430. Accessed 21 May 2025
16. Sawarkar S, Bapat A (2022) Global regulatory frameworks and quality standards for stem cells therapy and regenerative medicines. In: Stem cell production: processes, practices and regulations. Springer, pp 69–111
17. Drago D et al (2021) Global regulatory progress in delivering on the promise of gene therapies for unmet medical needs. Mol Ther Methods Clin Dev. https://doi.org/10.1016/j.omtm.2021.04.001
18. Rosemann A et al (2016) Global regulatory developments for clinical stem cell research: diversification and challenges to collaborations. Regen Med 11(7):647–657. https://doi.org/10.2217/rme-2016-0072
19. Hunt T (2024) 2024 state of the industry briefing: Achieving a new normal. Alliance for Regenerative Medicine. Biotech Showcase, San Francisco. https://alliancerm.org
20. Li C (2022) Strengthening regulations, recent advances and remaining barriers in stem cell clinical translation in China: 2015–2021 in review. Pharmacol Res. https://doi.org/10.1016/j.phrs.2022.106304
21. Cherukuri VP, Balakrishna CS, Golla M, Konagala SGS, Penugonda VLD, Addanki S (2021) Regulatory framework for regenerative medicine in India. Int J Drug Regul Affair 9(3):25–31
22. Tiwari SS, Desai PN (2018) Unproven stem cell therapies in India: regulatory challenges and proposed paths forward. Cell Stem Cell 23(5):649–652. https://doi.org/10.1016/j.stem.2018.10.007
23. Gao J, Gao C (2022) Development and regulation of stem cell-based therapies in China. Office of Clinical Evaluation of Biological Products, Center for Drug Evaluation, National Medical Products Administration, Beijing, China
24. Lu J et al (2023) Advanced therapy medicinal products in China: regulation and development. Center for Drug Evaluation, National Medical Products Administration, Beijing, China. https://doi.org/10.1002/mco2.251

25. Alahmad G, Aljohani S, Najjar MF (2020) Ethical challenges regarding the use of stem cells: interviews with researchers from Saudi Arabia. BMC Med Ethics 21(1):35. https://doi.org/10.1186/s12910-020-00482-6
26. Chen Y-S (2016) Mesenchymal stem cell: considerations for manufacturing and clinical trials on cell therapy product. Int J Stem Cell Res Ther 3(1):1155–1163. https://doi.org/10.23937/2469-570X/1410029
27. Pepper MS, Slabbert MN (2015) Human tissue legislation in South Africa: focus on stem cell research and therapy. S Afr J Bioeth Law 8(2):12. https://doi.org/10.7196/SAJBL.8008
28. UNESCO Office Cairo and Regional Bureau for Science in the Arab States (2011) Ethics and law in biomedicine and genetics: an overview of the national regulations in the Arab States. https://unesdoc.unesco.org/ark:/48223/pf0000215207. Accessed 21 May 2025
29. Bauer G, Elsallab M, Abou-El-Enein M (2018) Concise review: a comprehensive analysis of reported adverse events in patients receiving unproven stem cell-based interventions. Stem Cells Transl Med 7(9):676–685. https://doi.org/10.1002/sctm.17-0282
30. U.S. Food and Drug Administration (FDA) (2021) Cell & gene therapy guidances. https://www.fda.gov/vaccines-blood-biologics/biologics-guidances/cellular-gene-therapy-guidances. Accessed 21 May 2025
31. Turner L (2018) The US direct-to-consumer marketplace for autologous stem cell interventions. Perspect Biol Med 61(1):7–24. https://doi.org/10.1353/pbm.2018.0024
32. Sinclair KA (2019) The comprehensive strategic partnership agreement between South Africa and China: a geopolitical case study considering African realism (2010–2015). Doctoral dissertation, North-West University (South Africa). https://dspace.nwu.ac.za/handle/10394/33594. Accessed 21 May 2025
33. Finkelstein LS (1995) What is global governance? Glob Govern 1:367
34. Pease KS (2019) International organizations: perspectives on global governance. Routledge
35. National Human Genome Research Institute (2017) What are the ethical concerns of genome editing? https://www.genome.gov/about-genomics/policy-issues/Genome-Editing/ethical-concerns. Accessed 21 May 2025
36. Lovell-Badge R et al (2021) ISSCR guidelines for stem cell research and clinical translation: the 2021 update. Stem Cell Rep 16(6):1398–1408. https://doi.org/10.1016/j.stemcr.2021.05.012
37. Baylis F (2019) Human genome editing: our future belongs to all of us. Issues Sci Technol 35(3):42–44. https://issues.org/our-future-belongs-to-all-of-us/
38. Hanna E et al (2017) Gene therapies development: slow progress and promising prospects. J Market Access Health Policy 5(1):1265293. https://doi.org/10.1080/20016689.2017.1265293
39. Bokek-Cohen Y, Gonen LD, Tarabeih M (2023) The Muslim patient and medical treatments based on porcine ingredients. BMC Med Ethics 24:89. https://doi.org/10.1186/s12910-023-00975-0
40. Qur'an 5:3, 6:145
41. Qur'an 87:14–15
42. Shomali MA (2008) Islamic bioethics: a general scheme. J Med Ethics Hist Med 1:1
43. Hebga M (2020) Logic in Africa. In: Chimakonam JO (ed) Logic and African philosophy: seminal essays on African systems of thought. Vernon Press, Wilmington, pp 5–14
44. Chimakonam JO (2017) African philosophy and global epistemic injustice. J Global Ethics 13(2):120–137. https://doi.org/10.1080/17449626.2017.1364660
45. Levy-Bruhl L (1923/2018) Primitive mentality. Routledge, London (reprint)
46. Horton R (2020) African traditional thought and Western science. In: Chimakonam JO (ed) Logic and African philosophy: seminal essays on African systems of thought. Vernon Press, Wilmington, pp 27–70
47. Serequeberhan T (1991) African philosophy: the point in question. In: Serequeberhan T (ed) African philosophy: the essential reading. Paragon House, New York, pp 3–28
48. Asouzu II (2007) Ibuanyidanda: new complementary ontology beyond world immanentism, ethnocentric reduction and impositions. Lit Verlag, Zurich

49. Oyekan AO (2017) Rethinking African bioethics in the age of global ethics. In: Aina A, Adefarasin V, Anifowose OA (eds) Discourse on issues in African philosophy. Olabisi Onabanjo University Press, Ago-Iwoye
50. Asouzu II (2023) Complementary ethical reflection in Ibuanyidanda philosophy. In: Chimakonam JO, Cordeiro-Rodrigues L (eds) African ethics: a key guide to key ideas. Bloomsbury, London, pp 121–140
51. Chimakonam AE (2023) A personhood-based theory of right action. In: Chimakonam JO, Cordeiro-Rodrigues L (eds) African ethics: a key guide to key ideas. Bloomsbury, London, pp 103–120
52. Chimakonam JO (2019) Ezumezu: a system of logic for African philosophy and studies. Springer, Cham
53. Chimakonam JO, Chimakonam AE (2023) Examining the logical argument of the problem of evil from an African perspective. Relig Stud 59:326–339. https://doi.org/10.1017/S0034412522000300

Integrating Regenerative Medicine into Medical Curricula in the Global South: A Tanzanian Perspective

4

Afadhali D. Russa and Majigo M. Mtebe

Abstract

One of the most pressing health concerns currently facing Tanzania is the rise in antibiotic-resistant individuals, coupled with the emergence of superbugs. Regenerative medicine (RM) offers innovative solutions to these concerns, as well as to other pressing health issues. However, despite its promise and potential, students, scientists, and medical professionals in Tanzania have limited exposure to this field, and thus its benefits are not translated to the nation's population. In particular, the lack of RM-specific medical training in Tanzania serves as a major barrier for both physicians and patients to access the advantages of RM. This chapter explores the prospects, limitations, required curricular content, and current landscape of RM in Tanzania. It also highlights the issues surrounding antibiotic resistance and emerging superbugs, both of which underscore the need for alternative disease management models. Furthermore, we summarize the likely obstacles to RM in Tanzania, particularly in the context of moral, ethical, and religious dilemmas that may hinder its integration into school curricula and healthcare systems.

Keywords

Regenerative medicine · Medical curricula · Tanzania · Antimicrobial resistance · Stem cell therapy · Global South · Ethical · Regulatory challenges

---

A. D. Russa (✉)
Department of Human Anatomy, School of Biomedical Sciences Muhimbili University of Health and Allied Sciences, Dar es Salaam, Ilala District, Tanzania
e-mail: russa.denis@muhas.ac.tz

M. M. Mtebe
Department of Microbiology and Immunology, School of Diagnostic Medicine, Muhimbili University of Health and Allied Sciences, Dar es Salaam, Ilala District, Tanzania
e-mail: mtebe.majigo@muhas.ac.tz

U. Ogbogu et al. (eds.), *Regenerative Medicine and the Bioethics of Globalization, Justice and Hegemony*, Collaborative Bioethics 5,
https://doi.org/10.1007/978-3-032-07289-4_4

Introduction

The modern public health landscape requires healthcare professionals to adapt and implement new technologies to meet the needs of patients. One current and pressing public health concern, particularly in Tanzania, is the increase in individuals who are antibiotic resistant, coupled with the emergence of superbugs. Regenerative medicine (RM), in particular, stands at the forefront of providing innovative treatments and solutions for these and other similarly urgent public health issues.

In developed countries, research centers and institutions offer clinical and laboratory apprenticeships and hands-on experiences to medical students. These experiences help stimulate the interest and exposure necessary for RM therapies, research, and foundational knowledge of the field. However, despite RM's promise and applicability, prospective physicians in Tanzania are not exposed to RM content during their medical training. This shortcoming is likely to affect the competence of Tanzanian medical professionals in a world where healthcare will increasingly rely on RM-based therapies and technologies.

Ironically, it is these less-equipped future professionals who have the greatest need for RM expertise, as the diseases requiring such interventions are on the rise in Tanzania and throughout much of the Global South. Non-communicable diseases (NCDs) are increasing globally [1] (see Fig. 4.1), and current infectious disease studies show rising trends in antimicrobial resistance and the emergence of superbugs [2–4]. These developments point to a need for comprehensive disease management frameworks, which are likely to include RM-based modalities.

While training programs for physicians and biomedical scientists in developing countries provide some level of exposure to RM, more effort is needed to establish and scale up this novel skill set within the Tanzanian healthcare training and practice systems, as well as in other nations across the Global South. This chapter explores the prospects, limitations, required curricular content, and current landscape of RM in Tanzania. It also highlights the issues surrounding antibiotic resistance and emerging superbugs, both of which call for alternative disease management models. Furthermore, we summarize the likely barriers to RM in Tanzania, particularly in the context of moral, ethical, and religious dilemmas that may negatively affect its adoption in school curricula and healthcare in general.

Tanzania's RM Landscape: Barriers to Expertise and Implementation

RM and cell therapy are poised to become the primary means of managing non-communicable diseases (NCDs) [5–7]. Reports show that current exposure to RM during medical training is generally insufficient worldwide, particularly in terms of contact hours, foundational biology, and clinical clerkship [8]. The situation is even more limited in countries of the Global South [9]. There is a pressing need to more fully integrate RM into these training programs.

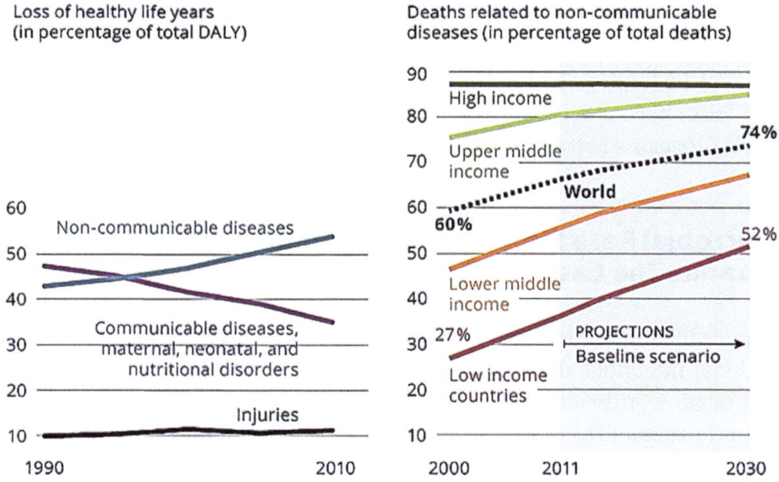

Fig. 4.1 The shift in global disease burden, and share of non-communicable diseases by world regions [1]

Like many countries in the Global South, Tanzania lags behind its northern counterparts in knowledge, technical skills, and clinical application of these emerging biomedical technologies [10, 11]. Incorporating RM and related technologies into Tanzanian medical school curricula could spark interest among future clinical practitioners and support a wider range of therapeutic applications. The knowledge, skills, and advocacy of these practitioners will serve as foundational competencies to guide them throughout their careers. Scientific studies and technical reports support the inclusion of this new area of biomedicine in both undergraduate and postgraduate curricula [12–14]. Doing so will help mainstream RM therapies and lay the groundwork for improved patient care in the foreseeable future [15].

The field of RM in Tanzania is still in its infancy. There are no RM centers of excellence (institutions that exemplify best practices) or dedicated research institutions, and overall awareness of the field among scientists, policymakers, and the general public remains low [7]. Currently, the country has only one RM-based therapy facility, which employs imported hematopoietic stem cells (HSCs) to manage sickle cell disease (SCD). However, this facility is likely insufficient to support sustainable development in stem cell technology and RM-based therapies in Tanzania [16, 17].

Although Tanzania cannot advance RM in isolation, its adoption across the Global South could help accelerate the application of novel technologies, including hematopoietic stem cells. Unsurprisingly, countries in the Global South vary widely in the pace of RM development when compared to those in the Global North. Nevertheless, some countries in the Global South have made considerable progress. For example, South Africa hosts several university institutions and private entities engaged in RM and related technologies such as genomics, biobanking, assisted reproductive technologies, and biomedical engineering. Brazil has also made advances in translating stem cell technologies into RM therapies, including the management of certain brain tumors using multipotent glial stem cells at the Universidad Federal do Rio de Janeiro. Despite these encouraging developments, RM still falls short of delivering tangible benefits in addressing the growing burden of complex diseases across the region.

Antimicrobial Resistance and Non-communicable Diseases in Tanzania: The Case for Regenerative Medicine

Increasing global antimicrobial resistance (AMR) poses a significant threat to the conventional infectious disease management model. AMR is currently the leading cause of death worldwide, with the highest toll occurring in low-resource settings, including Tanzania [18]. As of 2020, Tanzania recorded an estimated 1.29 million hospital-associated drug-resistant infections annually [19]. Additionally, there has been a notable rise in resistance to penicillin, ampicillin, ciprofloxacin, and third-generation cephalosporins, all of which are among the most commonly prescribed antibiotics [20]. The World Health Organization (WHO) reports over 40% resistance to some third-generation antibiotics, such as cephalosporins and quinolones [21]. Under the current One Health approach to disease management, these multidrug-resistant microbes, often referred to as "superbugs," threaten the health of both humans and animals [22]. Given the lack of effective treatment options, the outlook for public health is increasingly concerning.

Currently, data show that non-communicable diseases (NCDs) are the leading cause of mortality globally (see Fig. 4.1) [23, 24]. Moreover, projections indicate that by 2030, NCDs—including cancer, cardiovascular disease, and degenerative conditions—will surpass infectious diseases even in low- and middle-income countries (LMICs) of the Global South, including Tanzania [25].

In much of the Global South, current trends reflect a double burden, where both communicable and non-communicable conditions place significant strain on healthcare systems. A 2023 WHO report showed that in Tanzania, NCDs account for about 38.8% of deaths, while infectious diseases contribute approximately 51.4% [26]. Globally, NCDs are projected to cause more than three-quarters of all deaths by 2030 [23, 25]. As the burden of NCDs continues to grow, the ability to regenerate cells, tissues, and organs will become an essential, and ideally mainstream, treatment option.

While much of the developed world is prepared for this convergence of biomedical innovation and advanced healthcare, Tanzania and many other nations in the Global South remain at a distinct disadvantage. Governments, development partners, multilateral collaborators, agencies, academics, and other stakeholders in these countries must take the lead in promoting RM, stem cell research, and cell-based therapies within their research institutions, centers of excellence, universities, and health facilities.

To effectively incorporate RM in addressing these emerging healthcare challenges, many countries will need to undergo significant mindset shifts, followed by prompt and coordinated action. Given the alarming statistics on both NCDs and emerging superbugs, the medical profession and allied fields must pursue novel approaches to disease management, RM being among the most promising long-term solutions. In this context, RM has shown immense potential in treating diseases traditionally considered degenerative, incurable, or irreversible [27, 28]. These include prevalent conditions such as diabetes, heart disease, spinal cord injuries, burns, Parkinson's disease, and Alzheimer's disease [11, 28]. Since these diseases persist in both LMICs and developed nations, potential solutions should be viewed as matters of global importance.

Balancing Innovation and Integrity: Ethical Dimensions of RM in Tanzania

RM therapies are uncommon in Tanzania, and their introduction may raise concerns from social, moral, and legal standpoints. To properly introduce RM in Tanzania, the country should incorporate RM training into medical curricula and establish comprehensive legal frameworks to govern this emerging field. Legal documents, in the form of laws and policies, are essential to balance the interests of scientific advancement with the protection of public health.

The study and use of stem cells, in particular, have already led to ethical and moral dilemmas and skepticism surrounding stem cells persists despite longstanding awareness of their utility [29, 30]. The regulation of stem cells has served both as a safeguard and as a barrier to the advancement of the technology [31]. The debate over when life begins remains deeply contested across religious, scientific, political, and legal perspectives. Even in countries with relatively loose ethical and legal regulations, strong moral and religious sentiments often persist and can hinder the growth of RM more broadly.

Furthermore, in the pursuit of profit, private companies and agencies may market cell-based therapies without adhering to standardized safety and quality protocols [32–34]. The promotion of unproven treatments raises serious ethical and integrity issues and ultimately puts patients' safety at risk. In one particularly tragic case, a 13-year-old Israeli boy with ataxia-telangiectasia was treated in Russia with intrathecal and intracerebellar injections of human fetal neural stem cells. Four years later, he developed multifocal brain tumors that required neurosurgical intervention. A subsequent investigation revealed that some tumor cells came from the donor stem cells, suggesting that the transplanted neural stem/progenitor cells may have played a role in the development of the tumors [35].

While the use of embryonic stem cell therapies remains challenging globally, reports of fake therapies and dubious scientific breakthroughs are already widespread in countries such as India and China [36, 37]. Tanzania is no exception. Some companies, such as Nutech Mediworld in New Delhi, India, claim to have developed successful, clinically proven treatments, many of which are aggressively advertised on social networks and other media outlets in the region [38, 39]. Despite India's progress in RM, it should also serve as a cautionary tale for other developing nations like Tanzania. Although stem cell therapies in India are generally quite advanced, they are also loosely regulated, creating opportunities for some private health facilities to advertise fraudulent treatment outcomes and ineffective therapies, often under faked approval certifications from so-called regulatory bodies [40, 41].

Due to the unique nature of RM-based therapies, any new treatment regimen requires thorough and rigorous testing before it can be approved as a universally adopted and standard method of patient care.

In addition to expanding RM training for practitioners, Tanzania also needs to adopt existing and widely accepted standards of practice. The International Society for Stem Cell Research (ISSCR) has taken a global leadership role and strongly advises against the use of unproven stem cell therapies [42, 43]. In its efforts to standardize the clinical use of stem cell and regenerative medical procedures, the ISSCR has outlined best practices for the necessary procedures, facilities, expertise, outcomes, and regulations required to develop genuine and safe stem cell therapies [43].

Adopting and implementing these internationally recognized standards will be essential for Tanzania and the wider Global South as the region invests in RM development. To promote responsible and sustainable growth in this field, Tanzania should begin by establishing guiding laws and regulations, and by creating a national research and translational center of excellence. Tanzanian medical schools and healthcare institutions should prioritize collaboration by building institutions of learning and excellence to integrate RM training and practice into local communities. This collaboration depends on the inclusion of RM-focused curricula in basic, applied, and clinical programs so that healthcare professionals are well-equipped to adopt the emerging technologies and therapies in this field.

Due to the complexity of RM technology and its clinical translation, Tanzania should also organize quarterly to annual national forums involving professionals from academia, industry, government, patient and provider organizations, regulatory bodies, and other stakeholders. These forums should address the challenges facing RM, including barriers to scientific and therapeutic advances, as well as the discovery, development, and commercialization of RM therapies. Additional challenges include identifying and validating RM applications, and addressing the ethical, legal, and social issues that arise from the novelty and advancement of this field.

Across the African continent, efforts have been made to build capacity in cell therapy and RM. The African Academy of Sciences (AAS), based in Nairobi, Kenya, has led many of these initiatives. However, since these efforts are dependent on the budgets, regulations, and policies of individual nations, progress appears to have stalled following the onset of COVID-19 in 2020.

Integrating Regenerative Medicine into Medical Education: Capacity Building, Curriculum, and Core Concepts

Tanzania and the Global South nations in general have comparatively better chances of translating RM technology into clinical treatments, as there are many patients in need and many who cannot afford treatment options available locally or overseas. To ensure that RM is a viable and cost-effective option, governments, along with universities, development partners, and research and health institutions, must take a leading role in building greater capacity for RM in these regions. Strengthening this capacity will help ensure that RM becomes accessible to scientists, medical practitioners, patients, and the general public.

Until RM is successfully integrated into the mainstream—both for educational purposes and as a treatment option—the promise of RM therapies will remain out of reach. Early training exposure to RM not only builds the required competencies but also sparks interest and fosters opportunities for innovation in this evolving field.

Currently, RM and stem cell technology in Tanzania are primarily managed by senior scientists within various professional associations, networks, and councils. Furthermore, there is limited crosstalk between the few available facilities and the higher learning institutions responsible for preparing the future healthcare workforce. Medical curricula, facilitators, and training facilities must equip future health professionals with the necessary expertise to translate RM into patient care. Achieving this goal will require concerted efforts and consistent collaboration among all stakeholders.

Medical schools, research centers, government agencies, and all relevant ministries must lead the development of the necessary infrastructure—including laboratories and centers of excellence—for this emerging area of science and medicine. In addition, funding for students, lecturers, and other experts to acquire knowledge and skills from countries where RM is more advanced will be essential to stimulate and enrich Tanzania's capacity in this field. Collaborations with well-established RM centers abroad will also be fundamental to realizing RM's full potential within Tanzania.

A foundational curriculum for medical training in Tanzania should equip future medical professionals with a solid understanding of core RM concepts, beginning with the fundamental classification of cell types used in regenerative therapies. These have traditionally been divided into two main categories: embryonic stem cells (ESCs) and adult or mesenchymal stem cells (MSCs). A third category—induced pluripotent stem cells (iPSCs)—is a relatively recent discovery by Yamanaka and colleagues, who found that adult somatic cells could be reprogrammed into a pluripotent state. This groundbreaking work earned the 2012 Nobel Prize in Physiology or Medicine [44].

Cancer stem cells are a subpopulation of cancer cells that possess significantly enhanced self-renewal properties compared to ordinary stem cells. Although this category receives relatively little attention in stem cell research, it plays a substantial role in aggressive cancers such as glioblastomas and certain forms of leukemia [45, 46].

Beyond these basic categories, a more detailed classification system based on differentiation potential defines five levels of stemness: totipotency, pluripotency, multipotency, oligopotency, and unipotency [47]. Totipotency is generally defined as the ability of a single cell to divide and produce all the definitive cells of an organism, including extraembryonic tissues such as the placenta and amnion. Typically, the earliest stages of development, from the zygote to the morula, represent totipotency. The subsequent formation of the inner cell mass and the epiblast marks the beginning of pluripotency.

Pluripotent stem cells, which emerge from the gastrulation process, can differentiate into all cell types derived from the three germ layers: ectoderm, mesoderm, and endoderm. As differentiation continues through these pathways, cells develop into various tissues and are categorized as multipotent stem cells. The use of stem cells from the totipotent to multipotent stages usually involves the destruction of the embryo, and these are therefore classified as ESCs, a category that has triggered significant ethical and moral debate [48–50].

Several adult cellular and tissue models have been developed, including cardiomyocytes intended to regenerate heart muscle following infarction. However, there is currently no clinically proven application of ESCs in treatment, either in the Global South or elsewhere in the world [7].

One of the main challenges in implementing instruction on RM techniques, especially regarding stem cells, lies in the lack of a clear and universally accepted identification system [51]. Identification markers are limited, and those that have been established often overlap with markers found on tumor or otherwise defective cells.

For ESCs, in addition to strict ethical and moral regulations, there are also demanding in vitro culture and maintenance protocols. Moreover, embryonic stem cells carry the potential to clone an entire human, a possibility that opens the door to numerous moral and ethical controversies [52–54]. Because of these concerns, recent research has shifted toward adult stem cells, particularly MSCs. The identification of MSCs is well documented, their markers are well understood, and they are relatively easy to maintain under in vitro conditions. MSCs are also readily available in nearly all body tissues.

As noted earlier, RM and stem cell technology now emphasize adult stem cells, which may be multipotent, oligopotent, or unipotent. These progenitor cells have varying capacities to differentiate into specific, definitive cell types. Multipotent cells, for instance, can differentiate into one or two related body systems, and RM therapies involving MSCs are typically targeted toward these specific systems or organs. Although there is some variability in identity and markers, adult stem cells must meet minimal criteria: plasticity under standard culture conditions and expression of CD105, CD73, and CD90 as basic surface markers.

Stemness biomarkers are another important element of RM and should likewise be included in Tanzanian medical school curricula. The positive markers—CD105 (also called endoglin), CD73 (NT54 gene encoded), and CD90—are cluster of differentiation (CD) surface receptors consistently expressed by MSCs, although they may also appear on some cancer cells [7]. The minimal criteria for MSCs also include the absence of markers such as CD45, CD34, CD14, or CD11b, CD79alpha or CD19, and HLA-DR [55, 56]. Additionally, MSCs must be capable of differentiating into osteoblasts, adipocytes, and chondroblasts in in vitro culture.

A subset of adult stem cells includes cancer stem cells. Though rare, they can be dangerous if not properly identified. Accurate identification is important to monitor abnormal cell behavior and the development of various tumor types. Cancer stem cells from different tissues express antigens such as CD44, CD133, CD24, CD90, CD271, CD49f, and CD13 [57, 58]. Among these, CD44 is a key transmembrane glycoprotein used as a marker in cancers such as bladder, breast, pancreatic, gastric, prostate, and colon [58].

Another important subset of adult stem cells is hematopoietic stem cells (HSCs), which have the genetic potential to differentiate into all types of white blood cells. HSCs have been studied extensively, and their transplantation has become a standard-of-care in many hematological procedures, including bone marrow transplants (BTs). Therefore, the isolation, identification, and classification of these cells are critical skills that future medical professionals must acquire for RM therapies to successfully move from the bench to the bedside [6, 7].

To facilitate the integration of these foundational RM concepts into Tanzanian medical training, a structured, credit-based curriculum model can be adopted. As illustrated in Table 4.1, junior clerkship students can be introduced to core theoretical and practical components such as ESCs, adult stem cells, and iPSCs, along with a practicum in stem cell handling and application. This early exposure provides a strong conceptual grounding in pluripotency and cellular differentiation. At the senior clerkship level, regenerative medicine content is extended to clinical contexts through core modules in specialties with high patient volumes—orthopedics, oncology, general surgery, and internal medicine—while elective modules cover areas like hematology, dermatology, and ophthalmology (see Table 4.2). This modular approach ensures that students progressively connect theoretical knowledge with clinical relevance, aligning well with Tanzania's competency-based medical education framework.

Table 4.1 Key RM curricular content for junior clerkship

Serial number	Topic	Credits ratio[a]	Type
1	Embryonic stem cell	1	Core
2	Adult stem cells	2	Core
3	Induced pluripotent cells (IPSCs)	1	Core
4	Practicum in stem cells	2	Core

[a]For a basic medical degree of 4–5 years of training, a standard credit refers to 10 hours of theory and 15 hours of practical

Table 4.2 Key RM curricula content for senior clerkship[a]

SN	Topic	Credits ratio	Type
1	RM in orthopedics	1	Core
2	RM in oncology	1	Core
3	RM in general surgery	1	Core
4	RM in internal medicine	1	Core
5	RM in hematology	1	Elective
6	RM in dermatology	1	Elective
6	RM in ophthalmology	1	Elective

[a]Core vs elective based on the patient load by subspecialty

Building Clinical Impact Through RM Education: Addressing NCDs and Treatment Gaps in Tanzania

Due to the increasing prevalence of NCDs such as cancer and diabetes mellitus, RM-based therapies are expected to play a significant role in managing the burden of these diseases. Additionally, Tanzania faces its own unique set of health challenges, a double burden of both communicable and non-communicable diseases. The use of stem cell infusions to regenerate tissue in patients undergoing chemotherapy or radiotherapy is currently unavailable in Tanzania, but such therapies are already at the clinical stage in some Global South countries and other parts of the world [59].

Several stem cell-based treatments are possible, especially for patients who have very limited options. This includes individuals with conditions such as end-stage liver cirrhosis and heart failure, cases in which the potential benefits of stem cell treatment outweigh the risks associated with experimental therapies [7]. In Tanzania, where advanced procedures like liver transplants are not available, stem cell therapies could offer substantial relief to these critically ill patients. Use of stem cells in the treatment of chronic NCDs is already underway in some African healthcare institutions [60].

Easy and freely accessible methods for the isolation and characterization of stem cells are well-documented in scientific and medical literature [61–63]. Protocols for isolating adult stem cells from various tissues and organs have been tried and validated, including from umbilical cord and Wharton's jelly, spermatogonia and oogonia, and bone marrow [63]. Practical methods and routes for administering these cells to patients with liver cirrhosis have been reviewed previously, and many of these approaches are feasible in LMICs like Tanzania, as they require only minimal technology and infrastructure [60].

Commonly used markers for stem cell identification include CD family plasma membrane receptors, which can be characterized using commercially available monoclonal antibody kits in combination with confocal live-cell imaging, fluorescent in situ hybridization (FISH), or western blotting [63]. We have previously demonstrated the regenerative potential of fibroblast-like cells isolated from subcutaneous tissue [64], which can be easily applied to wound and injury management, conditions especially prevalent in LMICs like Tanzania.

Global Benchmarks and Regional Successes: Lessons for Advancing RM in Tanzania and the Global South

In technologically advanced countries, research centers and institutions provide clinical or laboratory attachments where medical students gain hands-on experience that helps stimulate the interest and exposure necessary for RM therapies, research, and foundational knowledge of the field. Advanced methods and techniques, including clinical trials and specialized patient care, are also regularly conducted by such facilities, offering trainees even greater exposure. These centers are currently nonexistent in Tanzania and are likely to remain so for the foreseeable future.

To illustrate the disparity between these countries and LMICs, in the state of California alone, there are over nine research and training centers—including the University of California (UC) San Francisco Alpha Clinic, Cedars-Sinai, City of Hope, UC Davis, UC Irvine, UC San Diego, UC San Francisco, the University of Southern California, and Stanford University—operating under the California Institute for Regenerative Medicine (CIRM) in South San Francisco. These institutions offer training in developmental biology, animal regeneration, stem cell science, and RM, along with exposure to laboratory techniques in stem cell research. Other centers, such as the University of Southern California (USC) Stem Cell: Eli and Edythe Broad Center for Regenerative Medicine at the Keck School of Medicine, offer academic minors designed for undergraduates interested in the scientific, ethical, legal, regulatory, and cultural dimensions of stem cell biology and RM. Students at these institutions may even participate in developing novel RM-based treatments for challenging conditions such as Huntington's disease, SCD, autism, muscular dystrophy, and Amyotrophic Lateral Sclerosis (ALS). Numerous similar centers operate across other technologically advanced countries. Additionally, the Stem Cell Network–Canada has trained thousands of students and professionals in RM therapies through university laboratories, hospitals, industry placements, and international exchanges [65].

There are also promising translational RM practices at the bedside level in several countries within the Global South, where infrastructure is comparable to that of Tanzania. The treatment of chronic NCDs such as cirrhosis, diabetes, and cardiac infarction relies heavily on adult stem cells with multipotent, oligopotent, or unipotent capabilities. Chemical treatments are used to induce differentiation into specific cell types using specialized media. In a series of studies conducted at the Hepatology and Tropical Medicine Institute in Cairo, over 150 post-hepatitis C end-stage liver disease patients treated with ex vivo culture-amplified autologous mesenchymal stem cells showed reversal of numerous severe disease indicators. These included improvements in albumin, bilirubin, INR, prothrombin concentration, and ALT levels, as well as stabilization of other clinical and biochemical variables. These outcomes were not replicated in the control group, which received standard palliative treatment for end-stage liver cirrhosis [31, 60, 66]. In Egypt alone, approximately eight stem cell centers were operational between 1997 and 2017, during which time over 4000 patients received hematopoietic stem cell transplants [67]. These treatments showed varying degrees of success, including improvements in anthropometry, lipid profiles, liver and renal function, hormonal balance, and inflammatory markers [68].

These examples demonstrate the growing influence of RM as an accepted medical practice across African countries. In a phase two clinical trial conducted by Tunisian researchers, eight patients with end-stage liver cirrhosis due to Hepatitis B, Hepatitis C, alcoholism, or cryptogenic causes experienced significant liver function restoration [69]. The treatment was well tolerated, with no serious side effects or instances of tissue rejection. A similar Egyptian study reported improvements in ascites and prothrombin time after a three-month treatment of 20 cirrhotic patients using autologous bone marrow-derived MSCs [60].

While many of the novel RM treatments to date focus on liver disease, several successful applications have also emerged for heart failure and other categories of NCDs, particularly autoimmune disorders ([70–72]). Countries across the Global South, including those in Sub-Saharan Africa, report high numbers of heart failure cases ([73–75]); thus, the availability of stem cell treatment could benefit a significant patient population. Promising research into the treatment of type 1 diabetes mellitus using MSCs from sources such as bone marrow, adipose tissue, umbilical cord blood, and Wharton's jelly is currently underway at the University of Khartoum in Sudan [76]. Moreover, the relatively simple infusion of allogeneic umbilical cord mesenchymal stem cells (UC-MSCs) via central or peripheral intravenous routes has shown positive outcomes in patients with heart failure and reduced ejection fraction, including improved ventricular function, better functional status, and overall enhanced quality of life [68, 70, 72].

Despite these encouraging achievements and case studies, budgetary constraints and technological limitations continue to hinder the adoption of RM in Tanzania and other parts of the Global South. Ultimately, the implementation of RM-based therapies in resource-limited settings will only be feasible if countries like Tanzania are able to take advantage of opportunities presented by more flexible ethical environments. Taken together, these scientific and clinical translational developments strongly support the need to integrate RM into medical curricula in Tanzania and across the Global South.

Concluding Thoughts

As the research and healthcare landscape in Tanzania currently stands, the country is not yet prepared to integrate RM into clinical practice. Deliberate and multifaceted efforts from scientists, clinicians, policymakers, and politicians are required, and these groups must work collaboratively to make the implementation of RM feasible. A critical first step is incorporating RM into medical training curricula. To achieve this, scientists working in RM and related fields must raise awareness of the discipline among policymakers in the ministries of health and education, as well as among university administrators, to support the introduction of RM as a medical specialty.

Medical schools, research centers, government agencies, and all relevant ministries must take the lead in initiating and investing in the necessary infrastructure—such as laboratories and centers of excellence—required to support this new area of science and medicine. It will also be essential for Tanzania to fund students, lecturers, and other professionals so they can acquire the knowledge and skills needed to apply RM in clinical and research settings.

In addition to developing internal capacity, Tanzania must invest in international partnerships to gain expertise from countries where RM is already well established. This investment and outreach will ultimately stimulate and enrich the country's knowledge base and practical capabilities in RM. Collaboration with the Global North will also be fundamental to the successful advancement of RM across the Global South.

References

1. European Environment Agency (2024) The shift in global disease burden, and share of non-communicable diseases by world region. European Environment Information and Observation Network. https://www.eea.europa.eu/en/analysis/maps-and-charts/the-shift-in-global-disease. Accessed 12 Jun 2025
2. Antimicrobial Resistance Collaborators (2022) Global burden of bacterial antimicrobial resistance in 2019: a systematic analysis. Lancet 399(10325):629–655. https://doi.org/10.1016/S0140-6736(21)02724-0
3. Painuli S, Semwal P, Sharma R, Akash S (2023) Superbugs or multidrug resistant microbes: a new threat to the society. Health Sci Rep 6(8):e1480. https://doi.org/10.1002/hsr2.1480
4. Salam MA, Al-Amin MY, Salam MT, Pawar JS, Akhter N, Rabaan AA, Alqumber MAA (2023) Antimicrobial resistance: a growing serious threat for global public health. Healthc (Basel) 11(13):1946. https://doi.org/10.3390/healthcare11131946
5. Hoang DM, Pham PT, Bach TQ, Ngo ATL, Nguyen QT, Phan TTK, Nguyen GH, Le PTT, Hoang VT, Forsyth NR, Heke M, Nguyen LT (2022) Stem cell-based therapy for human diseases. Sig Transduct Target Ther 7:272. https://doi.org/10.1038/s41392-022-01134-4
6. Kalra K, Tomar PC (2014) Stem cell: basics, classification and applications. Am J Phytomed Clin Therap 2:919–930. https://www.semanticscholar.org/paper/Stem-Cell%3A-Basics%2C-Classification-and-Applications-Kalra-Tomar/bc887910522365f0e2fcddf-c00c0b194c9e28733. Accessed 12 Jun 2025
7. Russa AD (2020) Stem cells: prospects and potential applications in Tanzania: a review. Tanzan J Health Res 21(2):1–8. https://doi.org/10.4314/thrb.v21i2.8
8. Pandya R, Talib Z (2022) A scoping review of regenerative medicine in medical education. BMC Med Educ 22:758. https://doi.org/10.1186/s12909-022-03816-7
9. Khasawneh RR, Rub EA (2022) Undergraduate medical students' perceptions regarding stem cells: is there a need for incorporating a stem cells elective course in the academic curriculum? Biochem Mol Biol Educ 50(6):620–626. https://doi.org/10.1002/bmb.21672
10. Greenwood HL, Singer PA, Downey GP, Martin DK, Thorsteinsdóttir H, Daar AS (2006) Regenerative medicine and the developing world. PLoS Med 3(9):e381. https://doi.org/10.1371/journal.pmed.0030381
11. Mao AS, Mooney DJ (2015) Regenerative medicine: current therapies and future directions. Proc Natl Acad Sci USA 112(47):14452–14459. https://doi.org/10.1073/pnas.1508520112
12. Alliance for Regenerative Medicine (2013) Annual report 2013. https://alliancerm.org/sites/default/files/ARM_Annual_Report_2013_Website.pdf. Accessed 12 June 2025

13. Ellythy L, Addani M, Master Z (2025) An analysis of stem cell training programs for physicians in the US—A mirage of credibility. Stem Cell Rep 20(6):102510. https://doi.org/10.1016/j.stemcr.2025.102510
14. Wyles SP, Terzic A (2019) Building the regenerative medicine workforce of the future: an educational imperative. Regen Med 14(7):613–615. https://doi.org/10.2217/rme-2019-0072
15. Wyles SP, Hayden RE, Meyer FB, Terzic A (2019) Regenerative medicine curriculum for next-generation physicians. NPJ Regen Med 4:3. https://doi.org/10.1038/s41536-019-0065-8
16. Mtenga J, Orf K, Zheng J, Chamba C, Chuwa H, Luoga F, Malangahe SW, Iversen PO, Makani J (2021) Haematopoietic stem cell transplantation in Tanzania. Br J Haematol 192(1):17–21. https://doi.org/10.1111/bjh.17106
17. Rwezaula S, Yonazi M, Panchal A, Dhoot A, Mathew J, Tony S, Rao S, Muhoka P, Mahfudh S, Budodi N et al (2024) Challenges and outcomes of the first stem cell transplant program in Tanzania, East Africa. Adv Hematol 1:1937419. https://doi.org/10.1155/2024/1937419
18. Mshana SE, Sindato C, Matee MI, Mboera LEG (2021) Antimicrobial use and resistance in agriculture and food production systems in Africa: a systematic review. Antibiotics 10(8):976. https://doi.org/10.3390/antibiotics10080976
19. Balasubramanian R, Van Boeckel TP, Carmeli Y, Cosgrove S, Laxminarayan R (2023) Global incidence in hospital-associated infections resistant to antibiotics: an analysis of point prevalence surveys from 99 countries. PLoS Med 20(6):e1004178. https://doi.org/10.1371/journal.pmed.1004178
20. Silago V, Oravcova K, Matthews L, Mshana SE, Claus H, Seni J (2024) Epidemiology and antimicrobial resistance trends of pathogens causing urinary tract infections in Mwanza, Tanzania: a comparative study during and after the implementation of the National Action Plan on Antimicrobial Resistance (2017–2022). Int J Infect Dis 147:107208. https://doi.org/10.1016/j.ijid.2024.107208
21. World Health Organization (2022) Global antimicrobial resistance and use surveillance system (GLASS) Report, 2022. https://www.who.int/publications/i/item/9789240062702. Accessed 12 June 2025
22. World Health Organization (2023) A one health priority research agenda for antimicrobial resistance, 2023. https://www.who.int/publications/i/item/9789240075924. Accessed 12 June 2025
23. Bigna JJ, Noubiap JJ (2019) The rising burden of non-communicable diseases in sub-Saharan Africa. Lancet Glob Health 7(10):e1295–e1296. https://doi.org/10.1016/S2214-109X(19)30370-5
24. Müller SA, Elimian K, Rafamatanantsoa JF, Reichert F, Mosala F, Böff TSF, Boone I, Ravaoarisoa, Nduenga S, Ortu G, Pozo-Martin F, Tomczyk S, Eckmanns T, Okwor T, Akoua-Koffi C, Diané B, Randriamanatany ZA, Ahuka S, Ochu CL, El-Bcheraoui (2024) The burden and treatment of non-communicable diseases among healthcare workers in sub-Saharan Africa: a multi-country cross-sectional study. Front Public Health 12:1375221. https://doi.org/10.3389/fpubh.2024.1375221
25. Ndumwa HP, Amani DE, Ngowi JE, Njiro BJ, Munishi C, Mboya EA, Mloka D, Kikula AI, Balandya E, Ruggajo P, Kessy AT, Kitambala E, Kapologwe N, Kengia JT, Kiologwe J, Ubuguyu O, Salum B, Kamuhabwa A, Ramaiya K, Sunguya BF (2023) Mitigating the rising burden of non-communicable diseases through locally generated evidence—lessons from Tanzania. Ann Glob Health 89(1):77. https://doi.org/10.5334/aogh.4111
26. World Health Organization (2025) Health data overview for the United Republic of Tanzania. World Health Organization Data. https://data.who.int/countries/834. Accessed 12 Jun 2025
27. Nelson TJ, Behfar A, Terzic A (2008) Strategies for therapeutic repair: the "R(3)" regenerative medicine paradigm. Clin Transl Sci 1(2):168–171. https://doi.org/10.1111/j.1752-8062.2008.00039.x
28. Zhidu S, Ying T, Rui J, Chao Z (2024) Translational potential of mesenchymal stem cells in regenerative therapies for human diseases: challenges and opportunities. Stem Cell Res Ther 15(1):266. https://doi.org/10.1186/s13287-024-03885-z

29. Charitos IA, Ballini A, Cantore S, Boccellino M, Di Domenico M, Borsani E, Nocini R, Di Cosola M, Santacroce L, Bottalico L (2021) Stem cells: a historical review about biological, religious, and ethical issues. Stem Cells Int 2021:9978837. https://doi.org/10.1155/2021/9978837
30. Volarevic V, Markovic BS, Gazdic M, Volarevic A, Jovicic N, Arsenijevic N, Armstrong L, Djonov V, Lako M, Stojkovic M (2018) Ethical and safety issues of stem cell-based therapy. Int J Med Sci 15(1):36–45. https://doi.org/10.7150/ijms.21666
31. Pepper SM, Slabbert MN (2015) Human tissue legislation in South Africa: focus on stem cell research and therapy. S Afr J Bioeth Law 8(2 Suppl 1):4–11. https://www.ajol.info/index.php/sajbl/article/view/125134. Accessed 12 Jun 2025
32. Lau D, Ogbogu U, Taylor B, Stafinski T, Menon D, Caulfield T (2008) Stem cell clinics online: the direct-to-consumer portrayal of stem cell medicine. Cell Stem Cell 3(6):591–594. https://doi.org/10.1016/j.stem.2008.11.001
33. Ogbogu U (2016) Combatting unlicensed stem cell interventions through truthful advertising law: a survey of regulatory trends. McGill J Law Health 9:311–328
34. Turner L, Knoepfler P (2016) Selling stem cells in the USA: assessing the direct-to-consumer industry. Cell Stem Cell 19:154–157. https://doi.org/10.1016/j.stem.2016.06.007
35. Amariglio N, Hirshberg A, Scheithauer BW, Cohen Y, Loewenthal R, Trakhtenbrot L, Paz N, Koren-Michowitz M, Waldman D, Leider-Trejo L, Toren A, Constantini S, Rechavi G (2009) Donor-derived brain tumor following neural stem cell transplantation in an ataxia telangiectasia patient. PLoS Med 6(2):e1000029. https://doi.org/10.1371/journal.pmed.1000029
36. Lv J, Su Y, Song L, Gong X, Peng Y (2020) Stem cell 'therapy' advertisements in China: infodemic, regulations and recommendations. Cell Prolif 53(12):e12937. https://doi.org/10.1111/cpr.12937
37. Tiwari SS, Raman S (2014) Governing stem cell therapy in India: regulatory vacuum or jurisdictional ambiguity? New Genet Soc 33(4):413–433. https://doi.org/10.1080/14636778.2014.970269
38. Master Z, Matthews KRW, Abou-El-Enein M (2021) Unproven stem cell interventions: a global public health problem requiring global deliberation. Stem Cell Rep 16(6):1435–1445. https://doi.org/10.1016/j.stemcr.2021.05.004
39. Takashima K, Minari J, Chan S, Muto K (2023) Hope for the best, but prepare for the worst: social media posted by participants in stem cell clinical trials. Regen Ther 24:294–297. https://doi.org/10.1016/j.reth.2023.07.009
40. Boopalan D, Pandian R, Kesavan G (2021) Prospects for stem cell-based regenerative therapies in India. Stem J 3(1):11–21. https://doi.org/10.3233/STJ-210002
41. Tiwari SS, Desai PN (2018) Unproven stem cell therapies in India: regulatory challenges and proposed paths forward. Cell Stem Cell 23(5):649–652. https://doi.org/10.1016/j.stem.2018.10.007
42. International Society for Stem Cell Research (ISSCR) (2021) ISSCR guidelines for stem cell research and clinical translation: 2021 update. https://www.isscr.org/guidelines. Accessed 12 June 2025
43. Lovell-Badge R, Anthony E, Barker RA et al (2021) ISSCR guidelines for stem cell research and clinical translation: the 2021 update. Stem Cell Rep 16(6):1398–1408. https://doi.org/10.1016/j.stemcr.2021.05.012
44. Yamanaka S (2012) Induced pluripotent stem cells: past, present, and future. Cell Stem Cell 10(6):678–684. https://doi.org/10.1016/j.stem.2012.05.005
45. Ayob AZ, Ramasamy TS (2018) Cancer stem cells as key drivers of tumour progression. J Biomed Sci 25:20. https://doi.org/10.1186/s12929-018-0426-4
46. Chu X, Tian W, Ning J, Xiao G, Zhou Y, Wang Z, Zhai Z, Tanzhu G, Yang J, Zhou R (2024) Cancer stem cells: advances in knowledge and implications for cancer therapy. Sig Transduct Target Ther 9:170. https://doi.org/10.1038/s41392-024-01851-y
47. Ntege EH, Sunami H, Shimizu Y (2020) Advances in regenerative therapy: a review of the literature and future directions. Regen Ther 14:136–153. https://doi.org/10.1016/j.reth.2020.01.004
48. Landry DW, Zucker HA (2004) Embryonic death and the creation of human embryonic stem cells. J Clin Invest 114(9):1184–1186. https://doi.org/10.1172/JCI23065

49. Lo B, Parham L (2009) Ethical issues in stem cell research. Endocr Rev 30(3):204–213. https://doi.org/10.1210/er.2008-0031
50. Peng YJ, Huang X, Zhou Q (2020) Ethical and policy considerations for human embryo and stem cell research in China. Cell Stem Cell 27(4):511–514. https://doi.org/10.1016/j.stem.2020.09.010
51. Dua HS, Joseph A, Shanmuganathan VA, Jones RE (2003) Stem cell differentiation and the effects of deficiency. Eye 17(8):877–885. https://doi.org/10.1038/sj.eye.6700573
52. Bonetti G, Donato K, Medori MC, Dhuli K, Henehan G, Brown R, Sieving P, Sykora P, Marks R, Falsini B, Capodicasa N, Miertus S, Lorusso L, Dondossola D, Tartaglia GM, Ergoren MC, Dundar M, Michelini S, Malcarne D, Beccari T, Connelly ST, Martin D, Bacu A, Herbst KL, Kapustin M, Stuppia L, Lumer L, Farronato G, Bertelli M (2023) Human cloning: biology, ethics, and social implications. Clin Ter 174 Suppl 2(6):230–235. https://doi.org/10.7417/CT.2023.2492
53. Collmann J, Graber G (2000) Developing a code of ethics for human cloning. Crit Rev Biomed Eng 28(3–4):563–566. https://doi.org/10.1615/critrevbiomedeng.v28.i34.370
54. Dinç L (2003) Ethical issues regarding human cloning: a nursing perspective. Nurs Ethics 10(3):238–254. https://doi.org/10.1191/0969733003ne603oa
55. Dominici M, Le Blanc K, Mueller I, Slaper-Cortenbach I, Marini F, Krause D, Deans R, Keating A, Prockop DJ, Horwitz E (2006) Minimal criteria for defining multipotent mesenchymal stromal cells. The International Society for Cellular Therapy position statement. Cytotherapy 8(4):315–317. https://doi.org/10.1080/14653240600855905
56. Renesme L, Pierro M, Cobey KD, Mital R, Nangle K, Shorr R, Lalu MM, Thébaud B (2022) Definition and characteristics of mesenchymal stromal cells in preclinical and clinical studies: a scoping review. Stem Cells Transl Med 11(1):44–54. https://doi.org/10.1093/stcltm/szab009
57. Saito S, Ku CC, Wuputra K, Pan JB, Lin CS, Lin YC, Wu DC, Yokoyama KK (2022) Biomarkers of cancer stem cells for experimental research and clinical application. J Pers Med 12(5):715. https://doi.org/10.3390/jpm12050715
58. Walcher L, Kistenmacher AK, Suo H, Kitte R, Dluczek S, Strauß A, Blaudszun AR, Yevsa T, Fricke S, Kossatz-Boehlert U (2020) Cancer stem cells—origins and biomarkers: perspectives for targeted personalized therapies. Front Immunol 11:1280. https://doi.org/10.3389/fimmu.2020.01280
59. Atewologun FA, Okesanya OJ, Okon II, Kayode HH, Ukoaka BM, Olaleke NO, Ogaya JB, Okikiola LA, Manirambona E, Lucero-Prisno DE III (2024) Examining the potentials of stem cell therapy in reducing the burden of selected non-communicable diseases in Africa. Stem Cell Res Ther 15:253. https://doi.org/10.1186/s13287-024-03864-4
60. Salama H, Zekri AR, Medhat E, Al Alim SA, Ahmed OS, Bahnassy AA, Lotfy MM, Ahmed R, Musa S (2014) Peripheral vein infusion of autologous mesenchymal stem cells in Egyptian HCV-positive patients with end-stage liver disease. Stem Cell Res Ther 5(3):70. https://doi.org/10.1186/scrt459
61. Bieback K, Schallmoser K, Klüter H, Strunk D (2008) Clinical protocols for the isolation and expansion of mesenchymal stromal cells. Transfus Med Hemother 35(4):286–294. https://doi.org/10.1159/000141567
62. Giordano A, Galderisi U, Marino IR (2007) From the laboratory bench to the patient's bedside: an update on clinical trials with mesenchymal stem cells. J Cell Physiol 211(1):27–35. https://doi.org/10.1002/jcp.20959
63. Mushahary D, Spittler A, Kasper C, Weber V, Charwat V (2017) Isolation, cultivation, and characterization of human mesenchymal stem cells. Cytometry A 93(1):19–31. https://doi.org/10.1002/cyto.a.23242
64. Russa AD (2014) Bromodeoxyuridine immunofluorescence and differential interference contrast imaging combination can precisely segregate adherent monolayer cells into specific cell-cycle phases. Tanzan J Health Res 16(2). https://doi.org/10.4314/thrb.v16i2.10
65. Stem Cell Network (2025) Training the next generation of Canadian Talent. 2025 Stem Cell Network. https://stemcellnetwork.ca/training/. Accessed 12 Jun 2025

66. Al Tayeb H, El Dorry A, Amer N, Mowafy N, Zimaity M, Bayoumy E, Saleh SA (2015) Autologous stem cells transplantation in Egyptian patients with liver cirrhosis on top of hepatitis C virus. Int J Stem Cells 8(2):209–218. https://doi.org/10.15283/ijsc.2015.8.2.209
67. Mahmoud HK, Fathy GM, Elhaddad A, Fahmy OA, Abdel-Mooti M, Abdelfattah R, Bokhary M (2020) Hematopoietic stem cell transplantation in Egypt: challenges and opportunities. Medditerr J Hematol Infect Dis 12(1):e2020023. https://doi.org/10.4084/MJHID.2020.023
68. Chin SP, Kee LT, Mohd MA, Then KY (2025) Umbilical cord-derived mesenchymal stem cells infusion in type 2 diabetes mellitus patients: a retrospective cytopeutics' registry study. Diabetes Metab Syndr Obes 18:1643–1659. https://doi.org/10.2147/DMSO.S507801
69. Kharaziha P, Hellström PM, Noorinayer B, Farzaneh F, Aghajani K, Jafari F, Telkabadi M, Atashi A, Honardoost M, Zali MR, Soleimani M (2009) Improvement of liver function in liver cirrhosis patients after autologous mesenchymal stem cell injection: a phase I–II clinical trial. Eur J Gastroenterol Hepatol 21(10):1199–1205. https://doi.org/10.1097/MEG.0b013e32832a1f6c
70. Bartolucci J, Verdugo FJ, González PL, Larrea RE, Abarzua E, Goset C, Rojo P, Palma I, Lamich R, Pedreros PA, Valdivia G, Lopez VM, Nazzal C, Alcayaga-Miranda F, Cuenca J, Brobeck MJ, Patel AN, Figueroa FE, Khoury M (2017) Safety and efficacy of the intravenous infusion of umbilical cord mesenchymal stem cells in patients with heart failure: a phase 1/2 randomized controlled trial (RIMECARD trial [randomized clinical trial of intravenous infusion umbilical cord mesenchymal stem cells on cardiopathy]). Circ Res 121(10):1192–1204. https://doi.org/10.1161/CIRCRESAHA.117.310712
71. Bartunek J, Terzic A, Davison BA, Filippatos GS, Radovanovic S, Beleslin B, Merkely B, Musialek P, Wojakowski W, Andreka P, Horvath IG, Katz A, Dolatabadi D, El Nakadi B, Arandjelovic A, Edes I, Seferovic PM, Obradovic S, Vanderheyden M, Jagic N, Petrov I, Atar S, Halabi M, Gelev VL, Shochat MK, Kasprzak JD, Sanz-Ruiz R, Heyndrickx GR, Nyolczas N, Legrand V, Guédès A, Heyse A, Moccetti T, Fernandez-Aviles F, Jimenez-Quevedo P, Bayes-Genis A, Hernandez-Garcia JM, Ribichini F, Gruchala M, Waldman SA, Teerlink JR, Gersh BJ, Povsic TJ, Henry TD, Metra M, Hajjar RJ, Tendera M, Behfar A, Alexandre B, Seron A, Stough WG, Sherman W, Cotter G, Wijns W, Program CHART (2017) Cardiopoietic cell therapy for advanced ischaemic heart failure: results at 39 weeks of the prospective, randomized, double blind, sham-controlled CHART-1 clinical trial. Eur Heart J 38(9):648–660. https://doi.org/10.1093/eurheartj/ehw543
72. Chin SP, Poey AC, Wong CY, Chang SK, Teh W, Mohr TJ, Cheong SK (2010) Cryopreserved mesenchymal stromal cell treatment is safe and feasible for severe dilated ischemic cardiomyopathy. Cytotherapy 12(1):31–37. https://doi.org/10.3109/14653240903313966
73. Gallagher J, McDonald K, Ledwidge M, Watson CJ (2018) Heart failure in Sub-Saharan Africa. Card Fail Rev 4(1):21–24. https://doi.org/10.15420/cfr.2018:4:1
74. Makubi A, Hage C, Sartipy U, Lwakatare J, Janabi M, Kisenge P, Dahlström U, Rydén L, Makani J, Lund LH (2016) Heart failure in Tanzania and Sweden: comparative characterization and prognosis in the Tanzania Heart Failure (TaHeF) study and the Swedish Heart Failure Registry (SwedeHF). Int J Cardiol 220:750–758. https://doi.org/10.1016/j.ijcard.2016.06.239
75. Siddikatou D, Mandeng Ma Linwa E, Ndobo V, Nkoke C, Mouliom S, Ndom MS, Abas A, Kamdem F (2025) Heart failure outcomes in Sub-Saharan Africa: a scoping review of recent studies conducted after the 2022 AHA/ACC/HFSA guideline release. BMC Cardiovasc Disord 25(1):302. https://doi.org/10.1186/s12872-025-04756-y
76. Babiker NE, Gassoum A, Abdelraheem NE, Arbab MA, Ahmed S, AlDeaf H, El-Sheikh MAA, Musa HH (2017) The progress of stem cells in the treatment of diabetes mellitus type 1. Prog Stem Cell 4(1):175–188. https://doi.org/10.15419/psc.v4i01.184

Part II

National Case Studies: Governance, Context, Constraint, and Endogenous Innovation

Translation and Commercialisation of Regenerative Medicine in South Africa: Navigating the Regulatory Landscape

5

Marietjie Botes

Abstract

Despite promises of improved diagnostic and therapeutic methods for diseases without effective existing treatments, disappointingly few regenerative medicines have entered the market, largely due to regulatory and ethical barriers. This chapter argues that South Africa must adopt an agile regulatory framework to balance innovation, commercialisation, and patient safety in the translation of regenerative medicine (RM). Although RM holds transformative potential for diseases lacking effective treatments, the country's current legislative instruments, the National Health Act and the Medicines and Related Substances Act, fail to adequately address the unique scientific, ethical, and commercial complexities of RM, creating a regulatory vacuum that hinders innovation and oversight. As a result, local innovation is stifled, patients risk exposure to unproven therapies, and South Africa remains tethered to Global North regulatory paradigms that do not align with its socio-economic and healthcare realities. Through a comprehensive analysis of South Africa's existing regulatory infrastructure, this chapter identifies key gaps, including ambiguous definitions, outdated approval processes, and ethical challenges around stem cell sourcing and clinical application. It further evaluates the benefits and limitations of adopting an agile regulatory framework informed by anticipatory, inclusive, and responsive governance practices. To address these gaps and limitations, the chapter recommends the development of RM-specific legislation, integration of international standards into local guidelines, public-private partnerships to fund innovation, and safeguards to prevent exploitative commercialisation. These actions aim to foster a regulatory environment conducive to responsible innovation while ensuring that regenerative treatments are accessible, affordable, and ethically governed.

M. Botes (✉)
University of KwaZulu Natal, Durban, South Africa

U. Ogbogu et al. (eds.), *Regenerative Medicine and the Bioethics of Globalization, Justice and Hegemony*, Collaborative Bioethics 5,
https://doi.org/10.1007/978-3-032-07289-4_5

Keywords

Regenerative Medicine · Stem Cell Therapies · South Africa · Ethical Governance · Agile Regulation

Introduction

Regenerative medicine (RM) is an interdisciplinary field that aims to repair, replace, or regenerate damaged or diseased tissues and organs using stem cells, tissue engineering, and biomaterials [1]. This field leverages the body's natural healing processes and often involves the use of cellular therapies, engineered tissues, and gene-editing techniques to restore normal function. These therapies represent novel classes of health products, promising to address various currently unmet medical needs. However, they differ from primary cells and tissues used for transplantation in several important ways, including: the complexity and variability of cellular starting materials, the use of raw materials and reagents of biological origin, disruptive manufacturing technologies, the scale of manufacture, living cells and competent vectors as active components, limited patient numbers with complex disease profiles, and the invasive nature of treatments [2].

The unique nature of RM also requires regulatory frameworks to accommodate their differences from other types of therapies. For this reason, countries such as the United States, European Union, Japan, Canada, Australia, South Korea, New Zealand, Singapore, China, and India have adopted regulatory frameworks specific to RM [3].

In South Africa, human biological materials are currently regulated by the National Health Act (NHA) of 2003 and the Medicines and Related Substances Act (MRSA) of 1965. Neither of these statutes has kept pace with scientific developments in the field of RM. Furthermore, South Africa does not have a dedicated legal or regulatory framework that specifically addresses RM and its unique features. This regulatory vacuum is hampering local research and development, manufacturing, and access to advanced cell-based products [4]. It also enables the provision of unproven therapies through the so-called "stem cell clinics," which exploit desperate and vulnerable patients [5, 6].

While countries like Japan and South Korea have developed accelerated pathways for RM approval [7], South Africa's static legal instruments have not addressed the regulatory status of RM therapies. The definitions of "medicine," "biological material," and "cell" in the MRSA and NHA are insufficient to cover the novel characteristics of RM products, which leads to regulatory ambiguity around classification and uncertainty for researchers. The inapplicability of statutory definitions to RM coupled with the lack of a dedicated legislative and policy framework suggests that the current system has yet to fully engage with the ethical complexity and commercial urgency of RM.

RM in South Africa is a growing field with significant potential to transform healthcare. With a robust regulatory framework, active research institutions, and a commitment to ethical and equitable practices, South Africa is poised to make substantial contributions to the global RM landscape. The ongoing challenges present opportunities for innovation and collaboration, ensuring that the benefits of RM can be realised by and for all South Africans. South Africa is actively involved in stem cell research, including the use of pluripotent and multipotent stem cells for therapeutic purposes [8], tissue engineering to develop biological substitutes that can restore, maintain, or improve tissue function [9], and gene therapy focused on correcting defective genes responsible for disease development [10].

Despite these scientific advances, South Africa also faces several challenges to the translation and commercialisation of RM, such as limited funding and resources for large-scale research and clinical translation, navigating a complex and yet underdeveloped regulatory environment, difficulty translating research findings into commercially viable therapies. However, these challenges are not unique to South Africa. Between 2009 and 2022, the European Medicines Agency (EMA) approved only 21 advanced therapeutic medical products (ATMPs) but withdrew or refused to renew market authorisation for 7 of them [11]. The European experience is an indication of the challenges faced when trying to translate RM for clinical use.

It has been observed that European countries as also face obstacles to clinical translation that include insufficient scientific knowledge, lack of dedicated funding, inadequate regulatory frameworks, ethical barriers, and intellectual property (IP) roadblocks [12]. These obstacles align with the most significant barriers to the global clinical translation of RM, including restrictive regulations, the lack of established protocols, and translational delays associated with international collaborations [13].

The foregoing illustrates how essential regulatory frameworks are in fostering innovation in RM by providing a structured pathway for developing, testing, and approving new therapies. These frameworks must ensure that innovative treatments meet rigorous safety and efficacy standards, which helps build public trust and confidence in new medical advancements. Clear guidelines and requirements can also empower regulatory bodies to facilitate the translation of cutting-edge research into clinical practice, enabling the timely delivery of advanced therapies to patients. Robust regulations help address issues of equitable access, ensuring that all segments of the population can benefit from breakthroughs in RM.

This chapter examines the laws and regulations governing the translation and implementation of RM in South Africa. Its objectives are to: provide an overview of the country's regulatory landscape in South Africa, including the key laws, guidelines, and regulatory bodies responsible for RM oversight; highlight ethical and professional standards relevant to RM; explore the implications of IP laws and consumer protection regulations for the commercialisation of RM products; and analyse both the opportunities and challenges posed by a more agile regulatory framework. In doing so, the chapter considers how such a framework could support the development and manufacturing of RM and expand patient access. Ultimately, it argues for reforms to South Africa's regulatory environment that would facilitate the ethical commercialisation of RM.

It is important to note that this chapter focuses solely on the clinical translation of RM and does not address research-related aspects of the field. In this regard, it is worth noting that the National Health Research Ethics Council (NHREC) has published the *South African Ethics in Health Research Guidelines: Principles, Processes and Structures*, which govern all research involving human participants. These guidelines require that such research be reviewed and approved by a registered Research Ethics Committee (REC) to ensure ethical compliance, meaningful community engagement, and that participants receive all relevant information in a language they understand to support informed decision making.

Regulatory Framework for RM in South Africa

South Africa's regulatory framework for RM comprises statutes, regulations, and guidance documents issued by health regulatory authorities. These instruments govern the classification, approval, and use of medicines, including biological products that fall within the broad ambit of RM. In this section, I discuss the primary regulatory instruments relevant to RM in South Africa and analyse their current limitations and applicability.

The Medicines and Related Substances Act 101 of 1965

The Medicines and Related Substances Act is the principal legal instrument for regulating medicines in South Africa. It governs the registration, regulation, and control of medicines and medical devices, including provisions for clinical trials. The MRSA is supported by the General Regulations (MRSA Regulations) issued by the Minister of Health. However, the MRSA was enacted in 1965, well before the advent of modern biotechnologies, and its regulatory design reflects the pharmaceutical landscape of the mid-twentieth century. As a result, it is ill-equipped to accommodate the complexity and novelty of contemporary regenerative medicine products, which involve cellular and genetic manipulation.

While the MRSA provides key definitions such as "medicine," "medical device," and "scheduled substance" it does not explicitly reference biological medicines or regenerative medicine. Section 1 defines "medicine" as:

> any substance or mixture of substances used or purporting to be suitable for use or manufactured or sold for use in—
>
> (a) the diagnosis, treatment, mitigation, modification, or prevention of disease, abnormal physical or mental state, or the symptoms thereof in man; or
> (b) restoring, correcting, or modifying any somatic or psychic, or organic function in man.

This broad definition has prompted debate among scholars and regulators about its adequacy in regulating advanced therapies such as RM, which typically involves complex, biologically derived substances not explicitly contemplated in the original statutory framework [14].

The National Health Act 61 of 2003 and Related Regulations

The National Health Act establishes a framework for the provision of health services and includes regulatory provisions on medical research, the use of human biological materials, and the functioning of health institutions. Section 68 of the NHA allows for the issuance of regulations governing the use of human biological material.

Under the "Regulations Relating to the Use of Human Biological Material" (the Regulations), "biological material" is defined expansively to include DNA, RNA, embryos, cultured cells, stem cells, growth factors, and categories of specific cell types, including embryonic stem cells, progenitor cells, and primordial germ cells. However, they limit the permitted uses of such materials to genetic testing, health research, medical education, and certain heritage-related studies.

The Regulations go further to specify various cell types, including cultured cells, defined as cells grown outside the body; embryonic stem cells, derived from the 30–200 inner cell mass of the blastocyst; primordial germ cells, stem cells found in the fetal gonad that can develop into ova or sperm; progenitor cells, stem cells that give rise to specific stem cell lines; and stem cells, which have the capacity for both self-renewal and differentiation into mature, specialized cell types. However, the Regulations limit the use of biological materials to genetic testing, health research, training of medical students, and studies of archaeological, medical, or heritage value. As a result, the definition in the NHA and the Regulations do not apply to the use of biological materials for RM.

Crucially, these policies do not extend to the clinical application of biological materials in regenerative therapies. Thus, while the NHA and its regulations fill certain definitional gaps left by the MRSA, they do not provide an operative framework for the ethical and clinical governance of RM.

South African Health Products Regulatory Authority (SAHPRA) Guidelines

The South African Health Products Regulatory Authority (SAHPRA) has taken steps to address regulatory gaps by issuing guidance documents and creating a Biological Medicines Sub-Unit within its Pharmaceutical Evaluation Management structure. SAHPRA is the national regulatory authority responsible for evaluating, registering, controlling, and continuously monitoring all health products in South Africa. This includes medicines, medical devices, and complementary medicines. Established under the MRSA, SAHPRA functions as an independent entity within the National Department of Health and plays a central role in ensuring that all health products meet the necessary standards of quality, safety, and efficacy.

SAHPRA defines biological medicines as products produced from or containing components of living organisms, including vaccines, biotherapeutics, monoclonal antibodies, and advanced therapies like gene-modified cells. In 2022, SAHPRA issued three key documents to guide the regulation of biological medicines, as follows:

1. General Guidance on Quality, Safety, and Efficacy Requirements for Biological Medicines (QSE Requirements, 2022)
2. Biological Medicines Stability Guidelines (2022)
3. Biological Medicines Amendment Guideline (2022)

These documents provide a framework aligned with the WHO's *Guidelines on the Quality, Safety and Efficacy of Biotherapeutic Protein Products Prepared by Recombinant DNA Technology* (Annex 4, TRS No. 987) and the ICH Q5C guidelines on *Stability Testing of Biotechnological/Biological Products*, both of which serve as foundational references that SAHPRA has adapted to the South African context. The QSE Requirements guide the registration of new biological medicines, with an emphasis on safety, efficacy, and quality. These requirements apply to all biological medicines and mandate the submission of a detailed dossier that includes risk management plans and stability data, with particular attention to South African health priorities such as HIV and tuberculosis. In addition, any material of animal or human origin must be certified as free from transmissible spongiform encephalopathies (TSEs), a group of rare prion-related neurodegenerative disorders that includes bovine spongiform encephalopathy (BSE or "mad cow disease"), and must be accompanied by a declaration from the final product manufacturer confirming its BSE/TSE-free status.

SAHPRA's definition of biological medicines is issued under its regulatory mandate under the MRSA, which empowers the authority to develop guidance for the evaluation, approval, and monitoring of medical products. However, this definition and the associated guidelines are not legally binding. They are merely intended to support industry compliance and inform regulatory decision-making.

In practical terms, SAHPRA provides a provisional regulatory pathway for the registration and post-market surveillance of RM products, particularly those classified as biological medicines. This includes the evaluation of product quality, safety, and efficacy through dossier submissions. However, in the absence of codified legal provisions, this guidance remains open to interpretation and does not carry the enforceability of statute. As a result, key aspects of regenerative medicine oversight—such as clinical trial governance, manufacturing protocols, and use-case classification—remain outside a clearly defined legal framework.

Comparative Insights from India and Brazil

While South Africa has taken preliminary steps through SAHPRA's guidance documents for biological medicines, it continues to lack a dedicated and legally codified regulatory framework for RM. This positions the country at a regulatory disadvantage compared to several other low- and middle-income countries (LMICs), most notably India and Brazil, that have instituted more targeted and responsive governance models for RM [15].

India, for example, issued the Guidelines for Stem Cell Research as early as 2007, which have since evolved into a comprehensive regulatory framework under the Central Drugs Standard Control Organization (CDSCO) [16]. These guidelines distinguish between permissible and restricted RM activities, mandate institutional ethics committee approvals, and specify clinical trial requirements. Moreover, India's Drugs and Cosmetics Rules (2019) formally regulate stem cell-based products as new drugs, requiring clinical evidence for safety and efficacy before market approval [16].

Similarly, Brazil, through its National Health Surveillance Agency (ANVISA), has adopted Collegiate Board Resolution No. 214 (2018), which provides a regulatory pathway for advanced therapy products (ATPs), including cell and gene therapies [17]. The Resolution outlines requirements for clinical trials, manufacturing, and post-market surveillance, reflecting a comprehensive life-cycle approach that is largely absent in the South African context.

These frameworks reflect a lifecycle-based regulatory approach, encompassing pre-clinical evaluation, ethical oversight, and commercial safeguards. By contrast, South Africa's reliance on the general provisions of the MRSA and the NHA, neither of which was designed with RM in mind, results in a fragmented and reactive system. Regulatory oversight is limited to the classification and stability of biological products, without substantive provisions for pre-clinical assessment, manufacturing protocols, or ethical clinical use of RM therapies. Thus, compared to India and Brazil, South Africa's regulatory lag is characterised by an absence of RM-specific legislation, a lack of codified approval processes for clinical application, limited ethical and safety governance for advanced therapies, and no distinction between experimental, approved, and unproven RM products.

The Case for Legal Reform in South Africa

These regulatory gaps have practical consequences. Skeen et al. [6] have emphasised the need for legal reform to respond to the risks posed by unproven and unregulated stem cell therapies. They cautioned against the premature commercialisation of RM in a study that explored the phenomenon of stem cell tourism among individuals with spinal cord injuries (SCI) in South Africa. Despite a lack of scientific evidence supporting their efficacy, commercial stem cell clinics (both local and abroad) continue to market these therapies directly to vulnerable patients (some of whom are South African citizens), often online. These procedures are typically expensive, ethically problematic, and frequently lack clinical validation or regulatory oversight.

Their study involved 11 South African participants who had undergone "stem cell therapy" in South Africa, India, China, Germany, and the Netherlands. Using validated assessment tools such as the Spinal Cord Independence Measure III

(SCIM III) and International Standards for the Neurological Classification of SCI (ISNCSCI), the researchers found no significant improvements in motor function, bladder or bowel control, or independence following treatment. In some cases, participants experienced worsened symptoms, including increased pain, and decreased mobility. Where improvements were observed, they were attributed to post-treatment rehabilitation rather than the stem cell intervention itself.

On average, participants spent ZAR 203,500 on these procedures, which often involved dubious cell sources, such as rabbit, sheep, fetal, or unspecified origin, and delivery methods including lumbar puncture and subcutaneous injections.

The regulatory lag in South Africa clearly undermines not only local innovation and investor confidence but also the country's ability to ensure patient safety and access to evidence-based, high-quality regenerative treatments. Learning from other LMICs, South Africa could benefit from a phased approach to RM regulation that integrates existing biomedical laws with adaptive, agile mechanisms.

Ethical and Professional Standards

In the absence of a cohesive legal framework, ethical guidance has taken on a vital compensatory role in shaping professional conduct and safeguarding patient welfare. As a result, the ethical and professional standards that govern medical research and clinical practice are essential for guiding the responsible development and use of regenerative medicine. This section explores how these ethical obligations function within the current regulatory gap and considers their role in addressing the risks posed by premature or unproven regenerative therapies.

The Ethical Guidelines from the Health Professions Council of South Africa (HPCSA) provide a comprehensive framework to ensure that medical research and practice are conducted according to the highest ethical standards. Several provisions are particularly relevant to RM. They include requirements for informed consent, confidentiality, and a rigorous risk-benefit analysis. The guidelines also emphasise scientific validity, the fair selection of participants, and the principle that the benefits of research should be accessible to all who need them, not just to those who can afford them.

Practitioners are expected to maintain a high level of competence in their field, which includes staying current with developments in RM through ongoing education and training. This ensures that they are equipped to provide the best possible care. The guidelines further underscore the importance of patient autonomy and the duty to provide high-quality care that meets established medical standards.

Given that RM involves the use of emerging and often complex technologies, RM techniques must be grounded on solid scientific evidence. Experimental treatments should only be offered within the context of ethically conducted clinical trials [18]. Practitioners are therefore required to comply with all applicable regulations and guidelines governing RM, including obtaining the necessary approvals before introducing new treatments [19].

A key consideration in the application of RM is that all innovative treatments must undergo rigorous ethical review before being offered to patients. This includes assessment by research ethics committees to ensure the scientific and ethical justification of the intervention, with a clear emphasis on prioritising patient welfare [20]. Clear communication with patients about the experimental nature of certain RM treatments is crucial.

Equity is another important concern: efforts should be made to ensure that RM treatments are accessible to all patients who may benefit, regardless of socio-economic status [21]. Practitioners are encouraged to advocate for fair pricing and insurance coverage to enhance affordability and access [22].

Finally, patients receiving RM treatments must be closely monitored to track clinical outcomes and manage potential adverse effects [23]. The collection of long-term data is critical to refining treatment protocols and improving the overall safety and efficacy of regenerative therapies over time.

In addition to the ethical guidelines issued by the HPCSA, several other ethical guidelines and regulations are relevant to medical research and practice in South Africa, particularly in the field of RM. The South African Medical Research Council (SAMRC) has published comprehensive ethical guidelines for medical research, which include provisions specific to clinical trials involving innovative treatments such as RM. The guidelines emphasise the ethical conduct of research and incorporate special protections for vulnerable populations to ensure they are not exploited in the research process [24].

The NHA further reinforces the ethical conduct of health research in South Africa. It provides for the establishment of health research ethics committees and mandates the ethical review of all research proposals. It also includes safeguards to protect the rights, safety, and well-being of research participants, thereby supporting the responsible development of novel medical technologies like RM.

South Africa has formally adopted several guidelines issued by the International Council for Harmonisation of Technical Requirements for Pharmaceuticals for Human Use (ICH). These guidelines stipulate standards for the design, conduct, performance, monitoring, auditing, recording, analysis, reporting, and ethical conduct of clinical trials. They also ensure that clinical trials respect and protect the rights, safety, and well-being of trial participants [25]. SAHPRA's Biological Medicines Unit, for example, has integrated these guidelines into its regulatory framework by adopting the ICH guidelines for quality, safety, and efficacy requirements [26]. Applicants are expected to refer to these guidelines when preparing applications for the registration of biological medicines.

However, none of these institutions currently exercise direct oversight over commercialised RM products and practices that fall outside of registered clinical trials. This gap constitutes a significant regulatory and ethical loophole, one that is increasingly being exploited.

Gaps in Oversight: The Rise of Unproven Stem Cell Therapies

Despite the existence of comprehensive ethical guidelines, commercial clinics in South Africa continue to offer unproven and unregistered cellular therapies, particularly to patients with spinal cord injuries and other chronic conditions. In addition, enforcement mechanisms remain focused primarily on research ethics, with limited reach into the clinical and commercial applications of RM. Ethical oversight likewise does not extend to post-approval surveillance, misleading advertising, or private sector compliance with ethical standards.

Consequently, these clinics often market their interventions as innovative or "experimental," thereby avoiding scientific and regulatory scrutiny. As discussed earlier, Skeen et al. [6] found that patients spent substantial sums on stem cell interventions, both in South Africa and abroad, without experiencing any clinically validated benefits. In some cases, patients even suffered harm because of these unproven treatments.

A simple Google search by a desperate patient will reveal numerous South African clinics advertising stem cell "cures" for conditions such as spinal cord injuries, stroke, diabetes, Alzheimer's disease, and heart conditions. These advertisements often overstate efficacy and downplay risks, contravening both the Consumer Protection Act (CPA) and established ethical marketing principles [27]. Although South Africa has statutory and ethical frameworks relevant to health research and medical products, many unproven regenerative therapies remain effectively unregulated in practice. This is due to a combination of outdated or non-specific laws, the absence of binding RM-specific regulation, and weak enforcement against commercial providers who exploit regulatory gaps by presenting their offerings as experimental or wellness-based interventions. As a result, the regulatory instruments discussed above are either not enforced or not directly applicable to many of these interventions, allowing their continued proliferation in the private sector.

In response to the growing interest among South African researchers in stem cells and their therapeutic potential, and in light of the regulatory vacuum described above, Moore [8] identified several ethically sound methods for deriving and creating stem cell lines that aim to advance the field without violating ethical norms. While these methods are not specific to South Africa, they draw on broader international debates and practices concerning ethically acceptable stem cell research. Nonetheless, they may offer South African regulators a valuable framework for guiding both the ethical use of stem cells in research and the oversight of activities within the country's expanding commercial stem cell sector. These include:

- Deriving adult stem cells from tissues or organs, such as hematopoietic stem cells from bone marrow or mesenchymal stem cells from fat, bone marrow, or umbilical cord tissue, which can be harvested without significant harm to the donor.
- Creating induced pluripotent stem cells (iPSCs) by reprogramming adult somatic cells (e.g., skin cells) using genes such as Oct4, Sox2, Klf4, and c-Myc, thus avoiding the ethical concerns linked to embryo destruction.

- Collecting umbilical cord blood, a rich and ethically acceptable source of stem cells, obtained non-invasively after childbirth.
- Harvesting stem cells from amniotic fluid or placental tissue, collected during routine medical procedures or post-delivery, again without harming the donor.
- Somatic cell nuclear transfer (SCNT) or therapeutic cloning, where a somatic cell nucleus is transferred into an enucleated egg to produce a blastocyst for stem cell extraction. Although controversial, efforts are ongoing to ensure this process does not result in viable embryos.

These methods reflect the progress made in aligning stem cell research with ethical standards, enabling scientific and medical advancement without compromising fundamental ethical principles.

In sum, the fragmented oversight system, which is lacking both legal clarity and institutional coordination, creates a regulatory vacuum that facilitates widespread evasion, especially within the private sector. The consequences include compromised patient safety, undermined scientific integrity, and erosion of public trust in the promise of RM.

Commercialising Regenerative Medicine

Commercialising RM in South Africa involves a comprehensive process encompassing research, regulatory approval, marketing, and distribution, all governed by strict regulations to ensure product safety, efficacy, and ethical conduct. Adherence to these steps is essential for the successful market entry and sustainability of RM products. Once a product has been approved and registered by SAHPRA, additional regulations apply. These cover marketing, advertising, consumer protection, customs and excise, data protection, adverse event reporting, and IP rights. In this context, the Intellectual Property Laws Amendment Act of 1997 (IPLAA 1997) and the CPA are particularly significant.

The IPLAA amends several existing IP laws to provide a more robust framework for protecting innovations, including those in biotechnology and RM. It broadens the scope of what can be patented, thereby allowing innovative RM products and processes to be protected under South African patent law [28]. Patent protection offers market exclusivity for a fixed period, encouraging companies to invest in research and development (R&D) by enabling them to recoup their investments and earn profits. This protection also fosters investor confidence and facilitates collaborations and licensing agreements among companies, researchers, and institutions, expanding the reach and revenue potential of new RM innovations.

While IP protection is vital for incentivising innovation, it must be balanced against public health needs. Excessive exclusivity can lead to high prices, limiting access to RM treatments for the broader population [29]. The World Trade Organisation (WTO) Agreement on Trade-Related Aspects of Intellectual Property Rights (TRIPS) [30] mandates patent protection for pharmaceutical products and processes, granting investors exclusive rights for a minimum of 20 years. However,

TRIPS also contains flexibilities that allow member states to protect public health. These include:

- Compulsory licensing, which permits the production or importation of generic versions of patented medicines without the patent holder's consent, under specific conditions such as public health emergencies or non-commercial public use [31]. Adequate remuneration must still be paid to the patent holder.
- Parallel importation (Article 6), which allows the importation of a patented product from another country where it is sold more affordably.
- Exceptions under Article 30, which permit limited use of a patented product (e.g., for experimental purposes), provided this does not unreasonably conflict with the normal exploitation of the patent or unreasonably prejudice the patent owner's legitimate interests.

To reinforce the use of these flexibilities, the WTO Doha Declaration on the TRIPS Agreement and Public Health (2021) affirmed that TRIPS should be interpreted and implemented in a manner that supports member states' right to protect public health and promote access to medicines. It emphasised the right to use all TRIPS flexibilities, including compulsory licensing and parallel importation.

For countries lacking sufficient domestic manufacturing capacity, a WTO General Council decision [32] created a mechanism allowing them to import generics produced under compulsory licenses. While originally designed for essential medicines in health crises such as HIV/AIDS, malaria, and tuberculosis, its potential application to RM remains unclear. Nevertheless, TRIPS does not restrict "medicines" to traditional pharmaceuticals. Stem cell and gene therapies may qualify, provided they address a significant public health need. In such cases, production or importation must be proportionate to the demand and duration of the health crisis.

Despite these progressive measures, pharmaceutical industry pressure and opposition from high-income countries often discourage the use of compulsory licenses and other TRIPS flexibilities [33]. International support and cooperation are essential to help countries leverage TRIPS flexibilities effectively. Notably, South Africa has never issued a compulsory licence for a pharmaceutical product, despite having enabling legislation [34].

Forere [35] notes a growing interest in South Africa's competition law regime to obtain compulsory licences. However, this bypasses procedural safeguards available in the Patents Act and TRIPS, potentially infringing patent rights. Forere recommends aligning the Competition Act with the compulsory licensing framework under the Patents Act and TRIPS to ensure coherence between competition law and IP regulation.

Rozek [36] warns that although compulsory licensing can improve access to existing treatments, indiscriminate use may discourage innovation. He suggests such licenses should be used sparingly and strategically, supported by complementary policies (e.g., public-private partnerships and government funding), to maintain incentives for research and innovation.

On a related note, Nagan [37] addresses the tension between IP rights and access to medicines in the human rights context. Citing the HIV/AIDS crisis in South

Africa, where high drug prices made treatment inaccessible, Nagan argues that the right to health should take precedence over IP rights. When South Africa sought to amend the MRSA to allow for compulsory licensing and parallel importation, the United States opposed the move on behalf of its pharmaceutical industry, claiming TRIPS violations. This dispute was partly resolved by the Doha Declaration, which reaffirmed member states' rights to protect public health and access to medicines.

The CPA also plays a critical role in regulating the commercialisation of RM. It protects consumers by ensuring fair business practices across all stages, from product development to marketing and post-sale services. Particularly relevant to RM are the rights to disclosure and information, and to fair and honest dealing. These provisions are crucial in a field where exaggerated claims about efficacy are common.

To safeguard consumers, the CPA requires that RM products be clearly labelled, with comprehensive information about usage, risks, side effects, and contraindications. Advertising must be truthful and evidence-based, and all efficacy claims must be scientifically substantiated. RM products must also meet established safety standards, and companies must implement recall procedures for unsafe products to protect consumer health [38].

While the IPLAA supports innovation by enabling patent protection for RM, this framework can also create monopolies over novel RM products. This has two major implications for affordability:

1. Patent holders may set prohibitively high prices to recover R&D costs.
2. Most regenerative therapies are biologics, which are difficult to replicate as generics, keeping prices high even after patent expiry.

Although TRIPS provides for compulsory licenses and other mechanisms, they are rarely used in practice, as seen in the HIV/AIDS therapies access debate [39]. The lack of precedent raises questions about their applicability to high-cost RM therapies.

Despite South Africa's strong research base in stem cell science and tissue engineering, few companies have successfully commercialised RM products [40]. One exception is NEXT Biosciences, based in Johannesburg, which produces biological products derived from placental tissues for use in applications such as eye surgery, wound care, and dental procedures. NEXT Biosciences is ISO 13485 accredited and licensed by the South African Department of Health. Its stem cell laboratory is accredited by the Association for the Advancement of Blood and Biotherapies (AABB), and its biological products are licensed by SAHPRA.

Other companies in South Africa that offer regenerative medicine services include R3 Stem Cell [41], an international company that has established clinics in Johannesburg and Umhlanga (near Durban). It provides regenerative procedures such as umbilical cord stem cell therapy, exosome therapy, and platelet-rich plasma (PRP) therapy for various conditions, including arthritis, diabetes, and neurological disorders. Another example is Altis Biologics [42], which developed the first injectable treatment for bone injuries requiring grafting, using a porcine-derived regenerative biological implant. This innovation has been approved by major South African medical aids and has been used in over 1000 patients.

These companies illustrate the growing presence of RM in South Africa, offering a variety of services and products in this emerging field.

Even if RM products are successfully commercialised, IP-driven pricing models may restrict access for the majority of South Africans who depend on public healthcare [43]. Public sector procurement is unlikely to cover high-cost RM therapies, and patent protections can delay or prevent local manufacturing of affordable RM alternatives. Moreover, insurance coverage remains rare, as most RM therapies are not yet formally recognised within medical schemes. Consequently, access to RM in South Africa is likely to remain limited to high-income patients who can pay out-of-pocket or travel abroad, deepening health inequities in access to cutting-edge therapies.

Agile Regulatory Frameworks

An agile regulatory framework refers to a dynamic, flexible approach to regulation that can adapt to evolving technologies, market conditions, and societal needs. In regulatory theory, Mandel [44] refers to this concept as "new governance," a model characterised by collaborative, flexible, and multi-stakeholder regulatory processes, as contrasted with traditional, top-down, command-and-control systems.

Soetemann-Hernandez [45] builds on this idea through the proposed "*Safe Innovation Approach*," developed to navigate the regulatory challenges posed by emerging fields like nanotechnology. This approach advocates for goal-based governance, incorporating anticipatory regulation and trusted environments that foster ongoing dialogue between regulators and innovators. It emphasises core values such as public interest, competence, respect, integrity, inclusion, fairness, and openness.

In a more systematic contribution to governance theory, Stilgoe et al. [46] propose a framework for *responsible innovation* built on four key dimensions: anticipation, reflexivity, inclusion, and responsiveness. These ideas prefigure the Safe Innovation Approach and support a move away from reactive regulation towards proactive engagement with ethical and social considerations.

The OECD's [47] Science, Technology, and Innovation Outlook echoes these concerns. Winickoff and Pfotenhauer [48] argue for a shift from post-hoc regulatory responses to more integrated, anticipatory models that embed governance within the innovation process. Emerging technologies like bioengineering pose unique complexities and risks, and governance must be capable of fostering innovation while addressing public concerns. They emphasise the importance of engaging with the innovation process itself to better balance innovation with precaution.

This challenge is further illustrated by the *Collingridge Dilemma* [49], which highlights a regulatory paradox: in the early stages of a technology's development, interventions are easier to implement but impacts are uncertain; at later stages, impacts are clearer, but the technology becomes more entrenched and difficult to modify due to its integration into broader social, economic, and technical systems.

In the context of RM, this dilemma underscores the risk that waiting for robust post-market data (e.g., on stem cell interventions) may result in missed opportunities for effective early intervention. However, acting too early and without sufficient evidence risks overregulation that may stifle innovation. Traditional regulatory instruments often fall short in addressing the long-term and systemic challenges posed by such emerging technologies.

To overcome this, scholars advocate for process-based governance that includes:

1. Anticipatory governance—engaging in foresight and planning to anticipate and address future impacts early in the development process.
2. Inclusive governance—involving diverse stakeholders, including the public, to align technological development with societal needs and values.
3. Directional governance—ensuring that innovation contributes to broader societal goals, such as the Sustainable Development Goals (SDGs).

These approaches call for collaborative innovation spaces where stakeholders can test new technologies under real-world conditions, and for policies that support participatory foresight, co-creation, and value-based standard-setting. International collaboration is also essential for sharing best practices, harmonising standards, and managing the global implications of emerging technologies.

Agile frameworks can accelerate the approval and adoption of novel therapies, particularly in biotechnology and RM [50]. Core features of such frameworks include:

- Iterative regulatory cycles that allow for continuous refinement based on feedback and new evidence [51].
- Stakeholder engagement to ensure regulations are practical, inclusive, and aligned with public and industry needs [52].
- Real-time data monitoring to support evidence-informed decision-making [53].
- Contextual adaptability, enabling rules to evolve with changing circumstances, thus enhancing their relevance and impact [54].

These features can significantly reduce the time to market for RM therapies, facilitating quicker patient access [55].

However, agile regulation is not without risks. Critics argue that flexibility may come at the cost of regulatory rigour, potentially leading to the premature approval of therapies without sufficient safety or efficacy data [56]. This is particularly concerning in RM, where long-term outcomes and risks are often unknown. Agile frameworks also require robust institutional capacity, continuous stakeholder engagement, and real-time data infrastructure, resources that many LMICs, including South Africa, may lack.

Frequent updates or overly flexibly interpretations of regulatory rules may also introduce uncertainty for investors and developers. Additionally, agility may be co-opted by commercial interests, resulting in regulatory capture or soft governance that prioritises market access over patient protection [57].

Although South Africa has strong constitutional and ethical commitments to healthcare access and innovation, its current regulatory system lacks the institutional agility needed to govern RM effectively. Frameworks such as the MRSA and the NHA remain largely static and reactive. They rely on risk-averse paradigms that require comprehensive data before approval and do not accommodate adaptive licensing, rolling reviews, or real-time safety monitoring.

To implement an agile regulatory model, South Africa would need to introduce several key reforms:

1. Conditional or time-limited approvals for early access to promising RM products.
2. Rolling reviews and living guidance documents to reflect evolving scientific consensus.
3. Integrated post-market surveillance systems to collect real-world data, safety, and efficacy.
4. Multi-stakeholder foresight exercises to anticipate ethical, social, and clinical risks.

International examples show that while the idea of agility is gaining traction, implementation varies significantly. Japan, for instance, has adopted a conditional approval system for RM that has accelerated patient access and spurred industry growth. However, it has also faced criticism for lower evidentiary standards and weak post-market monitoring [58]. In contrast, Canada's regulatory approach to "Advanced Therapeutic Products" was conceived as a flexible and adaptive framework but has largely remained bound to conventional models [59]. Despite promising language, it has yet to deliver meaningful agility in practice, highlighting how institutional inertia can undermine reform.

For South Africa, the challenge is not merely drafting agile regulatory provisions, but developing the operational capacity, political will, and ethical safeguards to make such a system function in practice.

Conclusion and Future Outlook

RM holds immense promise for transforming healthcare by offering innovative therapies to address previously unmet needs. In South Africa, the field is rapidly advancing, with growing research capacity in stem cells, tissue engineering, and gene therapy. However, the existing regulatory framework, primarily governed by the MRSA and the NHA, has not evolved in step with these scientific developments. This regulatory lag hinders local innovation, limits access to advanced cell-based products, and underscores the need for a more comprehensive and agile framework tailored to the unique demands of RM.

While agile regulatory frameworks offer the potential for responsiveness and adaptability in managing emerging technologies, their successful implementation requires more than aspirational commitments. It necessitates concrete institutional structures, accountable governance mechanisms, and coordinated policy integration. To support South Africa's transition from a static to an agile regulatory environment, the following phased implementation strategy is proposed:

Step 1: Legal and Policy Reform

- Amend the MRSA to include a dedicated chapter on regenerative and advanced therapies, explicitly recognising RM as a separate class of products.
- Introduce provisions for conditional approval, rolling submissions, and adaptive trial designs.

Step 2: Institutional Design and Capacity Building

- Establish a dedicated Regenerative Medicine Office within SAHPRA or as an independent advisory entity.
- Train a multidisciplinary team, including clinicians, regulatory scientists, ethicists, and legal experts, to evaluate RM applications using risk-based, stage-gated criteria.

Step 3: Infrastructure for Adaptive Oversight

- Develop a real-time post-market surveillance platform for RM therapies, possibly integrated with the National Health Laboratory Service or public health data systems.
- Create a living guidance document process to continually update RM protocols based on new clinical data and international best practices.

Step 4: Stakeholder Engagement and Public Education

- Launch transparent stakeholder consultation forums with patients, researchers, and industry actors to shape regulatory priorities and identify ethical concerns early.
- Develop public education campaigns to counter misinformation and stem cell tourism.

To ensure ethical and scientific integrity in this adaptive system, regulatory responsibilities must be clearly distributed, as shown in Table 5.1:

Table 5.1 Distribution of regulatory responsibilities for RM among South African Regulatory Agencies

Institution	Oversight Role
SAHPRA	Regulatory authority for product classification, approval, and post-market surveillance; lead agency for enforcement.
Regenerative Medicine Office	Specialised unit within SAHPRA (or a new entity) to assess RM-specific risks and benefits, develop RM-specific guidelines, and coordinate with ethics bodies.
NHREC	Oversight of ethics compliance in clinical trials; ensures patient protection during adaptive licensing.
HPCSA	Ensures compliance with ethical norms by health professionals offering RM therapies.
Competition Commission and Consumer Commission	Monitor pricing, advertising, and patient exploitation under the Consumer Protection Act (CPA) and Competition Act (CA).

To institutionalise agility and accountability, South Africa could adopt a model structured around three core pillars:

A. *Dedicated Oversight: Regenerative Medicine Office*
Embedded in SAHPRA or created as an independent, multi-stakeholder body with regulatory, clinical, ethical, and patient representation. Responsible for:

- Reviewing RM-specific applications.
- Publishing guidance documents.
- Leading international harmonisation efforts with entities like WHO, EMA, and ICH.
- Coordinating post-market safety reviews.

B. *Fast-Track Conditional Approval System*

- Introduce tiered approval categories based on therapy risk and novelty.
- Enable conditional time-limited authorisations for early patient access, especially for life-threatening or rare conditions.
- Require mandatory real-world evidence collection post-launch and allow for market withdrawal or modification based on evolving data.
- Adopt adaptive trial designs that allow smaller, quicker studies with the option to expand as data accumulates.

C. *Public-Private Innovation Partnerships.*
Create an RM Innovation Fund through matched public-private investments to:

- Support proof-of-concept studies and local clinical trials.
- Promote technology transfer from university labs to biotech start-ups.
- De-risk early-stage development, especially for therapies aimed at public health burdens.
- Establish a national Biotech Incubator Network, coordinated by the Department of Science and Innovation, to provide IP support, regulatory coaching, and business development services for RM start-ups.

For South Africa, embracing regulatory agility is not merely about accelerating processes—it is about moving smarter by embedding mechanisms for real-time learning, ethical accountability, and inclusive governance. Through the integration of dedicated oversight, conditional approval pathways, and innovation-enabling partnerships, an agile regulatory framework could foster both responsible innovation and equitable access. Such a framework would help ensure that the benefits of regenerative therapies extend to all South Africans—not only those with the means to afford them.

References

1. Mason C, Dunnill P (2008) A brief definition of regenerative medicine. Regen Med 3(1):1–5. https://doi.org/10.2217/17460751.3.1.1

2. Viljoen IM, Pepper MS (2021) When cells become medicines: a South African perspective. S Afr Med J 111(11):1055–1059. https://doi.org/10.7196/samj.2021.v111i11.15990
3. Qiu T, Hanna E, Dabbous M, Borislav B, Toumi M (2020) Regenerative medicine regulatory policies: a systematic review and international comparison. Health Policy 124(7):701–713. https://doi.org/10.1016/j.healthpol.2020.05.004
4. Pepper MS (2012) Partial relief from the regulatory vacuum involving human tissues through enactment of chapter 8 of the National Health Act and regulations thereto. S Afr Med J 102(9):736–737. https://doi.org/10.7196/samj.5940
5. McLean AK, Stewart C, Kerridge I (2015) Untested, unproven, and unethical: the promotion and provision of autologous stem cell therapies in Australia. Stem Cell Res Ther 6:1–8. https://doi.org/10.1186/scrt543
6. Skeen M, Eksteen C, Pepper M (2019) Stem cell tourism and spinal cord injury in South Africa. S Afr Med J 109(8b):17–23. https://doi.org/10.7196/samj.2019.v109i8b.13824
7. Jeong H, Purja S, Kim E (2024) Analysis and comparative evaluation of expedited programs for gene therapy products: insights from the United States, the European Union, Japan, and South Korea. Gene Ther 31(5):242–254. https://doi.org/10.1038/s41434-023-00437-7
8. Moore TJ (2007) Stem Cell Q and A-an introduction to stem cells and their role in scientific and medical research. Med Technol SA 21(1):3–6
9. Dzobo K, Senthebane DA, Pillay M, Ssemakalu C, Mkhumbeni N, Motaung KSCM (2017) The future of tissue engineering and regenerative medicine in Africa. Tissue Eng A 23(19–20):1023–1025. https://doi.org/10.1089/ten.tea.2017.0375
10. Pepper MS, Alessandrini M, Pope A, Van Staden W, Green RJ (2019) Cell and gene therapies at the forefront of innovative medical care: implications for South Africa. S Afr Med J 109(1):20–22
11. Piemonti L, Scholz H, de Jongh D, Kerr-Conte J, van Apeldoorn A, Shaw JAM, Engelse MA, Bunnik E, Mühlemann M, Pal-Kutas K, Scott WE, Magalon J, Kugelmeier P, Berishvili E (2023) The relevance of advanced therapy medicinal products in the field of transplantation and the need for academic research access: overcoming bottlenecks and claiming a new time. Transpl Int 36(11633):10.3389/ti.2023.11633
12. Seyhan AA (2019) Lost in translation: the valley of death across preclinical and clinical divide–identification of problems and overcoming obstacles. Transl Med Commun 4(1):1–19. https://doi.org/10.1186/s41231-019-0050-7
13. Da Silva RGL, Au L, Blasimme A (2024) Organizational aspects of tissue engineering clinical translation: insights from a qualitative case study. Transl Med Commun 9(17). https://doi.org/10.1186/s41231-024-00179-7
14. Viljoen IM, Pepper MS (2025) Addressing the limitations of the regulatory landscape in South Africa regarding advanced cell and gene therapies and related sectors involving human cells, tissues and organs. S Afr Med J 115(1):12–16. https://doi.org/10.7196/SAMJ.2024.v115i1.2629
15. McMahon D (2012) Regenerative medicine innovation in emerging economies: a case study comparison of China, Brazil and India (Doctoral dissertation). http://hdl.handle.net/1807/31864
16. Jose J, George T, Thomas AM (2020) Regulation of stem cell-based research in India in comparison with the US, EU and other asian countries: current issues and future perspectives. Curr Stem Cell Res Ther 15(6):492–508. https://doi.org/10.2174/1574888X15666200402134750
17. Saute JAM, Picanço-Castro V, de Freitas Lopes AC, da Silva Júnior JB, Nehm JH, dos Santos Maciel F, Schwartz IVD, Sallum JMF, Ozelo MC, Giugliani R, Bonamino M (2025) Clinical trials to gene therapy development and production in Brazil: a review. Lancet Reg Health Am 43. https://doi.org/10.1016/j.lana.2025.100995
18. Subbiah V (2023) The next generation of evidence-based medicine. Nat Med 29(1):49–58. https://doi.org/10.1038/s41591-022-02160-z
19. Sawarkar S, Bapat A (2022) Global regulatory frameworks and quality standards for stem cells therapy and regenerative medicines. In: Stem cell production: processes, practices and regulations. Springer Singapore, Singapore, pp 69–111. https://doi.org/10.1007/978-981-16-7589-8_4

20. Zocchi ML, Vindigni V, Pagani A, Pirro O, Conti G, Sbarbati A, Bassetto F (2019) Regulatory, ethical, and technical considerations on regenerative technologies and adipose-derived mesenchymal stem cells. Eur J Plast Surg 42:531–548. https://doi.org/10.1007/s00238-019-01571-5
21. Erikainen S, Couturier A, Chan S (2020) Marketing experimental stem cell therapies in the UK: biomedical lifestyle products and the promise of regenerative medicine in the digital era. Sci Cult 29(2):219–244. https://doi.org/10.1080/09505431.2019.1656183
22. Qiu T, Pochopień M, Hanna E, Liang S, Wang Y, Han R, Toumi M, Aballéa S (2022) Challenges in the market access of regenerative medicines, and implications for manufacturers and decision-makers: a systematic review. Regen Med 17(3):119–139. https://doi.org/10.2217/rme-2021-0083
23. Lukomska B, Stanaszek L, Zuba-Surma E, Legosz P, Sarzynska S, Drela K (2019) Challenges and controversies in human mesenchymal stem cell therapy. Stem Cells Int 2019(1):9628536. https://doi.org/10.1155/2019/9628536
24. South Africa. South African Medical Research Council (SAMRC) (2017) The South African Medical Research Council Guidelines on the Responsible Conduct of Research. https://www.samrc.ac.za/sites/default/files/attachments/2022-12/ResponsibleConductResearchGuidelines.pdf. Accessed 8 July 2024
25. International Council for Harmonisation of Technical Requirements for Pharmaceuticals for Human Use (ICH) (2024) ICH Guidelines. https://www.ich.org/page/ich-guidelines. Accessed 8 July 2024
26. South African Health Products Regulatory Authority (SAHPRA) (2022) General Guidance Document on Quality, Safety and efficacy requirements for Biological Medicines. https://www.sahpra.org.za/wp-content/uploads/2022/04/SAHPGL-PEM-BIO-06_v3_General-guidance-document-on-QSE-for-Biological-Medicines.pdf
27. Meissner-Roloff M, Pepper MS (2013) Curbing stem cell tourism in South Africa. Appl Transl Genom 2:22–27. https://doi.org/10.1016/j.atg.2013.05.001
28. Mnisi FM (2017) A normative case for the patentability of human embryonic stem Cell technologies. Safety, Ethics and Regulations, pp 121–153. http://link.springer.com/chapter/10.1007/978-3-319-59165-0_5
29. Tenni B, Moir HV, Townsend B, Kilic B, Farrell AM, Keegel T, Gleeson D (2022) What is the impact of intellectual property rules on access to medicines? A systematic review. Glob Health 18(1):40. https://doi.org/10.1186/s12992-022-00826-4
30. World Trade Organisation (WTO) TRT/WTO01/002. Agreement on Trade-Related Aspects of Intellectual Property Rights (as amended on 23 January 2017) (TRIPS). https://www.wipo.int/wipolex/en/text/500864. Accessed 9 July 2024
31. Wartini S (2018) The legal implication of compulsory Licence pharmaceutical products in the TRIPS agreement to the protection of the right to health in developing countries. Jurnal Dinamika Hukum 18(1):1–11. https://heinonline.org/HOL/LandingPage?handle=hein.journals/jwip6&div=14&id=&page=
32. World Trade Organisation. General Council Decision of 30 August 2003 on paragraph 6 of the Doha Ministerial Declaration on the TRIPS Agreement and Public Health. https://20152445.s21i.faiusr.com/61/ABUIABA9GAAgy8Ka7gUomejp3QE.pdf. Accessed 9 July 2024
33. Urias E, Ramani SV (2020) Access to medicines after TRIPS: Is compulsory licensing an effective mechanism to lower drug prices? A review of the existing evidence. J Int Bus Policy 3(4):367. https://doi.org/10.1057/s42214-020-00068-4
34. Vawda YA (2019) Analysing South Africa's compulsory licensing jurisprudence: is there room for the public interest (PI) in intellectual property (IP)? S Afr Intell Prop Law J 7(1):182–198. https://hdl.handle.net/10520/EJC-1b72a4d839
35. Forere M (2019) The compliance of South Africa's patents compulsory licensing regime to the TRIPs agreement. Queen Mary J Intell Property 9(2):156–177. https://doi.org/10.4337/qmjip.2019.02.02

36. Rozek RP (2000) The effects of compulsory licensing on innovation and access to health care. J World Intell Prop 3:889. https://heinonline.org/HOL/LandingPage?handle=hein.journals/jwip3&div=52&id=&page=]In
37. Nagan WP (2001) International intellectual property, access to health care, and human rights: South Africa v. United States. Fla J Int'l L 14:155. https://heinonline.org/HOL/LandingPage?handle=hein.journals/fjil14&div=16&id=&page=
38. Astvansh V, Antia KD, Tellis GJ (2024) What is (and isn't) a product recall? J Public Policy Mark:07439156241242419. https://doi.org/10.1177/07439156241242419
39. Papa Tchofou YC (2011) Compulsory licensing and access to medicines in the TRIPS-compliant world: policy options for anti-HIV drugs in sub-Sahara Africa (Master's thesis, Hanken School of Economics). https://helda.helsinki.fi/server/api/core/bitstreams/78c4ec42-91ed-4f85-bf6b-9aa6f7b162c3/content
40. Dzobo K, Thomford NE, Senthebane DA, Shipanga H, Rowe A, Dandara C, Pillay M, Motaung KSCM (2018) Advances in regenerative medicine and tissue engineering: Innovation and transformation of medicine. Stem Cells Int 8(1):2495848. https://doi.org/10.1155/2018/2495848
41. R3 Stem Cell: Stem Cell and Exosome Therapy South Africa (2025) https://r3stemcell.com/south-africa/
42. Altis Biologics (2025) https://www.altisbiologics.com/
43. Gurgula O, Lee WH (2021) COVID-19, IP and access: will the current system of medical innovation and access to medicines meet global expectations? J Gener Med 17(2):61–70. https://doi.org/10.1177/1741134321993182
44. Mandel GN (2009) Regulating Emerging Technologies. Law Innov Technol 1:75–92. https://doi.org/10.1080/17579961.2009.11428365
45. Soeteman-Hernández LG, Sutcliffe HR, Sluijters T, van Geuns J, Noorlander CW, Sips AJ (2021) Modernizing innovation governance to meet policy ambitions through trusted environments. Nano Impact 21:100301. https://doi.org/10.1016/j.impact.2021.100301
46. Stilgoe J, Owen R, Macnaghten P (2013) Developing a framework for responsible innovation. Res Policy 42(9):1568–1580. https://doi.org/10.1016/j.respol.2013.05.008
47. OECD Science, Technology, and Innovation Outlook (2018) Adapting to technological and societal disruption. https://doi.org/10.1787/sti_in_outlook-2018-en
48. Winickoff DE, Pfotenhauer SM (2018) Technology governance and the innovation process. In: OECD Science, Technology, and Innovation Outlook. Adapting to Technological and Societal Disruption. https://doi.org/10.1787/sti_in_outlook-2018-15-en
49. Collingridge D (1980) The dilemma of control. The Social Control of Technology, pp 13–22
50. Munsie M, Hyun I, Sugarman J (2017) Ethical issues in human organoid and gastruloid research. Development 144(6):942–945. https://doi.org/10.1242/dev.140111
51. Mansour Y, Singh S (2013) On the complexity of policy iteration'. arXiv preprint arXiv:1301.6718. https://doi.org/10.48550/arXiv.1301.6718
52. Vargo J, Seville E (2011) Crisis strategic planning for SMEs: finding the silver lining. Int J Prod Res 49(18):5619–5635. https://doi.org/10.1080/00207543.2011.563902
53. Guston DH, Sarewitz D (2002) Real-time technology assessment. Technol Soc 24(1–2):93–109. https://doi.org/10.1016/S0160-791X(01)00047-1
54. Black J, Baldwin R (2010) Really responsive risk-based regulation. Law Policy 32(2):181–213. https://doi.org/10.1111/j.1467-9930.2010.00318.x
55. Lysaght T, Kerridge IH, Sipp D, Porter G, Capps BJ (2017) Ethical and regulatory challenges with autologous adult stem cells: a comparative review of international regulations. J Bioethic Inquiry 14:261–273. https://doi.org/10.1007/s11673-017-9776-y
56. Bolislis WR, De Lucia ML, Dolz F, Mo R, Nagaoka M, Rodriguez H, Woon ML, Yu W, Kühler TC (2021) Regulatory agilities in the time of covid-19: overview, trends, and opportunities. Clin Ther 43(1):124–139. https://doi.org/10.1016/j.clinthera.2020.11.015

57. Geraci G, Bernat J, Rodier C, Acha V, Acquah J, Beakes-Read G (2023) Medicinal product development and regulatory agilities implemented during the early phases of the COVID-19 pandemic: experiences and implications for the future—an industry view. Ther Innov Regul Sci 57(5):940–951. https://doi.org/10.1007/s43441-023-00536-y
58. Jokura Y, Yano K, Yamato M (2018) Comparison of the new Japanese legislation for expedited approval of regenerative medicine products with the existing systems in the USA and European Union. J Tissue Eng Regen Med 12(2):e1056–e1062. https://doi.org/10.1002/term.2428
59. Koukoulas E (2013) Post-approval drug safety: moving from passive to active pharmacovigilance in Canada. Master of Science: Thesis, McGill University. https://escholarship.mcgill.ca/concern/theses/rb68xg320

Regenerative Medicine at the Tip of Africa: Far-Reaching Consequences of Slow Progress

6

Usri H. Ibrahim, Mari van de Vyver, Yvonne Holt, Carl T. Montague, and Carola U. Niesler

Abstract

The application of regenerative medicine (RM) in the clinic has gained significant momentum in countries of the Global North. In contrast, in the Global South, countries like South Africa have struggled to develop and integrate regenerative technologies within mainstream medicine. This is, in part, due to South Africa's unique quadruple burden of disease, which presents significant challenges to a healthcare system already under strain, as well as the alarming disparity in equitable access to therapeutic interventions, exacerbated by persistent socioeconomic factors. Translating RM research outcomes to clinical practice is further limited by insufficient supportive infrastructure, scarcity of designated

U. H. Ibrahim (✉)
Discipline of Biochemistry, School of Agriculture and Science, University of KwaZulu-Natal (UKZN), Pietermaritzburg, KwaZulu-Natal, South Africa
e-mail: IbrahimU@ukzn.ac.za

M. van de Vyver
Experimental Medicine Research Group, Department of Medicine, Faculty of Medicine & Health Sciences, Stellenbosch University, Cape Town, Western Cape, South Africa
e-mail: vandevyverm@sun.ac.za

Y. Holt
Next Biosciences, Midrand, Gauteng, South Africa
e-mail: yvonne.holt@nextbio.co.za

C. T. Montague
Shook Services, Durban, KwaZulu-Natal, South Africa
e-mail: carl@shookservices.com

C. U. Niesler
Discipline of Biochemistry, School of Agriculture and Science, University of KwaZulu-Natal (UKZN), Pietermaritzburg, KwaZulu-Natal, South Africa
e-mail: niesler@ukzn.ac.za

U. Ogbogu et al. (eds.), *Regenerative Medicine and the Bioethics of Globalization, Justice and Hegemony*, Collaborative Bioethics 5,
https://doi.org/10.1007/978-3-032-07289-4_6

research funds, and lack of adequate human healthcare resources. This has consequences not only for advancing healthcare in African countries but also in countries with more developed economies, advanced healthcare systems, and established regulatory frameworks. These consequences occur because a complete understanding of clinical responsiveness to developing therapies can only be achieved when armed with knowledge generated from studies that included populations rich in biological diversity, and many of these populations are found in the Global South. The current chapter highlights key challenges facing the development and clinical translation of RM approaches in South Africa, emphasizing how the lack of data from diverse populations during the development of regenerative modalities in the clinic has negative consequences not only with regard to clinical outcomes on a local level but also on a global scale.

Keywords

Regenerative medicine · South Africa · Cinical translation · Healthcare inequality · Research and development · Regulatory challenges · Cell therapies · Genetic diversity

Introduction

Over the last few decades, regenerative medicine has played an increasingly prominent role in delivering more effective healthcare in the Global North, especially in clinical areas where conventional pharmacological treatments may fail to achieve reasonable outcomes [1, 2]. Analysis of the stem cell market from 2015 to 2022 reflects a compound annual growth rate (CAGR) of more than 25% [3]. This growth in the contribution of RM modalities to modern healthcare systems means that RM is now considered one of the most important healthcare sectors, comprising approximately 2% of the global pharmaceutical market with a predicted CAGR of 16.79% from 2024 to 2030 [4–6]. However, for countries in the Global South, like South Africa, the status of RM is far removed from this global trend in both developmental research activities as well as integration within healthcare systems. As a result, access to these innovative therapies remains limited [7]. South Africa, despite being classified as an upper-middle-income country by the World Bank, can more accurately be described as a dual economy where two parallel economies exist, one being capitalist and advanced, while the second reflects the developing countries' poverty and underdevelopment [8].

South Africa is one of the most unequal societies in the world; individuals at the top 0.1% own a quarter of the country's total wealth and have access to quality private world class healthcare, while the majority rely on a public health system that struggles to deliver appropriate health services due to significant resource constraints and infrastructural challenges [7, 9–11]. In fact, at least 50% of the population actually have a so-called negative wealth position, living within the upper bound poverty line and unable to access basic quality healthcare [12]. Given this significant wealth gap, even in cases where RM therapies are available to South Africans, accessibility of these therapies remains disproportionately skewed toward the wealthy.

There are two reasons for this lack of accessibility: first, internationally-developed RM therapies are expensive and difficult to introduce in local markets (in both private and public healthcare contexts); and second, the lack of significant and targeted support for RM development inhibits the growth of local translational research and development (R&D) capacity, forcing individuals to rely on expensive, internationally-sourced alternatives. Because of this cycle, the incorporation and development of indigenous RM-modalities within South Africa remains out of reach [13]. Societal inequality, divergent national R&D priorities, limited funding, and a lack of clear regulatory frameworks suitable for incorporating RM therapies among African countries further complicate accessibility [14]. Additionally, the country's focus on developing solutions for pressing challenges presented by communicable diseases such as HIV and TB also overwhelms the healthcare system and detracts from RM developments [14]. A recent report published by the Alliance for Regenerative Medicine [15], on the global distribution of approved medicines, clinical trials, and investment in the RM field revealed slow progress and a potential lack of national clinical trials and investment prioritization regarding the development of these therapeutic modalities on the African continent [2, 16].

This chapter reflects on the current status of RM in South Africa, with a focus on identifying the challenges and constraints that hinder the translation of scientific developments to clinical application of RM-related therapies. Furthermore, we highlight specific challenges related to the ability of the Global South to effectively translate RM-driven approaches from the bench to the bedside.

Regenerative Medicine in South Africa

It is clear that when it comes to the research, development, and distribution of advanced RM therapeutic technologies, the Global North dominates [17, 18]. Although a number of South African academic and clinical research groups focus on various aspects of RM, including tissue engineering, stem cells, as well as cell- and gene-based therapies [19–21], the translation and subsequent clinical impact of this research is limited. To set the context for discussing the challenges and constraints that have led to these gaps between the Global North and the Global South, it is important to first examine the current state of RM field in South Africa, relative to global trends.

Research Development and Innovation

Research development and innovation in South Africa benefit from a strong tertiary education system, including 26 universities spread across the country that provide a unique blend of cultural diversity and outstanding infrastructure (Table 6.1) [22,

Table 6.1 Public universities in South Africa, along with their abbreviations and classifications. This classification is based on the distinction between Traditional Universities (offering academic- and research-intensive programs), Universities of Technology (focused on vocational and practical-skills training), and Comprehensive Universities (offering a mix of both)

University	Abbreviation	Classification
University of Cape Town	UCT	Traditional University
University of Fort Hare	UFH	
University of the Free State	UFS	
University of KwaZulu-Natal	UKZN	
North-West University	NWU	
University of Pretoria	UP	
Rhodes University	RU	
Stellenbosch University	SU	
University of the Western Cape	UWC	
University of the Witwatersrand	WITS	
Cape Peninsula University of Technology	CPUT	University of Technology
Central University of Technology	CUT	
Durban University of Technology	DUT	
Mangosuthu University of Technology	MUT	
Tshwane University of Technology	TUT	
Vaal University of Technology	VUT	
Nelson Mandela University	NMU	Comprehensive University
University of Johannesburg	UJ	
University of Limpopo	UL	
University of Mpumalanga	UMP	
Sol Plaatje University	SPU	
Walter Sisulu University	WSU	
University of South Africa	UNISA	
University of Venda	UNIVEN	
University of Zululand	UNIZULU	
Sefako Makgatho Health Sciences University	SMU	

23]. In addition, a number of established research institutions, scientific funding support systems, innovation hubs, and entrepreneurship programs as well as investment opportunities exist to support research and innovation activities within these universities and affiliated research facilities (Table 6.2).

Through these varied support systems, a number of local initiatives have sought to specifically focus on the development of RM therapies in South Africa. These initiatives include R&D, academic and clinical partnerships, as well as some biotechnology and pharmaceutical activities (Table 6.3). For instance, the South African Medical Research Council (SAMRC) has directly supported RM focus areas. The Stem Cell Research and Therapy Unit based at the University of Pretoria was the result of such funding, while the current call for SAMRC Extramural Research Unit funding includes RM and tissue engineering as an explicit priority area (Table 6.3). Finally, Next Biosciences is one of the few South African companies that focuses on developing and delivering RM biotechnology applications (Table 6.3).

Table 6.2 Research facilities and innovation hubs, funding agencies, and investment opportunities that support innovation and entrepreneurship in South Africa

Category	Organization	Aim/Role
Research Facilities and Innovation Hubs	Africa Health Research Initiative (AHRI) https://www.ahri.org *Based at UKZN*	Advance cutting-edge biomedical and population health research in Africa to improve disease prevention, treatment, and health policies. Focus on major infectious and non-communicable diseases (NCDs).
	Centre for the Aids Programme of Research in South Africa (CAPRISA) https://www.caprisa.org *Based at UKZNAffiliated with UCT and UWC*	Conduct innovative HIV/AIDS research to improve treatment, promote prevention, and contribute to policy strategies in Southern Africa. Focus on vulnerable populations including adolescent girls and young women.
	Council for Scientific and Industrial Research https://www.csir.co.za *Affiliated with numerous South African universities*	Drive scientific and technological innovation to foster socioeconomic development and industrial competitiveness through multidisciplinary research and innovation. Government owned.
	The Innovation Hub (Gauteng) https://www.theinnovationhub.com *Subsidiary of Gauteng Growth and Development Agency*	Incubate artificial intelligence, internet of things, and green tech start-ups by building networks between researchers, businesses, and investors.
	Stellenbosch Launch Lab https://www.launchlab.co.za *Based at SU*	Accelerate start-ups by providing business incubation, funding access, and market opportunities. Offers affordable lab space, equipment, and specialized business support to biotech start-ups.
	South African National Space Agency https://www.sansa.org.za *Affiliated with the Department of Science and Innovation*	Promote aerospace science and satellite applications through research, training, scholarships, and partnerships with national universities and international collaborations.
	WITS health Consortium https://www.witshealth.co.za *Based at WITS*	Advance public health research, training, and intervention programs in Southern Africa. Focus on HIV, TB, and maternal/child health.

(continued)

Table 6.2 (continued)

Category	Organization	Aim/Role
Funding Agencies	Technology Innovation Agency https://www.tia.org.za *Department of Science and Innovation*	Fund early-stage tech innovations in biotech, energy, and manufacturing sectors.
	South African Medical Research Council https://www.samrc.ac.za *Parastatal medical research organisation*	Fund health research including at the innovation, development, and technology transfer stages.
	South African National Research Foundation https://www.nrf.ac.za *Independent government agency*	Funding instrument that promotes scientific research, career development, public science engagement and research platforms.
	Department of Science, Technology and Innovation Fundhttps://www.dsti.gov.za *Public-private investment initiative*	Funds high-tech R&D with commercial potential, including AI, machine learning, biotechnology, nanotechnology, cybersecurity and space science.
	Small Enterprise Development Agency https://www.dsbd.gov.za/small-enterprise-development-agency-seda *Government agency*	Support tech start-ups with funding, incubation, and patent assistance.
	University of Cape Town Graduate School of Business—Bertha Centre https://www.gsb.uct.ac.za/berthacentre *Based at UCT*	Support entrepreneurship and innovation that solve social and environmental issues.
Industry/ Private investment	South African Small and Medium Enterprises Fund https://smesouthafrica.co.za *Collaboration between government, labour, and business*	Invest in high-growth start-ups that scale rapidly due to innovative products, scalable business models, and strong market demand.
	Launch Africa Ventures https://www.launchafrica.vc *Based in Mauritius*	Pan-African venture fund backing scalable tech startups.
	Naspers Labs https://www.naspers.com *Social impact programme*	Equips South Africa's unemployed youth with digital skills training, preparing them for careers in technology and entrepreneurship

A comparative analysis of articles published worldwide in the field of RM versus studies involving South African researchers in the field of RM was carried out to determine the relative contribution of South African research within the field of RM. To achieve this, data was retrieved from Clarivate [24] using the terms "*regenerative medicine,*" "*stem cell therapy,*" or "*regeneration*" to search within topics (referred to as the *Global RM Research Output*—see Fig. 6.1). This search was subsequently limited to South Africa (referred to as *South African RM Research Output*—see Fig. 6.1). The search revealed that RM outputs were active globally around the 1940s, while outputs only began emerging from South African research in the late 1970s. Furthermore, since 1975, a total of 1651 articles on RM involving South African scientists have been published, representing 0.5% of the more than 324,000 global outputs (Fig. 6.1). A significant body of work related to RM is therefore being produced locally.

Table 6.3 Research facilities, funding agencies, and industries that specifically support RM initiatives in South Africa

Category	Organization/ Initiative	Aim/ Role
Research Facilities	*African Stem Cell Initiative* https://neuroscience.uct.ac.za/research-research-platforms/african-stem-cell-inititative *Based at UCT*	Develops organoid models of neurological diseases from African populations to improve diagnosis and treatment
	Stem Cell Research and Therapy Unit (SAMRC) https://www.samrc.ac.za *Based at UP*	Investigates ways to lower the South African infection rate of communicable and NCDs through stem cell research
Funding Agencies	South African Medical Research Council https://www.samrc.ac.za	Supports RM research through various funding calls. Current call for Extramural Research Units includes Regenerative Medicine and Tissue Engineering as a priority area
	South African National Research Foundation https://www.nrf.ac.za	Supports RM research through broader programs. Past research support programs include "Unlocking the Future"
Industry	Next Biosciences https://nextbio.co.za	Focuses on the commercial application of cutting-edge RM biotechnology

Clinical Translation

South Africa, and Africa as a whole, is known for its genetically and socioeconomically diverse, and relatively drug-naïve population. Despite this diversity, less than 5% of clinical trials globally were hosted on the African continent in 2023 [25, 26]. The lack of clinical research conducted on African populations not only inhibits improved health outcomes for this region but also inhibits the ability of these trials to cater to diverse populations generally. Increased diversity within clinical trials not only strengthens the kind of data collected but also provides a more accurate snapshot of how these trials affect a variety of individuals.

To determine the number of clinical trials published in South Africa specifically within the scope of RM, we restricted the search described in Fig. 6.1 to "*clinical trials,*" in addition, we also searched the Pan-African Clinical Trials Registry [27] (as well as the NIH clinical trials site [28]). We found only five published clinical trials that could be classified as concentrating on RM from within South Africa in the Global South (Table 6.4). These five trials focused on the use of extracellular matrix-based therapies and stem cells in the treatment of osteochondral lesions, bone defects, ovarian-related infertility, and myelodysplastic syndromes. Although this search demonstrates instances where clinical translation has been attempted at a local level, it also emphasizes the challenges experienced in making these

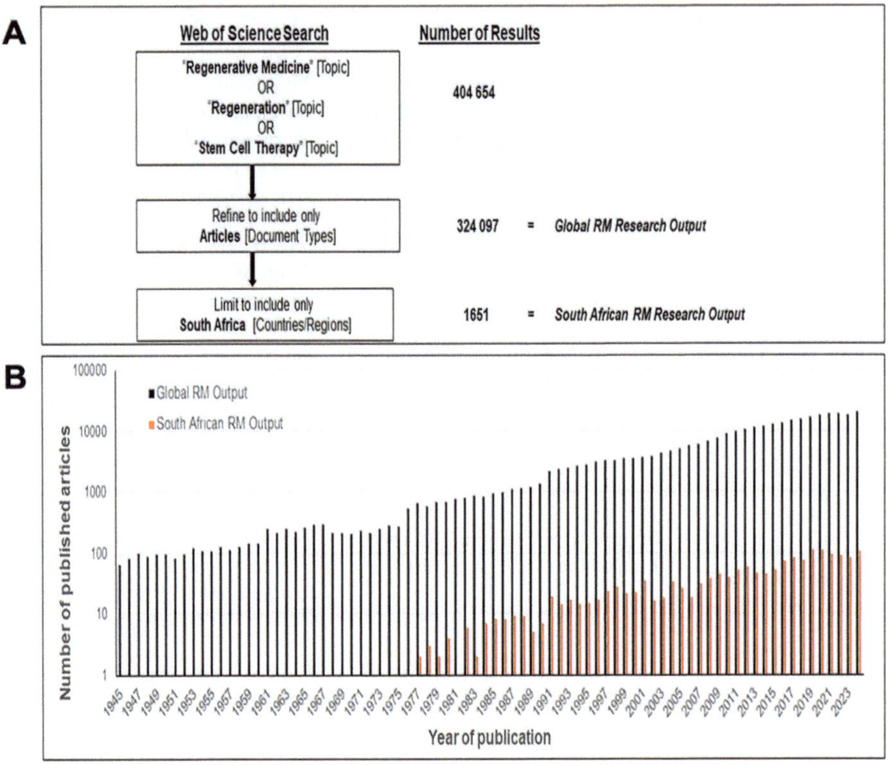

Fig. 6.1 Number of published articles in the field of RM involving South African vs. international researchers (**a**) The data was retrieved from Clarivate [24] using the search approach as indicated. (**b**) Graphical representation of the number of articles published from 1945 to 2024 [24]

Table 6.4 South African published clinical trials within the scope of RM [24, 27]

Condition/Disease	Approach	No. of trials	Reference
Bone defects & fractures	Extracellular matrix-based technologies	3	[29–31]
Ovarian fertility	Stem cell-based therapy	1	[32]
Myelodysplastic syndromes	Stem cell-based therapy	1	[33]

therapies available to the majority of individuals in the Global South, including South Africans. The reason for this low conversion is complex but could point to a combination of lack of funding and appropriate resources critical for trial management, suboptimal supply chain and data management systems, as well as insufficiently strong partnerships with key stakeholders in the private sector [25].

Challenges to the Healthcare System and the Disease Burdens Present in South Africa

Many of the progressive world economies consider RM to be critical for ensuring the future well-being of their populations. However, in South Africa the scientific and clinical activities in the innovation space are often geared toward more immediate local issues and needs, and as such there are less initiatives directed specifically at developing and understanding the long-term benefits of RM. South Africa faces a quadruple burden of disease which includes a combination of communicable diseases (mainly tuberculosis (TB) and HIV/AIDS), NCDs (including diabetes and cardiovascular diseases), high rates of maternal and child mortality, and related violence and injuries [34]. As a result, national health research priorities are shifted toward reducing this disease burden and toward the related need for cost-effective interventions [35]. Despite the fact that RM could address morbidities associated with a range of diseases, these types of implementation are low on the list of priorities due to these other, more pressing, health challenges. This trend of prioritization significantly impacts the research interest and governmental support of RM academic R&D activities, and further results in limited resources and research outcomes for diseases and related ailments for which RM may otherwise provide solutions.

The SAMRC has developed a Strategic Health Innovation Partnership program that aims to provide financial support to cutting-edge health research innovations, including within the field of RM [36]. For example, the Stem Cell Research and Therapy Unit (see Table 6.3) at the Institute for Cellular and Molecular Medicine (ICMM), University of Pretoria, is an extramural research facility funded by the SAMRC that focuses on RM research and applications. Other groups dedicated to RM research, funded by the SAMRC or other governmental bodies such as the National Research Foundation of South Africa (NRF), are distributed among the nation's universities [37]. Despite the availability of some resources to promote RM research within South Africa, the translation of this research remains limited [38]. One reason for this limitation is the substantial financial cost associated with RM therapies; this cost is a major setback when attempting to integrate these therapies within local healthcare systems, both public and private. This is further exacerbated by the lack of sizable local biotechnology industries in the RM space that could potentially engage in the translation of these therapies to the clinical sector.

Infrastructure, Funding, and Regulatory Constraints in South Africa

RM is one of the most rapidly evolving fields of healthcare, and effective partnerships and technology transfers are of paramount importance to achieve clinically desired outcomes [39]. Greater collaboration between institutional, industrial, and governmental sectors could provide a solid foundation for building a national RM society and for creating a system aimed at overcoming all obstacles that hinder RM's clinical applications. Similarly, effective technology transfer would support the translation of local RM research outputs into clinical applications, ultimately benefiting patients due to the availability of advanced treatment modalities and lower costs [40]. Furthermore, the transfer of such emerging cutting-edge technologies coupled with approaches such as decentralized manufacturing could be a cost-effective strategy that reduces costs and minimizes risks of supply chain interruptions [1]. Collaboration between industry and academia is likely a mandatory step to both overcome these challenges and to boost the translation abilities of academic research. This will be challenging given the evidence that when one analyzes the worldwide distribution of RM companies (Fig. 6.2), less than 3% of these companies are based on the African continent. RM's lack of visibility on the African continent thus limits the region's ability to foster local, private, and public partnerships in this critical field.

The regulation of RM in South Africa is governed by the *Medicines and Related Substances Act* (previously the *Drugs Control Act*) 101 of 1965 under the authority of the South African Health Products Regulatory Authority (SAHPRA), which oversees the approval and regulation of all healthcare products, including RM therapies. The current regulatory frameworks responsible for controlling and approving RM-related therapies in South Africa are still relatively immature and inefficient at guiding both clinical trials and the registration process related to RM therapies [7].

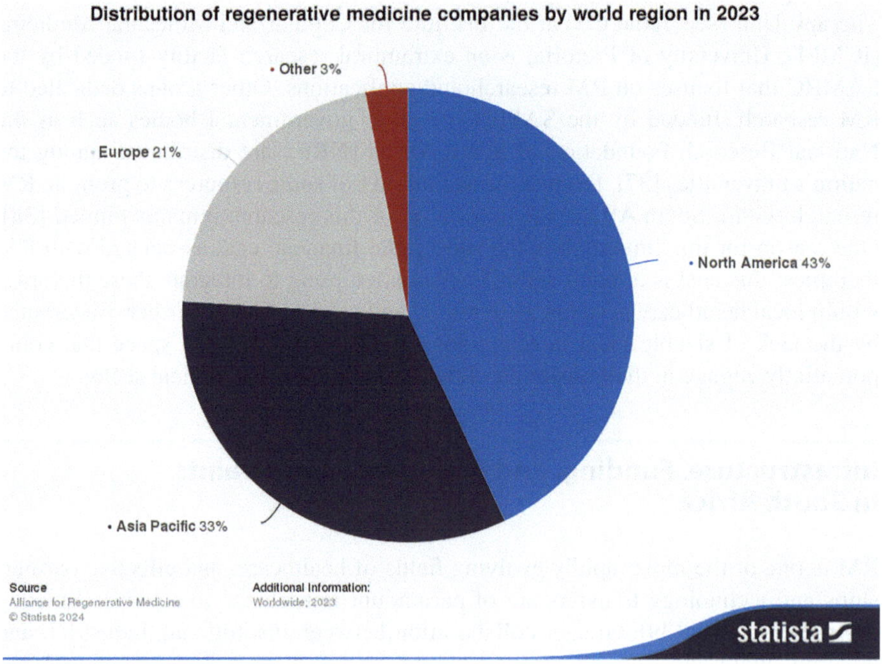

Fig. 6.2 Distribution of RM companies by World Region [41]

The lack of RM-specific regulatory frameworks has negatively impacted the clinical use of related technologies and, as a result, applications are often managed via hospital exemptions that allow the use of RM-based treatments under strict guidelines in particular medical settings [42]. RM therapies should be safe, effective, and high-quality, so it is unfortunate that the current regulatory framework does not take the complexity and technical specificity of RM products into account, therefore jeopardizing these crucial aims and limiting the positive impact that these therapies could have within the clinical sector in South Africa.

There is a dire need to develop a comprehensive, supportive regulatory framework that accommodates the unique features of RM therapies to ensure patient safety. Such a framework would need to address the control of gene treatments and cell-based products, while also considering the active inclusion of all involved stakeholders in this process. The unique features of South Africa, including the country's genetic diversity, must also be a key focus to promote accessibility and equity in offering these treatments and products. The importance of developing and improving current guidelines and rules that govern RM in South Africa cannot be overstated; legislative action must be taken to accommodate changes in this fast-growing field to expedite the acceleration of RM progress among the region and to encourage associated R&D activities. All of these steps are crucial to overcome national and regional health challenges.

Therapeutic and Economic Potential of RM in South Africa

RM possesses considerable potential for adding value to the South African health sector by providing both therapeutic and economic advantages to the region [39], as well as by providing advanced global understanding of biological diversity's impact on clinical responses [43, 44] and therefore the development of new pharmaceutical products [7, 45]. RM also offers new opportunities to expand the pool of innovative research within South Africa [45]. In this section, we will highlight the possible benefits that RM could provide and also shed light on the promising future of RM in South Africa.

RM offers advanced treatment options for chronic diseases, injuries, and congenital disorders by addressing the root causes of diseases and injuries and ultimately reducing long-term health burdens, including costs associated with chronic treatment and hospitalization. These treatment options improve patient outcomes, reduce recovery time, and enhance patients' quality of life [1]. These advancements are of particular relevance given the ever-increasing burden of NCDs, such as type 2 diabetes mellitus, and associated co-morbidities, including non-healing wounds [46]. Furthermore, RM could potentiate the South African personalized medicine efforts that aim to adapt treatment options according to various aspects of South African-wide genetic diversity. This focused adaptation could also have beneficial implications for the Global North, where the importance of genetic diversity in terms of pharmacogenomics has become especially clear [47]. More specifically, genetic variations affect how drugs are both absorbed and metabolized, and by extension affect the response observed by individuals within the population; this information is critical to maximize therapeutic efficacy. Besides the health-related cost savings, growth in the RM sector would boost the South African economy and initiate sustained economic development by attracting local and international investment, creating jobs within pharmaceutical and clinical applications, and ultimately establishing a solid bioeconomy based on innovative RM. Therefore, significant attention should be paid toward ensuring that RM facilities meet international standards (such as those embodied by the European Medicines Agency, the Canadian Food and Drugs Act, and the US Food and Drug Administration) to fully utilize both the economic and healthcare benefits of RM.

The unique biologic and genetic diversity in South Africa offers a profound opportunity to advance RM research through the investigation of the role of this diversity on clinical responses, and therefore the development of new pharmaceutical products [48]. This type of investigation would facilitate the design and manufacturing of personalized medication customized to distinct genetic profiles and would also facilitate a better understanding of disease pathophysiology differences of various demographic groups. Similarly, these efforts could accelerate the identification of various genomic variants and assist in building global genomic reference panels, which speaks again to the importance of genetic diversity in improving the efficacy of clinical modalities, including RM. Furthermore, unique South African genomic variants can be used to identify new therapeutic targets to advance local drug discovery research and industry, and to improve biotechnology innovation and clinical application in both the Global South and the Global North.

A Case Study: Amniotic Membrane as an RM Approach for Wound Healing

To demonstrate some of the challenges faced by developers of RM products in South Africa, this section will discuss a case study of an RM product that several of the authors of this chapter have been involved in developing and launching on the South African market. The case study will highlight the barriers to access for an effective RM wound therapy product.

Adjunctive wound healing therapies, such as amniotic membrane grafts, are used for the treatment of non-healing chronic wounds, including diabetic foot ulcers, open sores, or wounds that occur in approximately 15% of patients with diabetes and that are commonly located on the foot. Of those patients who develop a foot ulcer, 6% will be hospitalized due to infection or other ulcer-related complications. Approximately 14–24% of patients with diabetes who develop a foot ulcer will require an amputation. Diabetic foot ulceration precedes 85% of all non-traumatic related amputations, making it the leading cause of lower extremity amputations globally [49–51].

Globally, there have been at least seven randomized clinical trials conducted on the use of amniotic membrane products for the treatment of diabetic ulcers. These studies demonstrate the increased efficacy of amniotic membranes in healing of diabetic ulcers compared to current therapies and standards of care [52–58]. These clinical trials stemmed from more than 100 pre-clinical studies characterizing the properties and specific molecular mechanisms by which amniotic membranes promote healing [59]. A single pre-clinical study has been conducted in South Africa, illustrating the efficacy of a specific amniotic membrane product (AmnioRederm™) when used in combination with other treatment modalities [50].

AmnioRederm™ is a specialized wound recovery adjunctive, designed to naturally stimulate the recovery of non-healing soft tissue wounds. It is produced by the South African biotechnology company Next Biosciences, and is made from human amniotic membrane, a collagen-rich structure comprising the innermost layer of the fetal membrane of the placenta. Amniotic membrane has been shown to contain significant quantities of bioactive compounds and growth factors associated with the stimulation of wound healing. Amniotic membrane is an easily sourced human tissue, the manufacturing process is cost-effective and the product can be stored at room temperature. The product is easily applied to the wound's surface and does not require specialized application techniques or expertise. Given these attributes, amniotic membrane products, such as AmnioRederm™, are an ideal dressing that can be easily transported to outlying areas in South Africa, reaching rural populations that are serviced by basic primary healthcare facilities—facilities that sometimes suffer from both a lack of resources and specialized staff.

Amniotic membrane has been used to aid wound repair for over a century and is routinely used by ophthalmologists in South Africa, where its use is authorized and reimbursed by medical aids (medical insurance companies). Despite a long history of usage, excellent safety record, and numerous case reports illustrating its effectiveness [58], amniotic membrane products such as AmnioRederm™ are not used routinely for the treatment of non-healing diabetic ulcers in South Africa. One of the reasons for this is that the product is not reimbursed by medical aids. As part of the approval process to enable medical aids to pay for new treatments such as AmnioRederm™, the producers of the products need to submit an application to the medical aid, and the application must contain the results of clinical trials conducted in South Africa *specifically*—results from international trials are not acceptable. However, the cost of conducting human clinical trials in South Africa acts as a significant barrier, such that local biotechnology companies, including Next Biosciences, do not have the resources to conduct these trials independently. This impasse has created a situation where, although an efficacious and safe therapy for non-healing diabetic ulcers is approved and available in the South African market—at a fraction of the cost of similar products in the United States—such therapy is only accessible to those who are fortunate enough to purchase it privately. Therefore, despite the existence of an effective treatment option, many patients with non-healing diabetic ulcers will go on to need lower extremity amputations.

Strategies to Increase the Availability of RM Therapies

The South African government has recognized the promising potential of RM to address national healthcare challenges and to reduce the socioeconomic impact of diseases which are untreatable through pharmacological means [14, 49]. Therefore, the South African healthcare landscape has shifted toward an understanding that RM research, development, and technology transfer are needed. Several strategies could be employed to encourage and support South African R&D activities in the RM field. These strategies include flexible regulatory frameworks that promote RM investments and innovations while maintaining the safety, efficacy, and quality of RM-based therapies, as well as updating health policies and procedures to accommodate RM benefits and equitable access within this sector. Furthermore, promoting collaborative efforts, including strengthening public and private partnerships and cooperation with philanthropic entities, is also a promising avenue to investigate.

This AmnioRederm™ case study highlights several barriers related to the use of RM therapies by the general South African public, including:

- Lack of reimbursement by medical aids—rendering the product unaffordable for many.
- Requirements for locally conducted clinical trials—and associated high costs.
- Limited access in public healthcare settings—despite proven efficacy.

A possible solution to overcome these challenges is the provision of funding, either from the state or via medical aids, for the conduct of clinical trials in South African populations to inform new uses of currently available health products. Additional approaches could include adaptive regulatory pathways for local biotech products, state-subsidized clinical trial schemes, risk-sharing agreements between biotechnology companies and medical aids, and conditional reimbursement models. Furthermore, the SAHPRA could play an expanded role in facilitating trials designed for small biotechnology entities.

Finally, public and private partnerships could create solid bases for continuous funding, given the attractive investment and commercialization opportunities within the field of RM. These partnerships could then produce systems and channels for sharing infrastructure, expertise, and other resources to then develop suitable practical regulatory frameworks. Significant progress can be made by promoting collaborative research projects between public and commercial organizations. Through these partnerships, and utilizing the resources of both sectors, complicated RM-related challenges can hopefully be addressed. Sharing facilities, data, and knowledge can minimize duplication of efforts and can maximize the use of available resources. Moreover, increasingly effective partnerships would advocate for regulations that encourage and support the growth of the field of RM. Similarly, R&D surrounding RM in South Africa can be greatly enhanced by utilizing philanthropic entities. Applying the following strategies could effectively engage philanthropic support:

- Enhancing public awareness in a responsible way (potentially through carefully selected social media platforms) about the promise and potential of RM and its positive impact on healthcare.
- Engaging with authorities such as the Department of Health and the SAHPRA to clarify and emphasize the need for regulatory authority engagement.
- Creating endowments or special funds for translational studies (including clinical trials).
- Drawing attention to profitable projects and breakthroughs in RM by increasing visibility and attendance at local and international medical and scientific conferences.
- Recognizing and honoring philanthropic contributions to encourage continuous support through various media platforms.

Summary and Conclusions

Advancements in RM have already proven incredibly valuable with respect to their ability to manage common NCDs. The value and increasing potential presented by cutting-edge stem cell, gene, and cell therapies in particular cannot be overstated, and the implementation of RM in South Africa's healthcare landscape must become a matter of national priority. Despite the promise presented by RM, accessibility remains a pressing issue and this issue must be at the forefront when developing and implementing RM in South Africa and the Global South as a whole. The cost of products and therapies, especially those that originate from the Global North, puts the Global South at a significant disadvantage in these populations' ability to obtain proven, safe, and affordable RM technologies. Therefore, to promote further adoption of RM and RM technologies, the Global South must prioritize development in this sector. South Africa, in particular, can position itself as a leader in the Global South by establishing a strong environment for RM R&D growth through investment in both the private and public healthcare sectors. This investment should also include encouraging philanthropic entities to enter the market. By focusing on these concentrated investment efforts, South Africa will not only pave the way for economic expansion but will also promote greater accessibility and improved healthcare outcomes for all South Africans.

References

1. Altyar AE, El-Sayed A, Abdeen A, Piscopo M, Mousa SA, Najda A, Abdel-Daim MM (2023) Future regenerative medicine developments and their therapeutic applications. Biomed Pharmacother 158:114131. https://doi.org/10.1016/j.biopha.2022.114131
2. Ntege EH, Sunami H, Shimizu Y (2020) Advances in regenerative therapy: a review of the literature and future directions. Regen Ther 14:136–153. https://doi.org/10.1016/j.reth.2020.01.004
3. Bahari M, Mokhtari H, Yeganeh F (2023) Stem cell therapy, the market, the opportunities and the threat. Int J Mol Cell Med 12(3):310–319. https://doi.org/10.22088/IJMCM.BUMS.12.3.310

4. Alanazi BN, Alanazi YW, Al Otaibi MN, Alsuwairi HM, Al Sayegh AA, Alotaibi ZT (2023) Assessing the impact: a critical review of biomedical engineering's contributions to innovative healthcare solutions. J Surv Fish Sci 10(5):40–45. https://doi.org/10.53555/sfs.v10i5.2191
5. Atala A (2024) Thirty years of tissue engineering. Tissue Eng Part A 30(1–2):5–13. https://doi.org/10.1089/ten.tea.2023.0322
6. Grand View Research (2024) Regenerative medicine market size, share & trends analysis report by product (therapeutics, tools, banks), by therapeutic category (dermatology, musculoskeletal, immunology & inflammation), by region, and segment forecasts, 2025–2030. Grand View Research, Inc. wwwgrandviewresearchcom/industry-analysis/regenerative-medicine-market. Accessed 6 June 2025
7. Viljoen IM, Hendricks CL, Malherbe HL, Pepper MS (2022) Regenerative medicines: a new regulatory paradigm for South Africa. Biochimie 196:123–130. https://doi.org/10.1016/j.biochi.2022.02.010
8. May J, Meth C (2007) Dualism or underdevelopment in South Africa: what does a quantitative assessment of poverty, inequality and employment reveal? Dev South Afr 24(2):271–287. https://doi.org/10.1080/03768350701327178
9. Bhengu M, Chagwiza W (2023) The impact of illicit financial flows on South Africa's economic growth. SSRN. https://doi.org/10.2139/ssrn.4533652
10. Valodia I (2023) South Africa can't crack the inequality curse. Why, and what can be done. The Conversation. https://theconversation.com/south-africa-cant-crack-the-inequality-curse-why-and-what-can-be-done-213132
11. World Bank (2025) The world bank in South Africa. World Bank Group. www.worldbank.org/en/country/southafrica/overview#:~:text=South%20Africa%20remains%20a%20dual,coefficient%20of%200.67%20in%202018. Accessed 6 June 2025
12. Mabugu RE (2024) South Africa's fight against extreme poverty needs a new strategy - model shows how social grants could work. The Conversation. https://theconversation.com/south-africas-fight-against-extreme-poverty-needs-a-new-strategy-model-shows-how-social-grants-could-work-241694?utm_source=clipboard&utm_medium=bylinecopy_url_button. Accessed 6 June 2025
13. Corbett MS, Webster A, Hawkins R, Woolacott N (2017) Innovative regenerative medicines in the EU: a better future in evidence? BMC Med 15(1):49. https://doi.org/10.1186/s12916-017-0818-4
14. Qiu T, Hanna E, Dabbous M, Borislav B, Toumi M (2020) Regenerative medicine regulatory policies: a systematic review and international comparison. Health Policy 124(7):701–713. https://doi.org/10.1016/j.healthpol.2020.05.004
15. Alliance for Regenerative Medicine (2024) The sector snapshot: August 2024. www.alliancerm.org/sector-snapshot-august-2024. Accessed 6 June 2025
16. Ramezankhani R, Torabi S, Minaei N, Madani H, Rezaeiani S, Hassani SN, Gee AP, Dominici M, Silva DN, Baharvand H, Hajizadeh-Saffar E (2020) Two decades of global progress in authorized advanced therapy medicinal products: an emerging revolution in therapeutic strategies. Front Cell Dev Biol 8:547653. https://doi.org/10.3389/fcell.2020.547653
17. Pai M, Bandara S, Kyobutungi C (2024) Shifting power in global health will require leadership by the global south and allyship by the global north. Lancet 404(10464):1711–1713. https://doi.org/10.1016/S0140-6736(24)02323-7
18. Taylor-Robinson SD, Spearman CW, Suliman AAA (2021) Why is there a paucity of clinical trials in Africa? QJM 114(6):357–358. https://doi.org/10.1093/qjmed/hcab010
19. Cornetta K, Bonamino M, Mahlangu J, Mingozzi F, Rangarajan S, Rao J (2022) Gene therapy access: global challenges, opportunities, and views from Brazil, South Africa, and India. Mol Ther 30(6):2122–2129. https://doi.org/10.1016/j.ymthe.2022.04.002
20. Dzobo K, Senthebane DA, Pillay M, Ssemakalu C, Mkhumbeni N, Motaung KSCM (2017) The future of tissue engineering and regenerative medicine in Africa. Tissue Eng Part A 23(19–20):1023–1025. https://doi.org/10.1089/ten.tea.2017.0375

21. van de Vyver M, Idensohn PJ, Niesler CU (2022) A regenerative approach to the pharmacological management of hard-to-heal wounds. Biochimie 194:67–68. https://doi.org/10.1016/j.biochi.2021.12.016
22. Department of Higher Education and Training of South Africa (2024) Annual report. Government of South Africa. https://nationalgovernment.co.za/department_annual/506/2024-department-of-higher-education-and-training-(dhet)-annual-report.pdf. Accessed 6 June 2025
23. Sebola MP (2022) South Africa's public higher education institutions, university research outputs, and contribution to national human capital. Hum Resour Dev Int 26(2):217–231. https://doi.org/10.1080/13678868.2022.2047147
24. Clarivate (2025) Web of Science, Philadelphia. https://www.webofscience.com/wos/woscc/analyze-results/91beb4b1-3602-4aee-8263-f351a7a819cd-015d7b6a1f. Accessed 6 June 2025
25. Burgess LJ, Sulzer NU (2010) Clinical trials in South Africa: need for capacity building and training. S Afr Med J 100(7):402. https://www.scielo.org.za/scielo.php?script=sci_arttext&pid=S0256-95742010000700003. Accessed 6 June 2025
26. Centre for Health and Healthcare (2024) How scaling up clinical research in Africa can benefit society and the economy. World Economic Forum https://www.weforum.org/stories/2024/08/africa-scaling-up-clinical-research-benefit-society-economy/. Accessed 6 June 2025
27. Pan African Clinical Trials Registry (2025) The European & Developing Countries Clinical Trial Partnership. https://www.edctp.org/pan-african-clinical-trials-registry/. Accessed 6 June 2025
28. ClinicalTrials.gov (2025) U.S. National Library of Medicine, Bethesda. https://clinicaltrials.gov/. Accessed 6 June 2025
29. Ferretti C, Ripamonti U (2002) Human segmental mandibular defects treated with naturally derived bone morphogenetic proteins. J Craniofac Surg 13(3):434–444. https://doi.org/10.1097/00001665-200205000-00014
30. Kon E, Filardo G, Brittberg M, Busacca M, Condello V, Engebretsen L, Marlovits S, Niemeyer P, Platzer P, Posthumus M, Verdonk P, Verdonk R, Victor J, van der Merwe W, Widuchowski W, Zorzi C, Marcacci M (2018) A multilayer biomaterial for osteochondral regeneration shows superiority vs microfractures for the treatment of osteochondral lesions in a multicentre randomized trial at 2 years. Knee Surg Sports Traumatol Arthrosc 26(9):2704–2715. https://doi.org/10.1007/s00167-017-4707-3
31. Murdoch M, Wittstock C, Psaras G, Widgerow A, Lukhele M, Ramokgopa MT, Snyman J, Hutchings J, Marcos E, Grisillo Biscardi A, Cromarty D, Zheng X, Duneas N, Govender S (2025) Use of osteogenic bone matrix in patients with traumatic long bone defects: an open label, single center study. J Orthop 60:159–166. https://doi.org/10.1016/j.jor.2024.08.018
32. Cassim MI, Mohamed T, Adam JK, Niesler C, Chikandiwa A (2023) Use of autologous adipose-derived mesenchymal stem cells for ovarian rejuvenation in poor responder IVF patients: a phase 1 randomized placebo-controlled double-blind crossover study. Glob Reprod Health 8(2):e68. https://doi.org/10.1097/GRH.0000000000000068
33. Novitzky N, Mohamed R, Finlayson J, du Toit C (2000) Increased apoptosis of bone marrow cells and preserved proliferative capacity of selected progenitors predict for clinical response to anti-inflammatory therapy in myelodysplastic syndromes. Exp Hematol 28(8):941–949. https://doi.org/10.1016/s0301-472x(00)00489-6
34. Davies MT, Duckett ME (2024) The global burden of disease. In: Fiander A, Fry G (eds) A healthcare students introduction to global health: low-and middle-income countries. Springer, New York, pp 11–24
35. Tollman S, Price J (2021) Public health priorities in countries undergoing economic transition: the middle-income countries. In: Detels R, Karim QA, Baum F, Li L, Leyland A (eds) Oxford textbook of global public health. Oxford University Press, Oxford, pp 47–62
36. South African Medical Research Council (2020) Annual performance plan. South African Department of Health. https://www.samrc.ac.za/corporate-articles/annual-performance-plan. Accessed 6 June 2025

37. Hendricks CL, Alessandrini M, Pepper MS (2023) Equitable access to cell and gene therapies in South Africa: opportunities and hurdles. Gene Ther 30(1):180–186. https://doi.org/10.1038/s41434-021-00309-y
38. Atewologun FA, Okesanya OJ, Okon II, Kayode HH, Ukoaka BM, Olaleke NO, Ogaya JB, Okikiola LA, Manirambona E, Lucero-Prisno Iii DE (2024) Examining the potentials of stem cell therapy in reducing the burden of selected non-communicable diseases in Africa. Stem Cell Res Ther 15(1):253. https://doi.org/10.1186/s13287-024-03864-4
39. Takahashi T, Donahue RP, Nordberg RC, Hu JC, Currall SC, Athanasiou KA (2023) Commercialization of regenerative-medicine therapies. Nat Rev Bioeng 1:906–929. https://doi.org/10.1038/s44222-023-00095-9
40. Yamada S, Behfar A, Terzic A (2021) Regenerative medicine clinical readiness. Regen Med 16(3):309–322. https://doi.org/10.2217/rme-2020-0178
41. Mikulic M (2024) Distribution of regenerative medicine companies by world region in 2023. Statista. www.statista.com/statistics/687416/regenerative-medicine-companies-by-world-region/. Accessed 6 June 2025
42. Viljoen IM, Pepper MS (2025) Addressing the limitations of the regulatory landscape in South Africa regarding advanced cell and gene therapies and related sectors involving human cells, tissues and organs. SAMJ 115(1):12–16. https://doi.org/10.7196/SAMJ.2025.v115i1.2629
43. Ghosh S, Nehme R, Barrett LE (2022) Greater genetic diversity is needed in human pluripotent stem cell models. Nat Commun 13(1):7301. https://doi.org/10.1038/s41467-022-34940-z
44. Sirugo G, Williams SM, Tishkoff SA (2019) The missing diversity in human genetic studies. Cell 177(4):26–31. https://doi.org/10.1016/j.cell.2019.02.048
45. Singh V (2023) Emerging technologies and technological catch-up in the pharmaceuticals sector in South Africa. Dissertation, University of the Witwatersrand Johannesburg. https://wiredspace.wits.ac.za/items/4a174ad8-0ef4-4698-aec3-69216f4bbbb7. Accessed 6 June 2025
46. van de Vyver M (2017) Intrinsic mesenchymal stem cell dysfunction in diabetes mellitus: implications for autologous cell therapy. Stem Cells Dev 26(14):1042–1053. https://doi.org/10.1089/scd.2017.0025
47. Zhang H, De T, Zhong Y, Perera MA (2019) The advantages and challenges of diversity in pharmacogenomics: can minority populations bring us closer to implementation? Clin Pharmacol Ther 106(2):338–349. https://doi.org/10.1002/cpt.1491
48. Sibomana O (2024) Genetic diversity landscape in African population: a review of implications for personalized and precision medicine. Pharmacogenomics Pers Med 17:487–496. https://doi.org/10.2147/PGPM.S485452
49. Idensohn PJ (2023) Human amniotic membrane accelerates healing in a hard-to-heal diabetic foot ulcer. Wound Health S Afr 16(1):27–29. https://ojs.sabinet.co.za/index.php/whsa/article/view/44/13. Accessed 6 June 2025
50. Montague C, Holt Y, Vlok M, Dhanraj P, Boodhoo K, Maartens M, Buthelezi K, Niesler CU, van de Vyver M (2024) Combined therapeutic use of umbilical cord blood serum and amniotic membrane in diabetic wounds. Biochimie 227(Pt A):193–204. https://doi.org/10.1016/j.biochi.2024.07.012
51. Niesler CU, van de Vyver M (2022) Editorial: regeneration in health and disease. Biochimie 196:121–122. https://doi.org/10.1016/j.biochi.2022.03.009
52. DiDomenico LA, Orgill DP, Galiano RD, Serena TE, Carter MJ, Kaufman JP, Young NJ, Zelen CM (2016) Aseptically processed placental membrane improves healing of diabetic foot ulcerations: prospective, randomized clinical trial. Plast Reconstr Surg Glob Open 4(10):e1095. https://doi.org/10.1097/GOX.0000000000001095
53. Mohajeri-Tehrani MR, Variji Z, Mohseni S, Firuz A, Annabestani Z, Zartab H, Rad MA, Tootee A, Dowlati Y, Larijani B (2016) Comparison of a bioimplant dressing with a wet dressing for the treatment of diabetic foot ulcers: a randomized, controlled clinical trial. Wounds 28(7):248–254. https://www.hmpgloballearningnetwork.com/site/wounds/article/comparison-bioimplant-dressing-wet-dressing-treatment-diabetic-foot-ulcers-randomized. Accessed 6 June 2025

54. Snyder RJ, Shimozaki K, Tallis A, Kerzner M, Reyzelman A, Lintzeris D, Bell D, Rutan RL, Rosenblum B (2016) A prospective, randomized, multicenter, controlled evaluation of the use of dehydrated amniotic membrane allograft compared to standard of care for the closure of chronic diabetic foot ulcer. Wounds 28(3):70–77. https://www.hmpgloballearningnetwork.com/site/wounds/article/prospective-randomized-multicenter-controlled-evaluation-use-dehydrated-amniotic-membrane. Accessed 6 June 2025
55. Zelen CM, Serena TE, Denoziere G, Fetterolf DE (2013) A prospective randomised comparative parallel study of amniotic membrane wound graft in the management of diabetic foot ulcers. Int Wound J 10(5):502–507. https://doi.org/10.1111/iwj.12097
56. Zelen CM, Serena TE, Snyder RJ (2014) A prospective, randomised comparative study of weekly versus biweekly application of dehydrated human amnion/chorion membrane allograft in the management of diabetic foot ulcers. Int Wound J 11(2):122–128. https://doi.org/10.1111/iwj.12242
57. Zelen CM, Gould L, Serena TE, Carter MJ, Keller J, Li WW (2015) A prospective, randomised, controlled, multi-Centre comparative effectiveness study of healing using dehydrated human amnion/chorion membrane allograft, bioengineered skin substitute or standard of care for treatment of chronic lower extremity diabetic ulcers. Int Wound J 12(6):724–732. https://doi.org/10.1111/iwj.12395
58. Zelen CM, Serena TE, Gould L, Le L, Carter MJ, Keller J, Li WW (2016) Treatment of chronic diabetic lower extremity ulcers with advanced therapies: a prospective, randomised, controlled, multi-centre comparative study examining clinical efficacy and cost. Int Wound J 13(2):272–282. https://doi.org/10.1111/iwj.12566
59. Guo X, Zhang W, Lu J, Zhu Y, Sun H, Xu D, Xian S, Yao Y, Qian W, Lu B (2024) Amniotic miracle: investigating the unique development and applications of amniotic membrane in wound healing. Skin Res Technol 30(7):e13860. https://doi.org/10.1111/srt.13860

The Kenyan Legal and Policy Framework for Stem Cell Research

7

Paul Ogendi

Abstract

Regenerative medicine (RM), particularly stem cell research, is an emerging field with significant promise for addressing complex and chronic health conditions. In Kenya, while stem cell research remains in the early stages of development, recent policy and institutional investments indicate growing governmental interest in advancing the sector. However, the legal and regulatory environments are still underdeveloped. Kenya does not yet have a dedicated legislative framework for stem cell research, and existing laws related to health, science, and innovation engage with the field only in general terms. Critical issues, such as the permissibility of embryonic stem cell research, remain unclear, leaving researchers to operate within a fragmented and ambiguous regulatory landscape. At the same time, other areas of law, such as criminal law, data protection, and intellectual property, may indirectly constrain scientific activity. In contrast, South Africa offers a more structured and explicit regulatory model that Kenya could consider and adapt to its own context. As the country moves toward regulating this evolving area of biomedical science, aligning national efforts with both regional models and international standards will be important.

Keywords

Regenerative medicine · Stem cell research · Kenya · Research ethics · Health law · Biomedical regulation · Science policy

---

P. Ogendi (✉)
Faculty of Law, University of Nairobi-Kisumu Campus, Kisumu, Kenya

U. Ogbogu et al. (eds.), *Regenerative Medicine and the Bioethics of Globalization, Justice and Hegemony*, Collaborative Bioethics 5,
https://doi.org/10.1007/978-3-032-07289-4_7

Contextualizing Regenerative Medicine and Stem Cell Research in Kenya: Scientific Promise, Ethical Complexity, and Regulatory Challenge

Regenerative medicine (RM), including stem cell research, is an emerging field in many countries, including Kenya [1]. RM innovations have promising implications for scientists, companies, and governments seeking to resolve pressing healthcare challenges [1]. RM is a diverse and interdisciplinary field, which makes it difficult to define precisely. An often-cited definition by Daar and Greenwood [2] describes RM as "an interdisciplinary field of research and clinical applications focused on the repair, replacement or regeneration of cells, tissues or organs to restore impaired function resulting from any cause, including congenital defects, disease, trauma and ageing," and as one that "uses a combination of several converging technological approaches, both existing and newly emerging, that moves it beyond traditional transplantation and replacement therapies," including "the use of soluble molecules, gene therapy, stem and progenitor cell therapy, tissue engineering and the reprogramming of cell and tissue types."

A central feature of RM is its focus on the regeneration of living tissues and organs through a range of approaches, including stem cells, cell transplantation, tissue engineering, and nuclear transfer [3]. As Oerlemans et al. [4] explain, "the goal of regenerative medicine is to induce the regeneration of tissues and organs, either following disease or injury, or when congenital or developmental defects are present."

Among these approaches, stem cell research plays a particularly crucial role. Stem cells can be used in both biomedical research and therapeutic applications, to study disease mechanisms, test drugs, and generate cells and tissues for use in cell-based therapies and tissue engineering [5]. In addition to their clinical potential, stem cells provide a powerful tool for studying human development by offering insights into how tissues form, differentiate, and function, knowledge that is essential for designing effective regenerative treatments [5].

Building on its central role in RM, stem cell research has attracted significant interest from scientists, industry, and governments worldwide [1]. Between 2008 and 2012, global publications in the field grew by 7%, compared to an average growth rate of 2.9% across all scientific disciplines [5]. Leading contributors during this period included the United States (28.99%), China (11.41%), Italy (10.87%), Germany (6.88%), the UK (3.99%), Japan (3.80%), France (2.90%), India (2.72%), Canada (2.17%), and South Korea (2.17%) [6]. This growing interest is largely driven by the field's potential to treat previously incurable conditions, such as neurodegenerative diseases like Parkinson's and Alzheimer's [7].

While stem cell research has seen significant global growth, it has also faced ethical and moral challenges, particularly around the use of embryonic stem cells [8]. A major source of controversy stems from religious beliefs, especially the sanctity of life argument, which holds that human life begins at conception and that embryos possess inherent moral value, making their destruction for research purposes ethically unacceptable [9]. As Allum et al. [10] explain, this argument reflects a deontological view that treats the embryo as a human being with inviolable rights. Kantian moral philosophy adds another layer, arguing that if embryos possess moral status, then using them merely as a means to an end is unethical [11]. However, if embryos are not considered "persons" in the Kantian sense, then such research could be ethically permissible. This Kantian interpretation is contested, as the idea of personhood from conception has been criticized as problematic [12]. In contrast, utilitarian thinkers such as John Stuart Mill focus on outcomes; from this perspective, embryonic stem cell research is moral if it yields substantial medical benefits [12]. As Isasi and Knoppers [13] note, the wide range of perspectives suggests that there is no simple answer to the question of whether embryos should be regarded as potential persons.

A key development in stem cell research was the discovery of abundant adult mesenchymal stem cells (MSCs) in adipose tissue, which facilitated the expansion of cell-based therapies [14]. Unlike embryonic stem cells, adult stem cells have been used with less ethical controversy, particularly in autologous transplants to treat conditions such as leukemia and lymphoma [15].

The challenge in shaping public policy on stem cell research lies in the inherent blurring of boundaries between science, religion, ethics, and politics, particularly when embryonic stem cells are involved [12]. In Africa, religion is the most significant impediment to the advancement of stem cell research [8]. In Kenya, the Catholic Church remains a particularly formidable force, given its deep historical roots dating back to the late nineteenth century and its influential role in establishing missions, schools, and healthcare facilities. Its involvement in the 2010 Kenyan constitutional review process, especially in advocating for the protection of life from the point of conception, further underscores its influence on bioethical discourse [16]. These religious concerns are compounded by cultural beliefs. In some African contexts, the human body is regarded as sacred, and certain communities practice the burial of umbilical cords—a rich source of stem cells, particularly hematopoietic stem cells, which can be used to treat various blood and immune system disorders—to protect against spiritual harm [16].

Similar religious and cultural challenges to stem cell research exist outside of the African continent. In Mexico, for example, researchers have noted that "constant and forceful actions against abortion and stem cell research on the part of pro-life and religious groups" can obstruct efforts to regulate stem cells independently of broader moral debates [17]. This suggests the role of religion and culture in the field may be a global phenomenon. This shared difficulty presents an opportunity for international collaboration in developing strategies to address ethical concerns related to embryonic stem cells.

In Africa generally, the current climate surrounding human embryonic stem cell research regulation is fraught with opposition; many countries heavily regulate the area, while others ban it outright [18]. South Africa appears to be the most progressive nation as it is already using human embryo-derived stem cells for novel drug development and screening, while adult stem cells are being used in bone marrow transplants to cure diseases such as myeloma [8]. Moreover, as stem cell technology continues to advance and more clinical applications emerge, comparatively little attention has been paid to the regulation of products resulting from stem cell research. The primary focus has remained on the legal and ethical debates surrounding the technology [19]. This emphasis on regulatory and moral concerns has impeded stem cell research, in turn affecting clinical translation. Notably, Africa accounts for only 2.5% of clinical trials involving MSC-based therapeutic products [20].

Stem cell research is a relatively new but steadily growing field in Kenya. In July 2023, the Kenyan Medical Research Institute (KEMRI), the national body responsible for health research in Kenya, commissioned a state-of-the-art stem cell and regenerative medicine laboratory, marking a major milestone for the advancement of stem cell research in the country [21]. The laboratory received USD 542,600 in government funding to support research on non-communicable diseases (NCDs), including diabetes, cardiovascular diseases, cancer, road traffic injuries, burns, and oral, eye, and dental diseases [21]. This research-focused agenda demonstrates the Kenyan government's commitment to developing the sector and tapping into its full potential. It also underscores the importance of strong preclinical research as a foundation for clinical translation. Investing in stem cell research through KEMRI positions Kenya to become a leader in this field and to generate innovations that can ultimately benefit clinical practice.

In contrast to this research-focused trajectory, stem cell therapies have also emerged in the private, commercial healthcare sector. In 2020, Tia Clinics, based in Nairobi, became one of the first private clinics in the country to offer stem cell therapy for cosmetic and pain management purposes, including hair restoration, skin rejuvenation, and joint pain relief [22]. Tia Clinics markets its services as cutting-edge treatments designed to enhance natural beauty and confidence. Although the clinic appears to have begun offering services to the public in 2020, official records indicate that it was formally established in 2022 and only obtained government registration approval in 2024 [23]. This timeline raises questions about regulatory oversight and the status of stem cell therapy providers operating prior to formal recognition. While the services promoted by Tia Clinics may reflect the growing public interest in stem cell-based applications, they also raise important questions about the scientific validity, clinical evidence, and regulatory oversight supporting such therapies [24–28]. This juxtaposition between public-sector research and private-sector commercialization highlights the urgent need for a clear regulatory framework to distinguish between evidence-based interventions and potentially unproven treatments.

Against the backdrop of these global, regional, and ethical dynamics, the remainder of the chapter examines how Kenya governs stem cell research as an emerging aspect of RM. It adopts a doctrinal research methodology to assess relevant national laws, policies, and guidelines, with a focus on laboratory-based research rather than clinical applications or therapeutic products. The discussion is organized into four main sections. The section titled "Legal and Institutional Foundations for Stem Cell Research in Kenya" evaluates the enabling legal and institutional frameworks, such as the *Health Act* [29] and the *Science, Technology and Innovation Act* (*STI Act*) [30], along with key policies and guidelines that support stem cell research. The section "Legal Constraints and Regulatory Barriers to Stem Cell Research" examines limiting legal instruments, including criminal, intellectual property, and data protection laws, that may inhibit, constrain, or condition such research. The section titled "Uncertainty and Gaps in Kenya's Regulatory Framework for Stem Cell Research" identifies areas of legal uncertainty, regulatory overlap, or lack of policy guidance, particularly in relation to embryonic stem cell research. It also explores how these gaps might be addressed by drawing on comparative models, such as South Africa's, and by incorporating international ethical guidelines and best practices. Taken together, these sections show that while Kenya has made some progress, its regulatory environment remains underdeveloped. Further reform, grounded in both local context and international standards, is essential to support responsible and innovative stem cell research.

Legal and Institutional Foundations for Stem Cell Research in Kenya

Stem cell research falls under the umbrella of biomedical research, whose regulation in Kenya remained uncoordinated for many years and persists to date [31]. Currently, the legal instruments that could potentially support and facilitate stem cell research in Kenya include the *Health Act* and the *STI Act*. However, the controversial aspects of stem cell research, particularly the use of human embryos, remain unregulated, as they are not explicitly addressed within these instruments.

The Health Act and Related Policies and Institutional Frameworks

The *Health Act* is the primary legislative framework for implementing the constitutional right to health in Kenya. The reforms introduced by the 2010 *Constitution of Kenya* (the Constitution) [32] led to the devolution of health functions to county governments and explicitly recognized the right to health in the constitutional text. This, in turn, paved the way for the enactment of national health legislation [33]. Previously, the health sector was centrally managed by two health ministries, operating under the 1921 *Public Health Act* and other specialized statutes [33].

The *Health Act* now directly implements Section 43(1)(a) of the Constitution, which guarantees the right to the highest attainable standards of health, including the right to health care services. The *Health Act* and related health standards have enabled or led to the creation of key institutions to oversee and enhance governance of the health sector in Kenya, including the Office of the Director General, the Kenya Health Sector Inter-Governmental Consultative Forum, the Kenya Health Human Resource Advisory Council, the Kenya Health Professions Oversight Authority, and the National Research for Health Committee.

Strictly speaking, the provisions of the *Health Act* do not apply to stem cell research. However, the Act appears to contemplate the clinical application of stem cells. Section 2 of the Act defines "tissues" to include placenta, embryonic or fetal tissue, stem cells, and umbilical cord. As such, regulations governing tissues also apply to stem cells, as well as to the biological sources from which they are derived. Section 80 of the Act stipulates that the removal of tissue or gametes from a human being for transplantation into another person is prohibited unless specific conditions are met. Moreover, such transplantation must take place in a duly authorized health facility. In essence, stem cell therapy or treatment is prohibited unless certain prescribed conditions are fulfilled, including that the procedure is conducted in an authorized health facility. Currently, the facilities licensed to perform transplants in Kenya include Kenyatta National Hospital, Moi Teaching and Referral Hospital, Aga Khan University Hospital, M.P. Shah Hospital, Nairobi Hospital, St. Luke's Hospital–Eldoret, Nairobi West Hospital, and Mater Hospital [34].

Additional regulatory guidance comes from the 2022 *Kenya Policy on Donation, Transfusion and Transplant of Human Derived Medical Products* (2022 Policy), which defines a stem cell as an "undifferentiated cell that can divide to produce offspring cells in a human being" ([35], p. viii). The policy adopts a broad definition of "transplant" that includes the replacement of a "stem cell…with a healthy stem cell" ([35], p. viii) and aims to establish a comprehensive legislative, governance, and regulatory framework to ensure equitable access and accountability across blood, stem cell, tissue, and organ services. When read together with the *Health Act*'s inclusion of stem cells under the definition of tissues, the policy reinforces the view that stem cell therapy is, at least implicitly, contemplated within Kenya's broader clinical governance framework. However, significant ambiguity remains regarding the regulation of stem cell research itself, which the current instruments do not explicitly address. This gap underscores the need for the Kenyan government to clarify how stem cell research should be governed.

One of the strategies outlined in the 2022 Policy is to introduce legislation that would establish a semi-autonomous framework for the governance and regulation of blood, tissue, and transplant services. While stem cells are not explicitly mentioned in this context, it is possible they are included by virtue of the definition of tissues in the *Health Act*. However, this lack of specific reference contributes to the ongoing uncertainty surrounding the regulation of stem cells. The policy also calls for the development of national guidelines for transplant services, which may, in time, extend to stem cell therapy.

Thus, while the current legal and policy framework does contemplate stem cell therapy or treatment, it remains noncommittal with respect to stem cell research. This continued emphasis on clinical application, rather than scientific inquiry, suggests that the government is more inclined to adopt innovations developed abroad than to invest in building local research capacity in this emerging field.

Part XIV of the *Health Act* addresses the promotion and regulation of health research in Kenya and assigns responsibility for oversight and funding of health research to the National Health Research Committee (NHRC). The Act mandates the Cabinet Secretary for Health to establish the NHRC as a technical body and sets out its composition, which must include a distinguished biomedical science researcher. Among its functions, the NHRC is tasked with identifying emerging health issues as research priorities and setting standards for ethical clearance in health research.

Stem cell research is arguably one such priority in Kenya, given its potential to address complex medical conditions and contribute to scientific advancement. Under Section 97 of the *Health Act*, the Committee is required to collaborate with KEMRI and other research institutions in formulating national health policy. KEMRI is a state corporation established in 1979 and now governed under the *STI Act*. It serves as Kenya's national body for health research, with a mandate to carry out biomedical research aimed at improving the health of Kenyans. KEMRI conducts scientific investigations on a wide range of public health issues, including infectious and NCDs, and plays a central role in shaping national health policy through evidence-based research. It collaborates with local and international partners, including the World Health Organization (WHO), and operates multiple specialized research centers across the country.

Section 101 of the *Health Act* requires that no less than 30% of the National Research Fund (NRF) be allocated to health research, acknowledging the critical role of scientific and policy research in strengthening Kenya's health sector. The NRF is a statutory body established under Section 32 of the *STI Act*. Its primary mandate is to mobilize, allocate, and manage financial resources to support research and innovation across all fields of science and technology in Kenya [36]. The NRF is funded by an annual allocation from the national budget, amounting to 2% of Kenya's gross domestic product, with at least 30% of that allocation earmarked specifically for health research. The NRF supports a wide range of research activities, including basic and applied research, capacity building, and infrastructure development in research institutions.

The NRF plays a key role in strengthening the national innovation system by financing priority research areas identified by the government, including emerging fields such as regenerative medicine and stem cell research. A notable example of its work is the 2023 funding of KEMRI's state-of-the-art stem cell and regenerative medicine laboratory, demonstrating the NRF's commitment to advancing health research and positioning Kenya at the forefront of biomedical innovation.

Following the enactment of the Constitution and the subsequent establishment of county governments in 2013, Kenya began a process of aligning its national policies with the new constitutional and devolved governance framework. As part of this realignment, the *Kenya Health Policy 2014–2030* was developed to guide the country's health sector in the new constitutional era. The policy identifies research and development as a key action area in which all stakeholders are expected to invest in order to facilitate the achievement of national health objectives [37]. It emphasizes the importance of cultivating a culture where research plays a central role in guiding policy formulation and action to improve the health and development of the Kenyan population [37]. Notably, health research was included in the revised policy as the eighth component or building block, establishing it as a distinct regulatory element [33]. The policy also affirms its commitment to upholding ethical standards in health research in alignment with the provisions of the *STI Act*.

Prior to these reforms, two institutions were primarily responsible for health research: the National Council for Science and Technology (NCST), established in 1977, which provided oversight and stewardship, and KEMRI, which focused on resource creation through investment and training [33]. Following the enactment of the *STI Act* in 2013, the National Commission for Science, Technology and Innovation (NACOSTI) was established as the successor to the NCST [33].

Kenya's health research agenda has therefore evolved significantly over the past decade. It is now governed by a clearer and more coherent regulatory framework, underpinned by comprehensive laws and national policy commitments. This framework would likely be beneficial to RM research and development in Kenya, including stem cell research.

The Science, Technology, and Innovation Act (STI Act)

The *STI Act* is Kenya's principal legislation for the regulation of scientific research. According to its preamble, the Act aims to support the advancement of science, technology, and innovation in the country by promoting, coordinating, and overseeing their development [30]. While the Act does not expressly regulate stem cell research, its provisions could still apply to the conduct of such research in Kenya.

The *STI Act* could potentially govern two key aspects of stem cell research: the approval of research protocols and the establishment and regulation of the institutions where such research is conducted. With respect to research approvals, the *STI Act* empowers NACOSTI to regulate and ensure quality within the science, technology, and innovation sector, while also advising the government on related matters. Among its key functions are accrediting research institutions and approving scientific research activities across the country. Generally, all research projects in Kenya must be licensed by NACOSTI, except where the research is conducted within a university or an academic program. Notably, Section 12(5)(d) of the Act also prohibits the licensing of research that could negatively impact the lives of Kenyan citizens [30]. From the foregoing, two key points are evident as they relate to stem cell research approvals. First, all stem cell research in Kenya must obtain prior approval from NACOSTI, and a license must be issued before such research can proceed. Second, this license may be withheld if the proposed research is considered likely to pose adverse effects on the lives of Kenyan citizens. This requirement reflects a broader ethical imperative to prevent harm to human subjects, as illustrated by historical cases such as the Tuskegee Syphilis Study, which was unethically conducted in the United States from 1932 to 1972. In that study, African American men were deliberately left untreated for syphilis without their informed consent, leading to avoidable suffering and death [38]. By incorporating such ethical safeguards, the Kenyan regulatory framework seeks to prevent similar abuses and ensure that scientific progress does not come at the expense of human well-being. The licensing requirements also underscore the central role of NACOSTI in safeguarding public welfare while regulating emerging and potentially sensitive areas of scientific inquiry, such as stem cell research.

The *STI Act* also provides a framework for the establishment and regulation of research institutions in Kenya. Under this framework, the Cabinet Secretary for Health may establish a research institute through official notice in the Kenya Gazette, based on a recommendation from NACOSTI. It is an offence to operate a research institute without obtaining a certificate of registration from NACOSTI. Once registered, research institutions are required to comply with prescribed standards, codes of conduct, and any other applicable regulations. For institutions involved in stem cell research, registration is essential not only for legal compliance but also for ensuring adherence to ethical standards and preventing malpractice or abuse.

One of the regulations established by NACOSTI that is relevant to stem cell research is the *National Guidelines for Ethical Conduct of Biomedical Research Involving Human Participants in Kenya* (National Biomedical Research Guidelines) [39]. These guidelines address issues such as authorship, publication, and intellectual property rights, requiring that such matters be properly considered during the proposal development process to safeguard the interests of local researchers, institutions, and the country at large. The guidelines also emphasize that dissemination and utilization of research findings for educational purposes and their translation into policy, where appropriate, should be addressed at the proposal stage. This aligns with the African Union's Agenda 2063, which sets out aims and aspirations focused on developing Africa's human and social capital through an education and skills revolution with an emphasis on science and technology.

In addition to NACOSTI, other institutions established under the *STI Act* also have the potential to support stem cell research in Kenya. One such body is the Kenya National Innovation Agency (KeNIA), which is tasked with developing and managing the country's national innovation system by identifying and promoting strategic areas of innovation [40]. Health innovation is among its key priorities, and through the Department of Science and Innovation, KeNIA engages in partnerships to support initiatives such as nuclear medicine research, vaccine and pharmaceutical manufacturing, diagnostic test development, and the integration of indigenous knowledge into product innovation. KeNIA also seeks to build advanced research infrastructure, particularly in omics sciences, a field closely tied to stem cell research given its focus on genes and proteins relevant to treating non-communicable and degenerative diseases. As previously discussed, the NRF is another institution established under the *STI Act*.

Legal Constraints and Regulatory Barriers to Stem Cell Research

Laws that may impede, constrain, or limit the stem cell research agenda in Kenya include criminal laws, privacy laws, and intellectual property laws.

Criminal Laws

Stem cell research may fall under the purview of criminal law, particularly in cases where the law seeks to address extreme forms of human exploitation. Criminal law regulation is especially important in the context of stem cell tourism and its potential to exploit patients' desperation to find cures for complex diseases and conditions. Two key laws in this regard are the *Penal Code* [41] and the *Counter-Trafficking in Persons Act* [42].

The *Counter-Trafficking in Persons Act* is relevant to stem cell research because the definition of exploitation under Section 2 includes the "forcible or fraudulent use of any human being for removal of organs or any body parts." Given that stem cell research carries the potential for human exploitation, the state should strictly prohibit any such abuse. Additionally, where research is conducted in the context of organized crime, Section 10 of the Act imposes life imprisonment as the maximum penalty. A notable example of potential exploitation is the commercial procurement of human eggs from women, often under coercive or economically exploitative conditions, for use in embryonic stem cell research. As Widdows [43] highlights, the sale of eggs can raise serious ethical concerns similar to those in trafficking for prostitution, particularly when vulnerable populations are targeted. Such practices blur the lines between consent and coercion, underscoring the importance of a strong regulatory framework to prevent the commodification of human bodies in the name of scientific progress. Kenya's legal provisions, therefore, act as an important safeguard against these forms of exploitation, aligning with international ethical standards.

With regard to the *Penal Code*, fetal tissue may be subject to regulation if it is used in stem cell research. In this context, Sections 158–160, which address abortion, may apply. For example, in Hungary, there have been reports of illegal stem cell clinics charging exorbitant fees for treatments using embryonic and fetal stem cells obtained through unlawful means [44, 45]. Fetal stem cells raise similar ethical concerns as embryonic stem cells because they are derived from a five- to ten-week-old fetus obtained through abortion or miscarriage [15]. In Kenya, abortions conducted for the purpose of harvesting fetal tissue for stem cell research are strictly prohibited.

Intellectual Property Laws

Stem cell research and innovation fall directly under the purview of the *Industrial Property Act* (IPA) [46]. The Act is administered by the Kenya Industrial Property Institute (KIPI), which is established under Section 3 of the Act. Section 5 outlines the functions of KIPI, highlighting its role in granting industrial property rights and screening technology transfer agreements and licenses. Stem cell research undoubtedly involves aspects of both technology transfer and licensing at various stages.

With regard to licensing, Section 69 of the IPA imposes certain restrictions designed to protect the country's economic interests. For example, it permits the

Managing Director of KIPI to refuse registration of a contract if it includes terms that permit or require the importation of technology from outside Kenya when substantially similar or equivalent technology is available domestically on the same or more favorable terms. However, once all applicable conditions, including those under Section 69 are satisfied, the Managing Director is empowered under Section 70 to register the contract and issue a certificate of registration to the applicant. If a grievance arises, Section 71 provides for an appeal to the Industrial Property Tribunal, which must be filed within two months. Ultimately, approval for registration depends on whether the technology to be imported is unavailable in Kenya, among other criteria.

Innovations arising from stem cell research are also protected under the IPA, with patents being the most common form of protection. Part III of the IPA deals with patentability. Section 21(1) broadly defines an "invention" as "a solution to a specific problem in the field of technology," while Section 21(2)(c) includes within this definition "methods for treatment of the human or animal body by surgery or therapy, as well as diagnostic methods practised [sic] in relation thereto, except products for use in such methods." Under Section 22, an invention is patentable if it is new, involves an inventive step, and is industrially applicable or constitutes a new use. The meaning of these criteria is elaborated in Sections 23–25 of the IPA. However, Section 26(2) excludes from patentability any inventions that are contrary to public order, morality, public health and safety, principles of humanity, or environmental conservation. Therefore, certain aspects of stem cell research, particularly those involving embryonic stem cells, may face legal barriers under this provision, especially given the absence of specific legislation regulating stem cell technologies in Kenya.

Another relevant law in the area of intellectual property enforcement is the *Anti-Counterfeit Act* [47], which was enacted to combat the trade in counterfeit goods and to establish the Anti-Counterfeit Authority responsible for overseeing enforcement and related matters. The Act plays a key role in enforcing intellectual property rights, including those arising from stem cell research. Section 2 of the Act defines intellectual property to include any rights protected under the IPA. The functions of the Anti-Counterfeit Authority are detailed in section 5 and include combating counterfeiting and trade in counterfeit goods.

Privacy and Data Protection Laws

In addition to criminal and intellectual property laws, privacy and data protection laws are also relevant to the regulation of stem cell research in Kenya. This is particularly true where research involves the collection, storage, and sharing of sensitive personal data, such as genetic and health-related information. Although the risk of reidentifying individuals from such data is currently low, experts recommend a proportionate approach that distinguishes between perceived and actual risks [48]. This is especially important in biorepository-based research and in projects that involve data sharing across institutions and borders [48].

Kenya's primary data protection and safety law, the *Data Protection Act* [49], does not define the term "reidentification," leaving some ambiguity around how the concept should be addressed in the context of biomedical research. However, in the field of data protection more broadly, reidentification typically refers to the process by which anonymized or de-identified data is matched with other available information to reveal the identity of individuals [48]. This issue is especially relevant in biomedical research, including stem cell research, where sensitive personal and genetic data are often collected and stored.

Recent advances in data science and statistical analysis have increased the risk that supposedly anonymized datasets can be reidentified through inference, especially when combined with other data sources [48]. As a result, even if direct identifiers have been removed, the possibility of revealing a participant's identity still exists. Without appropriate legal and technical safeguards, stem cell research in Kenya could be vulnerable to claims of privacy violations, particularly in cases involving biorepositories or large-scale data sharing.

Under Section 2 of the *Data Protection Act*, genetic data is classified as a form of sensitive personal data. Because stem cell research often involves the collection and use of genetic information [50], such research activities fall within the scope of Kenya's data protection regime. Institutions conducting stem cell research must comply with specific legal obligations. First, they are required to register as either data controllers or data processors, depending on their role in handling personal data. The Data Commissioner, established under the Act, determines the thresholds for mandatory registration and assesses whether sensitive personal data is being processed ([49], Section 5).

As a precautionary measure, research institutions should proactively engage with the Data Commissioner to determine whether their activities involve the processing of sensitive data, particularly in joint research projects that involve cross-border data transfers. The Act imposes stricter conditions on the international transfer of personal data, including the requirement for adequate data protection safeguards or the informed consent of the data subject.

Data controllers and processors must also adhere to several core principles of data protection. These include processing personal data lawfully, fairly, and transparently; ensuring that data collected is relevant and limited to what is necessary; storing data in a manner that does not identify subjects longer than required; and avoiding cross-border data transfers unless the law permits it ([49], Section 25). These obligations are especially important for collaborative research projects that may involve international institutions or commercial entities.

Finally, Section 46 of the *Data Protection Act* introduces additional safeguards for health-related data. Such data may only be processed by a healthcare provider or by a person who is subject to a legal duty of professional confidentiality. The processing must also be necessary for the public interest in health or carried out under another lawful basis established by statute. These provisions are particularly significant for stem cell research, which often involves sensitive health and genetic information collected from human participants.

Uncertainty and Gaps in Kenya's Regulatory Framework for Stem Cell Research

The Kenyan legal framework remains largely uncertain, with significant gaps and ambiguities surrounding RM, and stem cell research in particular. Although the *Health Act* is Kenya's primary legislation governing the health sector, it fails to provide clear guidance on several contentious issues. Notably, it is either silent or vague about the permissibility of human embryonic stem cell research, one of the most ethically debated aspects of stem cell science. Given its status as the central statute, the *Health Act* should offer foundational guidance on the governance of stem cell research, yet this clarity is currently absent.

This lack of specificity is not limited to the *Health Act*. Other key national documents, including the *Kenya Health Policy* and the *Draft Science, Technology and Innovation Policy 2020–2030* [51], similarly omit any reference to stem cells or RM. While stem cell research may be implicitly permissible under the broader category of health research, as evidenced by the NRF's support for KEMRI's establishment of a state-of-the-art stem cell and regenerative medicine laboratory, this permissiveness is not firmly grounded in statute or regulatory policies, leaving the field exposed to regulatory uncertainty.

As previously noted, the National Biomedical Research Guidelines currently provide the most explicit acknowledgment of stem cell research within the national regulatory landscape. These guidelines were revised to reflect emerging issues in biomedical research, including biobanking and stem cell science. The guidelines describe stem cell research as "an emerging area of scientific research" ([39], p. 19) and call for the categorization of research activities to clearly indicate what is permissible, restricted, or prohibited. They also urge careful oversight of procurement, banking, and data sharing practices. However, the guidelines stop short of offering comprehensive regulation. Instead, they defer to both national and international frameworks and ethical principles, placing the burden on researchers to interpret and apply general norms in a highly specialized and evolving field.

This regulatory gap is compounded by the nature of stem cell science itself. As Wekesa [31] notes, regulating biomedical research in Kenya remains a challenge partly due to the constantly evolving nature of science. He identifies stem cell research as especially ethically fraught, particularly with respect to questions surrounding the destruction of embryos for the purpose of harvesting stem cells, or the use of placental-derived cells. These complexities make the need for clear, specific legislation even more pressing.

One approach to addressing this uncertainty is for Kenya to examine and adapt foreign models of regulation. South Africa presents an instructive case study. There, stem cell research is governed by both the *National Health Act of 2003* [52] and the *Medicine and Related Substances Act of 1965* [53]. Section 57(4) of the *National Health Act* permits research on stem cells and zygotes up to 14 days old, provided the researcher obtains written consent from the donor and documents the research appropriately. Therapeutic cloning involving adult or umbilical cord stem cells is also permitted under Section 57(2). However, South Africa has drawn a firm line against reproductive cloning, expressly banning the manipulation of genetic material for that purpose under Section 57(1). Violations of these provisions are punishable by fines or imprisonment of up to five years. While Kenya may need to tailor any imported model to its own sociopolitical and cultural context, the South African framework offers a working legislative example that could be presented to Parliament as a basis for deliberation. A similar level of specificity and legislative decisiveness would significantly advance Kenya's ability to govern this complex field.

In the absence of comprehensive domestic legislation, Kenya could look to established international guidelines to inform its regulatory approach to stem cell research. Although the African Union (AU) has not yet developed or adopted a unified framework for stem cell research across the continent, Kenya can still benefit by aligning with guidance from professional bodies such as the International Society for Stem Cell Research (ISSCR) and the International Stem Cell Forum (ISCF). The ISSCR *Guidelines for Stem Cell Research and Clinical Translation* [25], which cover human stem cell research, clinical translation, and related ethical issues, are based on well-known and widely accepted principles and international ethical frameworks, including the *Nuremberg Code* [54] and the *Declaration of Helsinki* [55]. The ISSCR Guidelines advocate for rigorous scientific and ethical oversight, particularly in the context of embryonic stem cell research, which they deem permissible in many jurisdictions under proper conditions. If Kenya were to adopt this approach, it would need to ensure the existence of oversight mechanisms robust enough to meet these standards while managing political and religious sensitivities.

Additional international instruments that Kenya might consider include the US National Institutes of Health *Guidelines for Human Stem Cell Research* [56], which regulate the procurement and use of biological materials, and the UNESCO *Universal Declaration on Bioethics and Human Rights* [57], which offers a general framework for bioethics that can be adapted for national legislation. Article 15 of the Declaration, which deals with benefit sharing, is particularly relevant for collaborative stem cell research. This provision highlights the ethical obligation of research institutions to ensure that communities and individuals who contribute to or are affected by stem cell research also share in its benefits, especially when such research leads to the development of therapies or commercial products. Benefit sharing reflects a commitment to fairness and justice in research, ensuring that vulnerable or marginalized populations are not merely used as sources of biological material but are recognized as stakeholders entitled to the fruits of scientific advancement [58, 59]. In the context of stem cell research, this could mean access to developed therapies, capacity building, or equitable participation in decision-making processes.

Kenyan policymakers and scientists may also benefit from engaging with the World Medical Association's ethical instruments, including the World Medical Association's *International Code of Medical Ethics* [60], the *Declaration of Helsinki* [55], and the *Declaration of Taipei* [61]. The *Declaration of Helsinki* outlines general ethical standards for research involving human subjects and allows for the use of unproven interventions in clinical settings, provided such interventions are subsequently studied for safety and efficacy. The *Declaration of Taipei* is especially relevant to stem cell research as it governs the ethical management of health data and biological materials, such as those stored in biobanks.

The absence of a binding international treaty regulating stem cell research highlights the broader regulatory vacuum that currently defines the global landscape. However, this also gives individual countries, including Kenya, the opportunity to define their own legal and ethical positions on stem cell science. The political will to do so, however, has often been constrained by religious influence and social conservatism. In Kenya, the close historical relationship between religion and the state has made it difficult to enact laws that engage directly with ethically contentious scientific fields. This tension is already evident in the cautious tone adopted by the National Biomedical Research Guidelines and will likely shape any future legislative or policy efforts.

Until comprehensive national legislation is formally adopted, international professional guidelines, such as those from ISSCR, will continue to play an influential role in shaping ethical conduct in stem cell research. Kenya has already incorporated many of these international standards into the National Biomedical Research Guidelines [31], and this trend should be deepened through deliberate legislative reform. The adoption of international best practices, adapted to Kenya's legal, cultural, and scientific context, offers a practical and effective path toward closing the current gaps in the regulation of stem cell research.

Conclusion

For Kenya to meaningfully develop its stem cell research sector, existing legislative frameworks require substantial revision, and complementary legal instruments must be designed and implemented. At present, the country's laws, policies, and guidelines only marginally engage with stem cell research. As a result, regulation largely falls under broadly framed statutes rather than a coherent, dedicated legal framework. While instruments such as the *Health Act*, the *STI Act*, and the National Biomedical Research Guidelines provide a foundation for oversight, they are neither sufficiently detailed nor equipped to address the specific demands of this complex and evolving field.

Although these instruments demonstrate a general openness to biomedical innovation, including regenerative medicine, their treatment of stem cell research is largely indirect. More significantly, there remains no formal recognition or guidance regarding the permissibility of embryonic stem cell research, leaving researchers and institutions to navigate a regulatory landscape marked by legal ambiguity and policy silence. Compounding this uncertainty is the influence of restrictive or limiting provisions found in other domains of law, particularly criminal, intellectual property, and data protection statutes, that may unintentionally constrain scientific activity.

In the absence of clear national legislation, Kenya risks falling behind in a rapidly advancing global field. Drawing on the regulatory example provided by South

Africa, where stem cell research, including embryonic stem cell research, is addressed with greater specificity, offers a practical model for Kenya to consider. Kenya should also look to international best practices and ethical frameworks to inform the design of context-specific rules that balance innovation, public interest, and cultural sensitivities. By doing so, the country can build a research environment that is both ethically grounded and scientifically progressive.

References

1. Beak C, Isasi R (2014) Regenerative medicine: socio-ethical challenges and regulatory approaches. In: Knoppers BM, Joly Y (eds) Routledge handbook of medical law and ethics. Routledge, Oxfordshire, pp 244–268. https://www.researchgate.net/publication/281714396_Regenerative_Medicine_Socio-ethical_challenges_and_regulatory_approaches. Accessed 17 Jul 2025
2. Daar AS, Greenwood HL (2007) A proposed definition of regenerative medicine. J Tissue Eng Regen Med 1(3):179–184. https://doi.org/10.1002/term.20
3. Atala A (2009) Engineering organs. Curr Opin Biotechnol 20(5):575–592. https://doi.org/10.1016/j.copbio.2009.10.003
4. Oerlemans AJM, Feitz WFJ, van Leeuwen E, Dekkers WJM (2013) Regenerative urology clinical trials: an ethical assessment of road blocks and solutions. Tissue Eng Part B Rev 19(1):41–47. https://doi.org/10.1089/ten.TEB.2012.0136
5. Barfoot J, Kemp E, Doherty K, Blackburn C, Sengoku S, van Servellen A, Gavai A, Karlsson A (2013) Stem cell research: trends and perspectives on the evolving international landscape. Elsevier. https://www.eurostemcell.org/system/files/documents/resources/Stem-Cell-Report-Trends-and-Perspectives-on-the-Evolving-International-Landscape_Dec2013.pdf. Accessed 17 Jul 2025
6. Liu M, Yang F, Xu Y (2022) Global trends of stem cell precision medicine research (2018–2022): a bibliometric analysis. Front Surg 9:888956. https://doi.org/10.3389/fsurg.2022.888956
7. Verginer L, Riccaboni M (2021) Stem cell legislation and its impact on the geographic preferences of stem cell researchers. Eurasian Bus Rev 11:163–189. https://doi.org/10.1007/s40821-021-00182-0
8. Gaobotse G (2018) Stem cell research in Africa: legislation and challenges. J Regen Med 7(1). https://doi.org/10.4172/2325-9620.1000142
9. Zarzeczny A, Caulfield T (2009) Emerging ethical, legal and social issues associated with stem cell research and the current role of the moral status of the embryo. Stem Cell Rev Rep 5:96–101. https://doi.org/10.1007/s12015-009-9062-4
10. Allum N, Allansdottir A, Gaskell G, Hampel J, Jackson J, Moldovan A, Priest S, Stares S, Stoneman P (2017) Religion and the public ethics of stem-cell research: attitudes in Europe, Canada and the United States. PLoS One 12(4):e0176274. https://doi.org/10.1371/journal.pone.0176274
11. Gulinson CL (2017) Embryonic stem cell tourism. Jurimetrics 58(1):17–58, 31
12. Manninen BA (2008) Are human embryos Kantian persons?: Kantian considerations in favor of embryonic stem cell research. Philos Ethics Humanit Med 3:4. https://doi.org/10.1186/1747-5341-3-4
13. Isasi RM, Knoppers BM (2006) Mind the gap: policy approaches to embryonic stem cell and cloning research in 50 countries. Eur J Health Law 13(1):9–25. https://doi.org/10.1163/157180906777036328
14. Volarevic V, Markovic BS, Gazdic M, Volarevic A, Jovicic N, Arsenijevic N, Armstrong L, Djonov V, Lako M, Stojkovic M (2018) Ethical and safety issues of stem cell-based therapy. Int J Med Sci 15(1):36–45. https://doi.org/10.7150/ijms.21666

15. Lo B, Parham L (2009) Ethical issues in stem cell research. Endocr Rev 30(3):204–213. https://doi.org/10.1210/er.2008-0031
16. Juma M (2024) The relationship between the church and state in Kenya. Presentation at the Apostolat Militaire International Conference 2024, Office of the National Security Advisor of the Republic of Kenya. https://www.apostolatmilitaire.com/wp-content/uploads/2024/09/00p-Relationship-between-the-Church-and-State-in-KENYA.pdf. Accessed 17 Jul 2025
17. Medina-Arellano MDJ (2011) Stem cell regulation in Mexico: current debates and future challenges. Stud Ethics L Tech 5(1). https://doi.org/10.2202/1941-6008.1142
18. Dhar D, Hsi-En Ho J (2009) Stem cell research policies around the world. Yale J Biol Med 82(3):113–115. https://pmc.ncbi.nlm.nih.gov/articles/PMC2744936/. Accessed 17 Jul 2025
19. Lowenthal J, Sugarman J (2015) Ethics and policy issues for stem cell research and pulmonary medicine. Chest 147(3):824–834. https://doi.org/10.1378/chest.14-1696
20. Chen Y, Chen Y, Tsai P, Chen C, Shaw S, Hsuan Y (2016) Mesenchymal stem cell: considerations for manufacturing and clinical trials on cell therapy product. Int J Stem Cell Res Ther 3(1):029. https://doi.org/10.23937/2469-570X/1410029
21. Kenya Medical Research Institute (2023) KEMRI commissions a centre of excellence in stem cell research. https://www.kemri.go.ke/kemri-commissions-a-centre-of-excellence-in-stem-cell-research/. Accessed 17 Jul 2025
22. Arum J (2020) Stem cell therapy now available in Kenya. The Star. https://www.the-star.co.ke/health/2020-09-07-stem-cell-therapy-now-available-in-kenya. Accessed 17 Jul 2025
23. Ministry of Health (n.d.) Kenya Master Health Facility Registry, Tia Clinics #32195. https://kmhfr.health.go.ke/public/facilities/e1000af4-ed47-494a-9e8f-705dbcc14d6d. Accessed 17 Jul 2025
24. Caulfield T, Murdoch B (2019) Regulatory and policy tools to address unproven stem cell interventions in Canada: the need for action. BMC Med Ethics 20:51. https://doi.org/10.1186/s12910-019-0388-4
25. International Society for Stem Cell Research (ISSCR) (2021) ISSCR guidelines for stem cell research and clinical translation: 2021 update. https://www.isscr.org/guidelines. Accessed 17 Jul 2025
26. Lau D, Ogbogu U, Taylor B, Stafinski T, Menon D, Caulfield T (2008) Stem cell clinics online: the direct-to-consumer portrayal of stem cell medicine. Cell Stem Cell 3(6):591–594. https://doi.org/10.1016/j.stem.2008.11.001
27. Master Z, Matthews KRW, Abou-El-Enein M (2021) Unproven stem cell interventions: a global public health problem requiring global deliberation. Stem Cell Reports 16:1435–1445. https://doi.org/10.1016/j.stemcr.2021.05.004
28. Munsie M, Hyun I (2014) A question of ethics: selling autologous stem cell therapies flaunts professional standards. Stem Cell Res 13(3 Pt B):647–653. https://doi.org/10.1016/j.scr.2014.04.014
29. Republic of Kenya (2017) The Health Act, 2017. https://kenyalaw.org/kl/fileadmin/pdfdownloads/Acts/HealthActNo.21of2017.pdf. Accessed 17 Jul 2025
30. Republic of Kenya (2013) Science, Technology and Innovation Act No. 28 of 2013. https://www.nacosti.go.ke/nacosti/Docs/Information%20Centre/Science-Technology-and-Innovation-Act-No.-28-of-20131.pdf. Accessed 17 Jul 2025
31. Wekesa M (2015) Challenges in regulation of biomedical research: the case of Kenya. Nigeria J Clin Pract 18(Suppl 1):S25–S30. https://doi.org/10.4103/1119-3077.170819
32. Laws of Kenya (2022) The Constitution of Kenya. https://kenyalaw.org/kl/fileadmin/pdfdownloads/TheConstitutionOfKenya.pdf. Accessed 17 Jul 2025
33. Mbindyo R, Kioko J, Siyoi F, Cheruiyot S, Wangai M, Onsongo J, Omwoyo A, Kisia C, Miriti K (2020) Legal and institutional foundations for universal health coverage. Kenya Bull World Health Organ 98(10):706–718. https://doi.org/10.2471/BLT.19.237297
34. Ndirangu W (2025, April 18) Mediheal hospitals barred from conducting organ transplant. Kenya News Agency. https://www.kenyanews.go.ke/mediheal-hospitals-barred-from-conducting-organ-transplant/. Accessed 17 Jul 2025

35. Ministry of Health Kenya, Kenya Blood Transfusion and Transplant Service (2022) Kenya Policy on Donation, Transfusion and Transplant of Human Derived Medical Products. https://www.scribd.com/document/778344102/Kenya-Policy-on-Human-Derived-Medical-Products-DonationsTransfus. Accessed 17 Jul 2025
36. National Research Fund (n.d.) National Research Fund Kenya. https://www.nrf.go.ke/. Accessed 17 Jul 2025
37. Ministry of Health (2014) Kenya Health Policy 2014–2030. https://arua-ncd.org/wp-content/uploads/2022/10/kenya-health-policy.pdf. Accessed 17 Jul 2025
38. Reverby SM (ed) (2012) Tuskegee's truths: rethinking the Tuskegee Syphilis study, Studies in Social Medicine. UNC Press Books, Chapel Hill
39. National Commission for Science, Technology and Innovation (2020) National Guidelines for Ethical Conduct of Biomedical Research Involving Human Participants in Kenya. https://www.nacosti.go.ke/nacosti/Docs/QUICK%20DOWNLOADS/National%20Guidelines%20for%20Ethical%20Conduct%20of%20Biomedical%20Research%20Involving%20Human%20Participants%20in%20Kenya.pdf. Accessed 17 Jul 2025
40. Otindo C, Kikuvi M (2021) Assessment of institutional structures governing science technology and innovation in Kenya. KIPPRA Discussion Paper No. 276. Kenya Institute for Public Policy Research and Analysis. https://www.academia.edu/download/119797807/DP276.pdf. Accessed 17 Jul 2025
41. Laws of Kenya (1930) Penal Code, Cap 63. https://new.kenyalaw.org/akn/ke/act/1930/10/eng@2023-12-11. Accessed 17 Jul 2025
42. Laws of Kenya (2010) Counter-Trafficking in Persons Act, No. 8 of 2010. https://kenyalaw.org/kl/fileadmin/pdfdownloads/Acts/Counter-TraffickinginPersonsAct_No8of2010.pdf. Accessed 17 Jul 2025
43. Widdows H (2009) Border disputes across bodies: exploitation in trafficking for prostitution and egg sale for stem cell research. Int J Fem Approaches Bioeth 2(1):5–24. http://www.jstor.org/stable/40339192
44. Hernandez HH (2009) Stem cell tourism on the rise. Stem Cell Institute. https://www.cellmedicine.com/tourism-caveat-emptor/. Accessed 17 Jul 2025
45. Than K (2009) Hungary detains 4 over illegal stem cell treatment. Reuters, July 29. https://www.reuters.com/article/healthNews/idUSTRE56S4PR20090729/. Accessed 17 Jul 2025
46. Laws of Kenya (2001) Industrial Property Act, No. 3 of 2001. https://kenyalaw.org/kl/fileadmin/pdfdownloads/Acts/Industrial_Property_Act__No._3_of_2001.pdf. Accessed 17 Jul 2025
47. Laws of Kenya (2008) Anti-Counterfeit Act, No. 13 of 2008. Republic of Kenya. https://www.aca.go.ke/images/downloads/Anti-Counterfeit-Act-No13-of-2008.pdf. Accessed 17 Jul 2025
48. Isasi RM, Andrews PW, Baltz JM, Bredenoord AL, Burton P, Chiu I, Hull SC, Jung J, Kurtz A, Lomax G, Ludwig T, McDonald M, Morris C, Ng HH, Rooke H, Sharma A, Stacey GN, Williams C, Zeng F, Knoppers BM (2014) Identifiability and privacy in pluripotent stem cell research. Cell Stem Cell 14(4):427–430. https://doi.org/10.1016/j.stem.2014.03.014
49. Republic of Kenya (2019) The Data Protection Act, No. 24 of 2019. https://kenyalaw.org/kl/fileadmin/pdfdownloads/Acts/2019/TheDataProtectionAct__No24of2019.pdf. Accessed 17 Jul 2025
50. Ahn SM, Simpson R, Lee B (2010) Genomics and proteomics in stem cell research: the road ahead. Anat Cell Biol 43(1):1–14. https://doi.org/10.5115/acb.2010.43.1.1
51. Ministry of Education (2020) Draft Science, Technology and Innovation Policy 2020–2030. https://www.nacosti.go.ke/wp-content/uploads/2021/Docs/Draft%20Science%2C%20Technology%20and%20Innovation%20%28STI%29%20Policy%202020-2030.pdf. Accessed 17 Jul 2025
52. Republic of South Africa (2004, July 23) National Health Act No. 61 of 2003. Government Gazette, vol 469, No. 26595. https://www.sahpra.org.za/wp-content/uploads/2020/02/National-Health-Act.pdf. Accessed 17 Jul 2025
53. Republic of South Africa (1965, July 7) Medicine and Related Substances Act No. 101 of 1965. Government Gazette Extraordinary, vol 68 No. 1171. https://www.gov.za/sites/default/files/gcis_document/201505/act-101-1965.pdf. Accessed 17 Jul 2025

54. Office of Research Integrity, U.S. Department of Health and Human Services (n.d.) Nuremberg Code: Directives for human experimentation. https://ori.hhs.gov/content/chapter-3-The-Protection-of-Human-Subjects-nuremberg-code-directives-human-experimentation. Accessed 17 Jul 2025
55. World Medical Association (2013) WMA Declaration of Helsinki—Ethical Principles for Medical Research Involving Human Subjects. https://www.wma.net/policies-post/wma-declaration-of-helsinki-ethical-principles-for-medical-research-involving-human-subjects/. Accessed 16 Jul 2025
56. U.S. National Institutes of Health (2009) Guidelines for Human Stem Cell Research. https://stemcells.nih.gov/research-policy/guidelines-for-human-stem-cell-research. Accessed 17 Jul 2025
57. UNESCO (2005) Universal Declaration on Bioethics and Human Rights. https://www.unesco.org/en/legal-affairs/universal-declaration-bioethics-and-human-rights?hub=387. Accessed 17 Jul 2025
58. Dauda B, Dierickx K (2013) Benefit sharing: an exploration on the contextual discourse of a changing concept. BMC Med Ethics 14:36. https://doi.org/10.1186/1472-6939-14-36
59. Kamau C, Prinsen L, Thaldar D (2025) Benefit-sharing with human participants in health research in South Africa: a call for clarity. Dev World Bioeth 25(2):144–153. https://doi.org/10.1111/dewb.12456
60. World Medical Association (2022) International Code of Medical Ethics. https://www.wma.net/policies-post/wma-international-code-of-medical-ethics/. Accessed 16 July 2025
61. World Medical Association (2016) WMA Declaration of Taipei on Ethical Considerations Regarding Health Databases and Biobanks. https://www.wma.net/policies-post/wma-declaration-of-taipei-on-ethical-considerations-regarding-health-databases-and-biobanks/. Accessed 16 Jul 2025

8 Stem Cell and Genetic Engineering for Regenerative Medicine: An Indian Perspective

Shambhavi Naik

Abstract

Regenerative medicine comprises technologies that can repair or replace damaged cells or tissues in order to address conditions caused by disease, injury, or aging. Among these technologies, stem cell therapy and genetic engineering hold particular promise for advancing regenerative medicine. While these innovations could significantly improve health outcomes, they also raise complex ethical and regulatory challenges. India, with its high disease burden, diverse population, and rapidly growing biomedical sector, provides a compelling case study for examining the governance of regenerative medicine. The country's dynamic healthcare landscape, which integrates both modern and traditional medicine, presents unique regulatory and ethical considerations. However, India's experience is not merely of local interest, it also offers insights relevant to other emerging markets navigating similar challenges. This chapter explores the governance of stem cell therapy and genetic engineering in India, identifying key opportunities and challenges in their adoption. It argues that while public acceptance of these technologies in India is likely to be high, significant efforts are needed to: (i) reduce the cost of therapies through local research and production, (ii) increase public awareness of their benefits and risks, and (iii) enhance regulatory capacity to ensure responsible oversight. Similar strategies may be valuable for other countries seeking to facilitate equitable and ethical access to regenerative medicine.

Keywords

Regenerative medicine · Stem cell therapy · Genome editing · CRISPR · Ethical governance · Equitable access · India · Emerging markets

S. Naik (✉)
Takshashila Institution, Bengaluru, Karnataka, India
e-mail: shambhavi@takshashila.org.in

U. Ogbogu et al. (eds.), *Regenerative Medicine and the Bioethics of Globalization, Justice and Hegemony*, Collaborative Bioethics 5,
https://doi.org/10.1007/978-3-032-07289-4_8

Introduction

Regenerative medicine (RM) encompasses a broad range of technologies aimed at repairing, replacing, or regenerating cells, tissues, or organs to restore function impaired by disease, injury, congenital defects, or aging. These technologies hold transformative potential for treating a wide array of conditions, many of which currently have no effective medical solutions. Among the various approaches to RM, stem cell therapy and genetic engineering have emerged as particularly promising, with stem cell treatments already in clinical use and genome editing technologies advancing rapidly toward medical applications. In December 2023, the first clustered regularly interspaced short palindromic repeats or CRISPR-based genome editing therapies were approved for human use, marking a significant milestone in the clinical adoption of genetic engineering [1].

While RM offers enormous opportunities for improving health outcomes, its adoption is fraught with ethical, regulatory, and accessibility challenges. These challenges include concerns about tampering with nature, inequitable access to therapies, patient exploitation, and the ethical implications of enhancement applications. In addition, the governance of RM must address data privacy, benefit-sharing, regulatory enforcement, and long-term monitoring of safety and efficacy. Unlike many other emerging technologies, RM carries the unique ability to alter biological systems at a fundamental level, necessitating careful oversight even when its application is restricted to therapeutic use.

The field of RM is advancing rapidly, with various stem cell-based therapies already in use. The most widely established stem cell treatment is hematopoietic stem cell transplantation, formerly known as bone marrow transplantation, which has been successfully used to treat blood disorders such as leukemia and thalassemia. However, research has also begun expanding into tissue regeneration for neurodegenerative diseases, including Parkinson's [2] and Alzheimer's [3], as well as efforts to restore vision through retinal regeneration [4].

Cardiovascular diseases, a leading cause of mortality worldwide, are another major target for regenerative therapies, with ongoing studies investigating the potential for stem cell-based repair of myocardial tissue following cardiac events [5]. Similarly, stem cell-based therapies for type 1 diabetes are exploring ways to regenerate insulin-producing beta cells, offering hope for long-term disease management, through a one-time intervention instead of continuous insulin replacement therapy [6]. Beyond disease treatment, RM is also being investigated for use in injury repair, including spinal cord regeneration, healing for chronic wounds, burns, surgical trauma, etc., and tissue reconstruction following surgical trauma [7]. One of the more speculative but highly anticipated applications of RM is its potential to combat aging-related degeneration and even extend the human lifespan [8].

Genome editing technologies, particularly CRISPR, have introduced new possibilities for RM. Unlike traditional stem cell therapies—which rely on the body's existing regenerative potential—CRISPR-based approaches enable precise genetic modifications, allowing for programming of cell differentiation and targeted correction of disease-causing mutations. The approval of CRISPR-based therapies for beta-thalassemia and sickle cell anemia demonstrates the growing feasibility of genome editing as a medical intervention of genetic origin [9].

While much of the research and policy agenda for RM has been driven by high-income countries and jurisdictions such as the United States, the European Union, and the United Kingdom, the demand for these therapies is expected to be highest in emerging markets, where diseases such as diabetes and cardiovascular conditions are prevalent. Although high-income countries have traditionally led scientific innovation and regulatory frameworks, their dominance is being challenged by the increasing scientific and economic influence of emerging economies. Reports have warned about declining American leadership in RM [10], and their recent formulation of critical technology lists to maintain technological superiority highlights this global shift [11].

India, with its large population, increasing economic strength, and high burden of non-communicable diseases, presents a valuable case study for understanding the governance and adoption of RM. While each country has its unique context, India's experience provides insights for emerging markets facing challenges relating to economic considerations, regulatory capacity, public trust in science, and religious perspectives.

Religious and cultural beliefs are particularly important in shaping public attitudes toward emerging biotechnologies. By 2030, India's population is projected to include 1.3 billion Hindus, 311 million Muslims, 37 million Christians, and 46 million adherents of other religious groups [12]. Religious traditions can influence the acceptance of biomedical innovations [13] such as stem cell therapy and genome editing, with studies suggesting that 47% of Indian Hindus and 44% of Indian Muslims viewing science and religion as conflicting [14]. However, unlike in the United States, where religious objections to genetic interventions are more pronounced [15], Indian public attitudes toward genome editing appear to be relatively accepting [14].

Economic factors will also play a crucial role in determining access to RM in India. While India's economic growth is projected to continue at a rapid pace, the high cost of these advanced therapies may render them inaccessible to large portions of the population. Emerging biomedical technologies tend to be expensive at the outset, raising concerns about equitable access and the potential to create a medical tourism market that prioritizes wealthier international patients over domestic healthcare needs. Additionally, India's diverse medical landscape, which includes the tenets of Ayurveda, Unani, Siddha, and Homeopathy—alongside modern Western medicine—raises questions about how RM will be integrated into existing healthcare practices.

Given these considerations, this chapter will examine the governance of RM in India, focusing on stem cell therapy and genome editing. It will explore key regulatory, ethical, and operational challenges, including concerns around patient safety, regulatory enforcement, equitable access, and public trust. This chapter will also consider lessons from international experiences, including those of other emerging markets, while maintaining an India-centric perspective. By providing a structured analysis of India's evolving policy landscape, this chapter aims to contribute to the broader discourse on how emerging economies can navigate the challenges of integrating RM into their healthcare systems.

Governance of Stem Cells and Genome Editing for Human Medical Use

India does not have a dedicated law to govern applications of stem cells or genome editing for human use. This lack of statutory oversight is intentionally designed to allow flexibility in rule-making—guidelines prescribed by governmental agencies are easier to amend in pace with technological developments, whereas formal laws require long Parliamentary processes.

While this flexibility has its advantages, certain aspects of stem cell and genome editing governance require legal enforcement to effectively protect human rights. This is especially relevant for private entities operating outside the scope of government funding, such as those offering direct-to-consumer products and services. Moreover, the benefits of opting for guidelines over legislation only materialize if those guidelines are effectively implemented. For example, India's stem cell guidelines were first proposed by the Indian Council of Medical Research (ICMR) and the Department of Biotechnology (DBT) in 2007 and have been periodically updated in 2013, 2017, and 2020. The ICMR and the DBT issued separate guidelines for genome editing for therapeutic use in humans in 2019, and these guidelines recommended the setting up of working expert groups to formulate disease-specific guidelines for gene therapies.

Recognizing that different diseases can have different relative risks and benefits and thus need tailored standards for approval is an excellent step. However, nearly 5 years after these recommendations, there has still been no update on the formation of working groups or the release of disease-specific guidelines. While it is difficult to pinpoint any one reason for the failure to update and standardize guidelines for stem cell use or genome editing, it does signal a weak bureaucratic process. While some of the delay can be attributed to the emergence of COVID-19, there is no clear reason on why the guidelines have not been formulated in the years since.

The current stem cell and gene therapy guidelines share some common features, as follows:

1. *Limits on the Clinical Application of Stem Cell/Gene Therapies*

   The stem cell guidelines prescribe the use of stem cells for certain hematologic diseases, while gene therapy guidelines limit the use of genetic engineering for therapeutic purposes. Stem cell therapies for other diseases can only be carried out as clinical trials and require necessary approvals from Indian authorities. The stem cell guidelines could benefit from the working group methodology employed by gene therapy guidelines in setting standards for approval for stem cell use for non-hematologic purposes. Such proactive guidelines will instill confidence in investors and researchers who are willing to put in the funds and efforts into creating stem cell therapies for those diseases.

The current gene therapy guidelines limit the use of genetic engineering technologies to therapeutic products. However, the guidelines do not define what therapeutic products are limited in use, and this ambiguity leads to confusion regarding the use of such therapeutic products in RM. Tissue repair could be considered a therapeutic product, but even this type of repair is dependent on both the type of injury and the extent of the repair required. The current guidelines lack clarity, making it difficult to distinguish between symptoms and disease. For example, they do not specify whether age-related muscle loss qualifies as a disease. This ambiguity highlights the broader challenge of classifying physical repairs—and under these definitions, it is unclear whether restoring muscle function would count as enhancement or regeneration. The need to clarify whether stem cell technology and genetic engineering for RM will be allowed to be commercialized cannot be overstated and is necessary for research to progress.

2. *They Prohibit Germline Genome Editing*

   Both sets of guidelines allow research on genetically edited embryos in vitro until day 14, after which embryos must be discarded. Implantation of genetically edited embryos is entirely prohibited. While some jurisdictions have adopted similar restrictions, there is no single internationally accepted standard for governing embryo editing. Ethical perspectives on this issue vary widely—some argue that embryo editing should not be attempted at all, while others believe it could be a moral obligation under certain conditions. India's approach aligns with a cautious regulatory stance observed in several countries, including the United States, the European Union, Australia, and Canada, though this area remains a topic of ongoing debate in global bioethics and policy.

3. *Both Guidelines Do Not Address Ethical Concerns with Using Stem Cell or Genetic Engineering Technologies*

   While both guidelines have sections on ethical considerations as a mainstay with these technologies, there is nothing concrete that explicitly addresses these concerns. The ICMR's ethical guidelines for biomedical research in human participants that govern clinical trials are generally referred to as the overarching ethical guidelines for any research involving humans. But these guidelines are also substantive, referring to protocols for consent and ethical committee approvals, and do not address normative questions about the use of genetic engineering or stem cell technologies in an Indian context [16]. Furthermore, a 2005 survey by the ICMR showed that in the absence of any powers of enforcement, only 22% of India's 179 institutional ethics committees followed the principles laid down by the ICMR [17].

The overall governance of stem cell and genetic engineering technologies has evolved in response to technological development in specific therapeutic use cases, and the guidelines do not clarify whether RM would fall under the ambit of existing policy. The current narrative of using such technologies examines their use for rectifying monogenic diseases such as thalassemia and sickle cell anemia or muscular dystrophy. Conversations around more complicated uses of these technologies to repair injuries are limited; however, the ethical considerations in grappling with therapeutic use cases of genetic engineering will also extend to RM. India will be a critical market for these technologies given its aging population and widespread endemic genetic disease, and it is necessary to understand these considerations and possible ways to mitigate them in light of these factors.

Ethical Considerations for Stem Cells and Genome Editing for Human Medical Use

Before addressing the ethical questions surrounding RM, it is important to recognize three distinct types of considerations: those grounded in individual ethics, those informed by scientific judgment, and those shaped by the global political context. For example, many countries—including India—prohibit the commercial use of germline genome editing because the technology is still in its early stages and requires further research to fully understand its long-term effects. Using it prematurely is widely considered unethical due to the potential harm to both individuals and society. However, such restrictions are primarily based on scientific caution rather than personal moral objections, such as the belief that editing the genome is "playing God." In contrast, arguments against germline editing based on the inability of future generations to consent reflect a deeper philosophical stance rooted in individual ethical beliefs.

Furthermore, inequitable access to healthcare is a consequence of the global economic order and persists regardless of the ethical frameworks applied to specific technologies. It is essential to distinguish between individual ethics, scientific considerations, and structural global forces because only the latter two can be addressed through public policy. For instance, policies can establish minimum safety standards for genome editing—similar to those in place for drugs and vaccines—or offer financial support to expand access to therapies, thereby reducing health inequities. However, public policy cannot substitute for the consent of future generations. If a society determines that such consent is ethically necessary for germline genome editing, and that parental consent is insufficient, then the application of the technology must be halted, regardless of its scientific readiness. Therefore, it is crucial that ethical foundations are clearly articulated and collectively agreed upon before developing policies that address scientific safety, efficacy, and access.

Ethical considerations can also be addressed through public engagement, but this approach has greater implications for an individual's ethical choices. The section below will first discuss the ethical acceptability of RM before exploring other considerations.

Are Regenerative Medicine-Based Therapies "Playing God"/ Tampering with Nature?

The idea that RM and genetic engineering "tamper with nature" or amount to "playing God" has been a persistent concern in bioethics. However, such arguments must be critically examined in light of philosophical reasoning rather than being taken at face value. John Stuart Mill, in his essay collection Nature, The Utility of Religion and Theism [18], challenged the assumption that what is natural is inherently good or that human intervention in nature is intrinsically immoral. Mill argued that human civilization has always sought to improve upon nature—from medicine to agriculture—and that rejecting an innovation solely because it alters nature is logically inconsistent. Therefore, the idea that RM is unethical simply because it modifies biological processes lacks philosophical grounding.

However, public concern about "playing God" continues to shape debate surrounding biotechnology. Historically, these anxieties surfaced as early as the 1975 Asilomar Conference, where scientists proposed self-imposed guidelines for the responsible development of recombinant DNA technologies [19]. More recently, concerns about genetic engineering and synthetic biology have been dismissed as a result of scientific illiteracy [20]. However, framing opposition as mere ignorance is inadequate—many of these concerns stem from deeply held personal values, religious beliefs, or broader philosophical positions on human intervention in life sciences.

Public engagement efforts have shown that concerns over "playing God" can sometimes be addressed through rational discussion and value-based dialogue. For instance, deliberative public consultations in the UK have demonstrated that when discussions focus on potential misuses, unintended consequences, and ethical safeguards, public attitudes toward genome editing and RM become more receptive [21, 22]. However, ethicists should be careful to separate public perception from rational ethical analysis—arguments should not be accepted or rejected simply because they are popular or unpopular [23].

Indian perspectives on the "playing God" argument have not been extensively explored in bioethics literature. However, available data suggest that public attitudes in India toward genome editing are more accepting than in many other countries. A Pew Research study found that 56% of Indian adults view genome editing for therapeutic purposes as appropriate—the highest level of support among 20 surveyed countries [14]. Notably, support for genome editing was equally high among Hindus (57%) and Muslims (59%), indicating that religion may not be a point of differentiation when it comes to public acceptance of genome editing in India. This acceptance aligns with the broader philosophical perspectives found in Indian traditions. In Hinduism, Jainism, and Buddhism, principles such as *ahimsa* (non-violence) and *karuna* (compassion) shape views on the sanctity of life, emphasizing collective well-being. Furthermore, unlike Christianity and Islam, these traditions do not prescribe a definitive moment of ensoulment, which may reduce ethical resistance to the use of embryos in research and medical therapies. This acceptance has sometimes been seen in the unsubstantiated claims of some scholars that stem cell research was developed in India by ancient Hindus [24]. Islamic traditions dictate that ensoulment happens in early pregnancy, and the use of embryos in research has generally seen low resistance from Islamic countries [25]. A detailed study of how ancient Indian philosophies would approach RM is lacking, but the attitude toward emerging technologies from the Indian public has generally been positive, suggesting that religious beliefs may outcompete the need for improved healthcare.

Unlike some other nationalities, Indian respondents were also more accepting of genome editing for enhancement purposes, such as increasing intelligence [26]. This suggests that at least a portion of the Indian public may not view human genetic modification as inherently problematic, whether for therapeutic or non-therapeutic purposes. However, further research is needed to understand whether this acceptance is based on a rejection of the "playing God" argument or if it arises from other socio-cultural factors, such as India's emphasis on scientific progress and medical advancements.

Given this data, it is unlikely that the "playing God" argument will significantly impede the adoption of regenerative medicine in India. However, public engagement efforts should not assume universal acceptance, and ethical frameworks should still address concerns about safety, access, and the potential for misuse of these emerging biotechnologies. If regenerative medicine is eventually extended to heritable genetic modifications in embryos, policymakers will need to ensure that ethical considerations and regulatory safeguards keep pace with scientific advancements.

Other considerations, driven largely by the socioeconomic context of India, may be more important to the adoption of regenerative medicine in India. The next section will explore some of the main concerns stemming from this context.

Inequitable Access to Benefits

The exorbitant cost of regenerative medicine treatments remains a significant barrier to access for millions of Indians from lower socioeconomic strata. For instance, at one center in Tamil Nadu, almost 71% of patients with acute myeloid leukemia were unable to undergo hematopoietic stem cell transplantation (HSCT), primarily due to financial constraints [27]. Similarly, the Indian Stem Cell Transplant Registry recorded only 960 transplantations between 2012 and 2016, despite an estimated 3600 thalassemia cases being eligible for transplantation annually [28].

However, financial constraints are not the only factor affecting access to regenerative medicine. Infrastructure-related challenges, such as the need for Good Manufacturing Practice (GMP) standard facilities and cold storage, may further restrict availability to specific regions [29]. This regional disparity could exacerbate existing inequities in healthcare access, meaning that access to regenerative medicine may be limited not just by affordability, but also by geography and infrastructure availability.

The concern around inequitable access to novel biomedical technologies is not unique to India. In South Africa, a deliberative public engagement study on heritable human genome editing (HHGE), conducted by Thaldar et al. [30], revealed that

South Africans view access to genome editing for serious health conditions as a matter of social justice. Participants emphasized that such technologies should not be exclusively available to the wealthy, as this would only worsen existing healthcare inequalities.

Building on these findings, Shozi and Thaldar [31] examined the philosophical and ethical implications of these public attitudes, particularly through the lens of Ubuntu, a central principle in African ethics. Ubuntu is an African philosophy meaning "I am because we are," emphasizing shared humanity, compassion, and community interconnectedness. Shozi and Thaldar argued that ensuring equitable access to emerging biomedical technologies should not merely be aspirational but a core policy priority requiring government intervention and targeted regulatory frameworks.

These insights from South Africa are highly relevant to India, given its similarly stratified healthcare system. The principle of social justice in healthcare—as articulated in the South African public engagement study—suggests that Indian policymakers should not only focus on safety and efficacy but also proactively ensure that regenerative medicine benefits a broad cross-section of society rather than an elite minority.

To address inequities in access to regenerative medicine, three main strategies could be considered:

1. *Promoting Domestic Production of Lower-Cost Therapies*

   Locally developed products can significantly reduce treatment costs. For example, India's indigenously developed CAR-T therapy, approved in 2023, costs roughly one-tenth of its US-made counterparts [32]. However, policy streamlining and infrastructure investment are essential to boosting indigenous research capacity. Disease-specific guidelines for stem cell and gene therapies could help channel local research efforts toward India's most pressing health needs. The Indian government should provide financial and regulatory support in the short term to boost manufacturing in order to help scale the domestic production of these therapies.
2. *Encouraging International Collaboration and Policy Incentives*

   Indian policymakers should work toward bilateral and multilateral agreements to incentivize affordable product development for emerging markets. As seen in South Africa's deliberative engagement study, the public expects governments to actively shape equitable access rather than leaving it to market forces. India should adopt a similar proactive approach, ensuring that regenerative medicine technologies are developed with accessibility in mind from the outset.
3. *Extending Insurance Schemes*

   India could include regenerative medicines under the AYUSHMAN Bharat scheme to facilitate access to low-income groups to these therapies. The AYUSHMAN Bharat scheme is a flagship health initiative of the Government of India that focuses on providing primary healthcare accessible to everyone and providing financial assistance to vulnerable communities for accessing secondary and tertiary healthcare. Alternatively, a tiered pricing model based on individual spending ability could be developed through the support of government subsidies to provide equitable access across the country.

By combining domestic innovation, strategic regulation, and global partnerships, India can move toward a more equitable model of regenerative medicine—ensuring that life-changing treatments are not restricted to the privileged few, but available to all who need them.

Impact on Human Rights

In India, gaps in regulatory oversight have created space for unscrupulous private practices offering untested stem cell therapies. India's first attempt at regulating stem cells came with the 2007 guidelines issued by the DBT and the ICMR. These guidelines prohibited the clinical use of stem cells unless approved through formal clinical trials, but they lacked statutory backing [33]. The primary authority for the approval of stem cell therapies is the Central Drugs Standard Control Organisation. A legal framework was eventually introduced through the *New Drugs and Clinical Rules, 2019*, which required all clinics offering stem cell-based treatments to obtain governmental licenses. However, the 2019 rules failed to adequately define which stem cell products fell under their ambit, leading to confusion and inconsistent enforcement.

This regulatory uncertainty has also led to significant disruptions in patient care. Since clinics were required to halt treatments until they obtain licenses, many patients were left without access to ongoing therapies. In some cases, judicial intervention is necessary. In *Shikha v. Union of India and Anr.*, the Delhi High Court ruled on May 28, 2019, that a clinic in New Delhi could continue providing embryonic stem cell therapy to two patients with cerebral palsy until the licensing process was completed [34]. This case exemplifies the incoherence in India's regulatory framework, the inconsistency of enforcement, and how judicial rulings may allow continued provision of unapproved therapies.

The lack of stringent oversight has also fueled the expansion of direct-to-consumer marketing for stem cell treatments. Many unregulated clinics claim to offer stem cell-based therapies for conditions ranging from neurological disorders to anti-aging treatments, despite little or no clinical evidence supporting their efficacy [35]. Moreover, bio-networking strategies are used to circumvent regulations, often involving foreign-sourced stem cell products which are marketed directly to patients [36, 37].

While these regulatory challenges are particularly acute in India, similar concerns have been raised elsewhere. In South Africa, Jordaan [38] discusses how autologous stem cell therapy is regulated under the *Medicines and Related Substances Control Act*, ensuring that experimental therapies must pass clinical trials before being legally offered to patients. Although this system has not eliminated all concerns, the stricter legal framework prevents the large-scale proliferation of unproven treatments seen in India. This comparison underscores the need for India to strengthen its enforcement mechanisms, ensuring that regulatory gaps do not allow unsanctioned experimentation on vulnerable patients.

Beyond stem cell therapies, genome editing presents an even more complex regulatory challenge, particularly regarding its potential impact on human rights. In India, public discussions on the human rights implications of genome editing are nascent, but insights from Thaldar et al. [39] provide a useful framework for considering its broader ethical and legal significance.

Thaldar et al. [30] argue that genome editing holds significant promise from a human rights perspective. They identify three key reasons why genome editing could enhance human rights rather than undermine them. First, the ability to edit disease-causing genes before birth aligns with the fundamental human right to health. If safe and effective, genome editing could eliminate hereditary conditions such as sickle cell disease, Huntington's disease, or cystic fibrosis, it could thereby prevent unnecessary suffering. Second, prohibiting safe and effective genome editing could infringe upon the reproductive rights of parents who wish to prevent their children from inheriting serious genetic conditions. If prospective parents have the freedom to select embryos via in vitro fertilization, then the extension of this autonomy to genetic modifications is a logical progression from a human rights standpoint. Third, a carefully regulated genome editing framework could reduce genetic inequalities rather than exacerbate them. Thaldar et al. emphasize that if genome editing remains available only to the wealthy, it risks widening health disparities. However, with proper public policy and accessibility measures, genome editing could instead be a tool for advancing health equity by ensuring that life-saving treatments are available to all.

For India, these insights underscore the need for a more proactive policy framework on genome editing. Without such regulation, India risks following the same trajectory as stem cell therapy, where a lack of oversight allows private clinics to commercialize unproven and even unsafe interventions. However, if India develops a coherent legal framework that prioritizes safety, accessibility, and ethical governance, genome editing could become a valuable tool for advancing healthcare and human rights rather than a domain of exploitation.

India's regulatory struggles with stem cell therapy highlight the risks of weak enforcement mechanisms, particularly when private clinics exploit patients with unproven treatments. The comparison with South Africa's stricter legal framework on autologous stem cell therapy shows that more robust regulatory oversight is necessary to prevent medical exploitation. However, genome editing presents an even greater challenge, requiring greater ethical and legal clarity before the technology becomes widely available.

Drawing from Thaldar et al. [39], it is clear that genome editing can be framed as a human rights-enhancing technology, provided that policy safeguards are in place. As India grapples with how to regulate regenerative medicine and genome editing, it must balance scientific innovation with regulatory integrity, ensuring that these technologies are used responsibly and equitably. Without such measures, India risks becoming a site for experimental medicine without adequate patient protections rather than a leader in biomedical innovation.

Informed Inclusion in Clinical Trials

India has a unique genetic makeup, characterized by over 4500 distinct genetic subpopulations [40]. This genetic makeup will likely play an important role in the applicability of emerging gene therapy products to the Indian population. Yet, current trials for gene therapies for diseases endemic to India, such as sickle cell anemia or thalassemia, do not adequately focus on Indian genotypes. A search on clinicaltrials.gov using the terms "gene therapy" and "sickle cell anemia" shows 37 studies, while "gene therapy" and "thalassemia" show 33 studies; none of these clinical trials are being conducted in India [41]. Most of the sickle cell anemia trials are based in the United States or the European Union, and the majority of thalassemia trials are based in the United States or China. This disparity in representation at clinical trials is not unique to genome editing or to India. Of the 324,854 randomized clinical trials that were registered from 2010 to 2019, only 17,777 (5%) of the trials were set in countries located in Asia, and only 5756 (2%) of the trials were set in sub-Saharan African countries [42]. This lack of inclusion of the Indian genotypes in clinical trials for products that are eventually meant for an Indian market raises significant ethical issues, and addressing context-specific questions is fundamental to designing interventions that improve health outcomes [43]. The lack of genetic representation, particularly in stem cell research, is not a uniquely Indian problem—earlier studies have shown that white Americans were poised to receive the benefits of embryonic stem cell research while ethnic minority groups are often excluded from such benefits. The study argued that creating panels of embryonic stem cells representative of the genetic diversity existing in the United States would be necessary to remedy this important concern [44].

Conversely, there are also concerns regarding possible exploitation of vulnerable populations through clinical trials. This issue arises in that some of the trials being conducted in developing countries are seeking to answer the questions of, and to benefit, the developed world; therefore, the clinical trial's effects are not considered through the lens of the needs of those vulnerable populations. A review indicated that about one-third of 509 clinical trials sponsored by US-based companies from 1995 to 2005 were conducted outside the United States, many in poor and low-income countries [45], without targeting diseases that are actually prevalent in those countries. A recent study has found that while the absolute number of clinical trials in India may be low, several foreign-sponsored clinical trials in India over-recruit Indian participants as compared with other clinical sites, in contravention of existing rules [46]. As discussed in the previous section, Indian guidelines on the ethics of human biomedical research are procedural in nature but do not account for the practical context in which they have to operate, making consent more like a ritual rather than an informed discussion. Socioeconomic factors such as class, gender, kinship, and education are also determinants of the patient–company–doctor dynamics, leading to questions of whether participants are truly consenting to undergo clinical trials [47].

Furthermore, stakeholders might not be even aware of the provisions of the ethical guidelines. A review of Indian policies around participant injury in clinical research found that researchers were largely unaware of their responsibilities [48]. Treatment provision was reported to be facilitated by India's unregulated healthcare and biotechnology industry [49]. Rosemann et al. [50] observed that two parallel processes operate in the Indian stem cell market—one led by the ICMR and other agencies to bring stem cell research in alignment with international standards, and the other by companies and physicians exploiting regulatory and capacity loopholes to bring in unapproved clinical applications.

Appropriate inclusion of Indian populations in clinical trials for regenerative medicine is necessary for the development of equitable therapies for this market. This requires clinical trials to be either held in India specifically, though this may delay product approvals or the inclusion of Indian data sets and participants in global trials. However, clinical trials also need to be ethical and ensure that participants are well-informed and adequately consenting to the benefits and risks of the trial. This would require formulating stronger rules for ethical approvals, educating the public on consenting procedures, and training doctors to assess the utility of technologies involved in regenerative medicine.

Conclusion

India has the potential to become a global leader in regenerative medicine, but achieving this status requires overcoming significant governance, financial, and ethical challenges. While public acceptance of stem cell and gene therapy appears to be high, the current regulatory framework remains fragmented and lacks enforceable legislation, which creates uncertainty for researchers, clinicians, and investors. Additionally, access to regenerative therapies is currently limited to affluent populations, as high costs place these innovations beyond the reach of many patients. Without targeted reforms, India risks becoming a passive consumer of foreign-developed regenerative therapies rather than an active innovator and leader in the field.

To fully capitalize on the promise of regenerative medicine, India should take five key steps:

1. *Establish a Clear, Enforceable Regulatory Framework*

   Current guidelines on stem cell therapy and genetic engineering lack statutory backing, leading to uncertainty and inconsistent enforcement. By enacting comprehensive legislation with clear approval pathways for regenerative therapies, India can create a stable environment for investment and innovation.
2. *Expand Government Funding for Domestic Biotech Research*

   RM in India is heavily dependent on imported technologies, making therapies expensive and limiting access. Increased public funding for research and development and incentives for private sector investment can strengthen India's local biotech ecosystem and reduce reliance on costly foreign alternatives.

3. *Develop Transparent Pricing and Subsidy Models*

   RM has the potential to treat high-burden diseases such as diabetes, cardiovascular disorders, and neurodegenerative conditions, but remains unaffordable for much of the population. Policymakers should explore tiered pricing models, government subsidies, and insurance integration to ensure equitable access to these life-changing treatments.
4. *Strengthen Clinical Trial Oversight and Ethical Review Processes*

   India has a history of exploitative clinical practices in the biotech sector, with concerns over inadequate informed consent and weak ethical oversight. Strengthening ethics committees, clinical trial transparency, and patient protections will be crucial in ensuring that regenerative therapies are developed and deployed responsibly.
5. *Engage in International Collaborations*

   India should actively participate in global regulatory dialogues on RM to align its policies with international best practices while ensuring that its own ethical, cultural, and economic concerns are addressed. Collaborations with leading biotech hubs in Europe, the United States, and Australia may accelerate the adoption of cutting-edge technologies and position India as a key player in global biotech innovation.

By implementing these measures, India can transform RM from an emerging scientific field into a cornerstone of its healthcare system. With the right mix of regulatory clarity, investment in local research, financial accessibility, and ethical oversight, India has the potential not only to serve its own population but also to emerge as a global hub for RM research and application.

References

1. U.S. Food and Drug Administration (2023) FDA approves first gene therapies to treat patients with sickle cell disease. FDA News Release. https://www.fda.gov/news-events/press-announcements/fda-approves-first-gene-therapies-treat-patients-sickle-cell-disease. Accessed 6 Jun 2025
2. Stoker TB (2018) Stem cell treatments for Parkinson's disease. In: Stoker TB, Greenland JC (eds) Parkinson's disease: pathogenesis and clinical aspects. Codon Publications, Brisbane
3. Qin C, Wang K, Zhang L, Bai L (2022) Stem cell therapy for Alzheimer's disease: an overview of experimental models and reality. J Anim Model Exp Med 5(1):15–26. https://doi.org/10.1002/ame2.12207
4. Van Gelder RN, Chiang MF, Dyer MA, Greenwell TN, Levin LA, Wong RO, Svendsen CN (2022) Regenerative and restorative medicine for eye disease. Nat Med 28:1149–1156. https://doi.org/10.1038/s41591-022-01862-8
5. Wernly B, Mirna M, Rezar R, Prodinger C, Jung C, Podesser BK, Kiss A, Hoppe UC, Lichtenauer M (2019) Regenerative cardiovascular therapies: stem cells and beyond. Int J Mol Sci 20(6):1420. https://doi.org/10.3390/ijms20061420
6. Brovkina O, Dashinimaev E (2020) Advances and complications of regenerative medicine in diabetes therapy. PeerJ 8:e9746. https://doi.org/10.7717/peerj.9746

7. Ashammakhi N, Kim HJ, Ehsanipour A, Bierman RD, Kaarela O, Xue C, Khademhosseini A, Sedlits SK (2019) Regenerative therapies for spinal cord injury. Tissue Eng Part B Rev 25(6):471–491. https://doi.org/10.1089/ten.TEB.2019.0182
8. Rosenthal N, Badylak S (2016) Regenerative medicine: today's discoveries informing the future of medical practice. npj Regen Med 1. https://doi.org/10.1038/npjregenmed.2016.7
9. European Medicines Agency (2023) First gene editing therapy to treat beta thalassemia and severe sickle cell disease. Sci Med Health. https://www.ema.europa.eu/en/news/first-gene-editing-therapy-treat-beta-thalassemia-and-severe-sickle-cell-disease. Accessed 6 Jun 2025
10. U.S. Department of Health and Human Services (2005) 2020: a new vision. A future for regenerative medicine. U.S. Department of Health and Human Services. https://www.scribd.com/document/37338866/2020-A-New-Vision-a-Future-for-Regenerative-Medicine. Accessed 6 Jun 2025
11. National Archives (2020) National strategy for critical and emerging technologies. Trump White House Archives. https://trumpwhitehouse.archives.gov/wp-content/uploads/2020/10/National-Strategy-for-CET.pdf. Accessed 6 Jun 2025
12. Pew Research Center (2021) Religious composition of India. https://www.pewresearch.org/religion/2021/09/21/population-growth-and-religious-composition/. Accessed 6 Jun 2025
13. Charitos IA, Ballini A, Cantore S, Boccellino M, Di Domenico M, Borsani E, Nocini R, Di Cosola M, Santacroce L, Bottalico L (2021) Stem cells: a historical review about biological, religious, and ethical issues. Stem Cells Int 2021(1). https://doi.org/10.1155/2021/9978837
14. Funk C, Tyson A, Kennedy B, Johnson C (2020) Biotechnology research viewed with caution globally, but most support gene editing for babies to treat disease. Pew Research Center. https://www.pewresearch.org/science/2020/12/10/biotechnology-research-viewed-with-caution-globally-but-most-support-gene-editing-for-babies-to-treat-disease/. Accessed 6 Jun 2025
15. Rainie L, Funk C, Anderson M, Tyson A (2022) Americans are closely divided over editing a baby's genes to reduce serious health risk. Pew Research Center. https://www.pewresearch.org/internet/2022/03/17/americans-are-closely-divided-over-editing-a-babys-genes-to-reduce-serious-health-risk/. Accessed 6 Jun 2025
16. Dixit S (2020) Ethical guidelines for clinical trials and research in India. Takshashila Institution. https://static1.squarespace.com/static/618a55c4cb03246776b68559/t/62f0e6a87ea7f916e9ebcc7e/1659954868961/10011-TDD-Ethical+Guidelines+for+Clinical+Trials+and+Research+in+India.pdf. Accessed 6 Jun 2025
17. Mudur M (2005) India plans to audit clinical trials. BMJ 331:1044. https://doi.org/10.1136/bmj.331.7524.1044-b
18. Mill JS (1874) Nature, the utility and religion of theism. Longmans, Green, Reader, and Dyer, London
19. Goodfield J (1977) Playing god: genetic engineering and the manipulation of life. Random House, New York
20. Carter L, Mankad A, Hobman EV, Porter NB (2021) Playing god and tampering with nature: popular labels for real concerns in synthetic biology. Transgenic Res 30:155–167. https://doi.org/10.1007/s11248-021-00233-2
21. Bhattachary D, Calitz JP, Hunter A (2010) Synthetic biology dialogue. Royal Soc. https://haseloff.plantsci.cam.ac.uk/resources/SynBio_reports/1006-synthetic-biology-dialogue.pdf. Accessed 6 Jun 2025
22. UK Biotechnology and Biological Sciences Research Council and the UK Medical Research Council (2017) BBSRC and MRC review of vulnerable skills and capabilities. BBSRC and MRC. https://www.ukri.org/wp-content/uploads/2021/10/BBSRC-13102021-BBSRC-MRC-VulnerableCapabilties-Report.pdf. Accessed 6 Jun 2025
23. Torgersen H, Schmidt M (2013) Frames and comparators: how might a debate on synthetic biology evolve? Futures 48:44–54. https://doi.org/10.1016/j.futures.2013.02.002
24. Osborne H (2019) Indian scientist claims ancient Hindus invented stem cell technology and test tube babies. Newsweek. https://www.newsweek.com/india-scientist-ancient-hindu-stem-cell-test-tube-babies-guided-missile-1281360. Accessed 6 Jun 2025

25. Fadel HE (2010) Developments in stem cell research and therapeutic cloning: Islamic ethical positions, a review. Bioethics 26(3):128–135. https://doi.org/10.1111/j.1467-8519.2010.01840.x
26. Naik S, D'Almeida P (2023) Views on heritable gene editing: surveys from India. Takshashila Institution. https://static1.squarespace.com/static/618a55c4cb03246776b68559/t/63bb9b99f0927a3da72c80fe/1673239452214/TSA_Heritable+gene+editing_SN_V.10_06012023.pdf. Accessed 6 Jun 2025
27. Philip C, George B, Ganapule A, Korula A, Jain P, Alex AA, Lakshmi KM, Sitaram U, Abubacker FN, Abraham A, Viswabandya A, Srivastava VM, Srivastava A, Balasubramanian P, Matthews V (2015) Acute myeloid leukaemia: challenges and real world data from India. Br J Haematol 170(1):110–117. https://doi.org/10.1111/bjh.13406
28. Kulkarni U, George B (2019) Access to hematopoietic stem-cell transplantation in India. J Postgrad Med 65(1):1–4. https://doi.org/10.4103/jpgm.JPGM_391_18
29. Meneghel J, Kilbride P, Morris GJ (2020) Cryopreservation as a key element in the successful delivery of cell-based therapies—a review. J Front Med 7. https://doi.org/10.3389/fmed.2020.592242
30. Thaldar D, Shozi B, Steytler M, Hendry G, Botes M, Mnyandu N, Naidoo M, Pillay S, Slabbert M, Townsend B (2022) A deliberative public engagement study on heritable human genome editing among south Africans: study results. PLoS One 17(11):E0275372. https://doi.org/10.1371/journal.pone.0275372
31. Shozi B, Thaldar D (2023) Promoting equality in the governance of heritable human genome editing through Ubuntu: reflecting on a South African public engagement study. AJOB 23(7):43–49. https://doi.org/10.1080/15265161.2023.2207524
32. Willyard C (2024) The effort to make a breakthrough cancer therapy cheaper. MIT Technol Rev. https://www.technologyreview.com/2024/04/12/1091161/car-t-cancer-therapy-treatment-expense/. Accessed 6 Jun 2025
33. Kapoor M (2022) Decoding the legality of stem cell therapy in India. The Viewpoint. https://www.barandbench.com/law-firms/view-point/the-viewpoint-decoding-the-legality-of-stem-cell-therapy-in-india. Accessed 6 Jun 2025
34. LawStreet News Network (2019) Stem cell treatment to continue until licences under new drugs and clinical rules, 2019 are processed: Delhi HC. Lawstreet J. https://lawstreet.co/judiciary/stem-cell-treatment-to-continue-until-licences-under-new-drugs-and-clinical-rules-2019/#google_vignette. Accessed 6 Jun 2025
35. Datta S (2018) An endogenous explanation of growth: direct-to-consumer stem cell therapies in Pr China, India and the USA. Regenerat Med 13(5):559–579. https://doi.org/10.2217/rme-2017-0144
36. Patra PK, Sleeboom-Faulkner M (2009) Bionetworking: experimental stem cell therapy and patient recruitment in India. Anthropol Med 16(2):147–163. https://doi.org/10.1080/13648470902940689
37. Sleeboom-Faulkner M, Patra PK (2011) Experimental stem cell therapy: biohierarchies and bionetworking in Japan and India. Soc Stud Sci 41(5):645–666. https://doi.org/10.1177/0306312711409792
38. Jordaan DW (2017) Autologous stem cell therapy: an analysis of the South African regulatory regime. S Afr J Hum Rights 28(1):31–54. https://doi.org/10.1080/19962126.2012.11865035
39. Thaldar D, Botes M, Shozi B, Townsend B, Kinderlerer J (2020) Human germline editing: legal-ethical guidelines for South Africa. S Afr J Sci 116(9/10). https://doi.org/10.17159/sajs.2020/6760
40. Mastana SS (2014) Unity in diversity: an overview of the genomic anthropology of India. Ann Hum Biol 41(4):287–299. https://doi.org/10.3109/03014460.2014.922615
41. ClinicalTrials.gov (n.d.) U.S. National Library of Medicine, Bethesda. https://clinicaltrials.gov/. Accessed 6 Jun 2025
42. Park JJH, Grais RF, Taljaard M, Nakimuli-Mpungu E, Jehan F, Nachega JB, Ford N, Xavier D, Kengne AP, Ashorn P, Socias ME, Bhutta ZA, Mills EJ (2021) Urgently seeking efficiency and

sustainability of clinical trials in global health. Lancet Glob Health 9(5):681–690. https://doi.org/10.1016/S2214-109X(20)30539-8
43. Barreto ML (2009) Health research in developing countries. BMJ 339:b4846. https://doi.org/10.1136/bmj.b4846
44. Faden RR, Dawson L, Bateman-House AS, Agnew DM, Bok H, Brock DW, Chakravarti A, Gao X, Greene M, Hansen JA, King PA, O'Brien SJ, Sachs DH, Schill KE, Siegel A, Solter D, Suter SM, Verfaillie CM, Walters LB, Gearhart JD (2003) Public stem cell banks: considerations of justice in stem cell research and therapy. Hast Cent Rep 33(6):13–27
45. Glickman SW, McHutchison JG, Peterson ED, Cairns CB, Harrington RA, Califf RM, Schulman KA (2009) Ethical and scientific implications of the globalization of clinical research. N Engl J Med 360(8):816–823. https://doi.org/10.1056/NEJMsb0803929
46. Mendiratta J, Vaswani RN, Saberwal G (2023) Representation from India in multinational, interventional, phase 2 or 3 trials registered in clinical trials registry-India: a cross-sectional study. PLoS One 18:9. https://doi.org/10.1371/journal.pone.0284434
47. Glasner P (2009) Cellular division: social and political complexity in Indian stem cell research. New Genet Soc 28(3):283–296. https://doi.org/10.1080/14636770903151976
48. Thatte UM, Kulkarni-Munshi R, Kalekar SA (2009) Review of policies for injuries to research participants in India. J Med Ethics 35(2):133–139. https://doi.org/10.1136/jme.2008.025155
49. Kandhari R (2021) Stem cell treatments in India: an ethnography of regular practice. Med Anthropol 40(4):348–360. https://doi.org/10.1080/01459740.2020.1857379
50. Rosemann A, Bortz G, Vasen F, Sleeboom-Faulkner M (2016) Global regulatory developments for clinical stem cell research: diversification and challenges to collaborations. J Regenerat Med 11(7):647–657. https://doi.org/10.2217/rme-2016-0072

Ethical Crossroads and Regulatory Pathways: Navigating Regenerative Medicine in Malaysia

9

Nishakanthi Gopalan

Abstract

Regenerative medicine (RM) is transforming modern healthcare due to its potential to repair or replace damaged tissues and organs, rather than merely managing the symptoms of ailments or diseases that cause this damage. This field includes treatment options in the form of stem cell therapies, tissue engineering, and gene editing technologies such as Clustered Regularly Interspaced Short Palindromic Repeats ("CRISPR"), all of which hold significant promise in addressing chronic and previously untreatable diseases. For countries in the Global South, particularly Malaysia, RM presents both opportunities and ethical challenges. This chapter examines Malaysia's evolving position in RM, tracing the country's development through government investments, research infrastructure, private sector engagement, and the growing medical tourism market. This chapter critically explores the ethical and regulatory implications surrounding consent, equity, commercialization, and cultural-religious sensitivities, all of which intersect with regulatory gaps and enforcement challenges. By highlighting the regulatory frameworks and existing guidelines relevant to stem cell and gene therapy, this chapter illustrates the complexities of ensuring patient safety, promoting responsible research, and balancing innovation with public trust. In doing so, the analysis offers broader insights for developing nations navigating similar concerns. Ultimately, this chapter argues that Malaysia's path forward lies in harmonizing international best practices with local ethical norms, strengthening governance mechanisms, and cultivating inclusive and culturally sensitive approaches to biomedical innovation.

N. Gopalan (✉)
Medical Humanities and Ethics Unit (MedHEU), Faculty of Medicine, Universiti Malaya, Kuala Lumpur, Malaysia
e-mail: nisha.gopalan@um.edu.my

U. Ogbogu et al. (eds.), *Regenerative Medicine and the Bioethics of Globalization, Justice and Hegemony*, Collaborative Bioethics 5,
https://doi.org/10.1007/978-3-032-07289-4_9

Keywords

Regenerative medicine · Ethical challenges · Regulatory framework · Stem cell therapy · Malaysia · Biotechnology policy · Medical tourism

Introduction

Regenerative medicine ("RM"), which encompasses cutting-edge therapies such as stem cell therapies, tissue engineering, and gene editing, is rapidly transforming modern healthcare on a global scale. By focusing on restoring damaged tissues and organs, RM offers the potential to cure diseases once considered untreatable, thereby significantly improving patient outcomes and quality of life. For instance, stem cell-based therapies are currently being developed for conditions like Parkinson's disease, spinal cord injuries, and various heart diseases, potentially shifting the paradigm from symptom management to long-term resolution [1].

The societal impact of RM is far-reaching as it addresses persistent healthcare challenges and reduces the burden of chronic illnesses. This advancement not only contributes to increased life expectancy but also promotes overall population health and economic productivity. In particular, RM supports economic growth by creating high-value jobs, reducing long-term healthcare costs, enhancing patient functionality and workforce participation, and possibly positioning countries like Malaysia competitively in the global biotechnology market. Additionally, RM plays a vital role in national development by fostering biotechnology industries, supporting innovation, and attracting both local and international investments. This chapter begins by outlining Malaysia's research and development ("R&D") surrounding RM, highlighting how strategic government investments have supported national capacity building while also allowing the country to keep pace with global advancements in biotechnology.

Developments of Regenerative Medicine in Malaysia

Malaysia is a rapidly developing nation known for its strategic investments in technology and healthcare, particularly in the emerging field of RM. This field aims to leverage biotechnological innovation to enhance national health outcomes and promote economic growth. As a multicultural and multilingual society, Malaysia's healthcare policies and practices are necessarily shaped by its pluralistic socio-cultural context. In recent years, RM in Malaysia has experienced notable growth propelled by proactive government initiatives, a surge in biomedical research, and the rising prominence of medical tourism [2–4].

Malaysia's first stem cell transplantation involved a hematopoietic stem cell transplant and was performed in 1987 at University Malaya Medical Centre (the "UMMC") for a pediatric patient; shortly after, in 1993, an adult hematopoietic stem cell transplant was performed at the same hospital [5, 6]. From 1993 to 2015, a total of 401 transplants were conducted across 13 hospitals nationwide [7]. Presently, bone marrow transplants for conditions such as leukemia, anemia, and thalassemia are well established and recognized by the Ministry of Health (the "MOH"); such transplants conform to local regulatory standards. Recognition by the MOH indicates that these procedures meet national clinical standards and are subject to oversight mechanisms that ensure patient safety, institutional accreditation, and adherence to treatment protocols. Conforming to local regulatory standards involves compliance with MOH-issued guidelines, including protocols on donor-recipient matching, laboratory procedures, quality control, infection control, and post-transplant monitoring. Hospitals performing these transplants are expected also to report outcomes to national registries and undergo periodic audits to maintain licensure and accreditation. To support bone marrow transplant efforts, the Malaysian Stem Cell Registry (the "MSCR") was established in 2000. The registry allows for voluntary registration of bone marrow or stem cell donors, increasing the number of possible matches available for those on transplant lists [5, 6].

Research interest in stem cells and tissue engineering research in Malaysia accelerated in the late 1990s. In 1999, the Tissue Engineering Laboratory at the National University of Malaysia's Faculty of Medicine evolved into the Centre for Tissue Engineering and Regenerative Medicine (the "CTERM"). This center laid the groundwork for the establishment of the Tissue Engineering and Regenerative Medicine Society of Malaysia ("TESMA"), now the country's leading professional network in the field. TESMA hosts regular international conferences and publishes the semi-annual *Regenerative Research Journal* to promote scholarly exchange and visibility. Beyond organizing conferences and publishing the semi-annual Regenerative Research Journal, TESMA is Malaysia's largest professional society in tissue engineering and regenerative medicine. It plays a broader role in advancing the field by facilitating cross-sector collaboration, supporting early-career researchers, and engaging with national stakeholders on education, policy, and innovation in regenerative medicine [8].

In the private sector, numerous biotechnology companies—both local and foreign (notably from India, China, and Japan) have been licensed to operate in Malaysia, engaging in R&D across genetics, tissue engineering, and stem cells. Malaysia has also established several biobanks to support biomedical research. The most prominent is "The Malaysian Cohort," a landmark program approved by the Malaysian government in 2005. The program stores hundreds of thousands of human samples for epidemiological and genomic research. Today, both public and private biobanks actively contribute to ongoing research in genetics, oncology, and RM [9, 10].

Strategic Investments in Malaysia

The Malaysian government has taken a proactive and strategic role in advancing RM through various initiatives led by the Ministry of Science, Technology, and Innovation ("MOSTI") and the MOH. Notably, in 2005, the MOH supported the

Malaysian Cohort study, while MOSTI established multiple funding mechanisms and research grants to stimulate innovation in biotechnology and RM. MOSTI's National Biotechnology Policy ("NBP") was first launched in 2005 and structured in the following three phases:

1. *Phase I (2005–2010)*: Capacity Building—focused on infrastructure and talent development.
2. *Phase II (2011–2015)*: Science to Business—emphasized commercializing scientific research.
3. *Phase III (2016–2020)*: Developing Global Business—aimed to position Malaysia as a regional biotechnology leader.

In 2022, Malaysia launched the NBP 2.0, extending the policy's vision to 2030. The NBP is guided by nine strategic thrusts, which include healthcare biotechnology development, human capital building, regulatory framework development, and government commitment—each designed to foster a robust biotechnology ecosystem across sectors, including regenerative medicine [11, 12]. Among these, the thrust on healthcare and well-being places strong emphasis on RM and tissue engineering, as well as bio-innovation. BiotechCorp, the implementing government agency for the policy, was rebranded as the Bioeconomy Corporation in 2016, signifying an expanded mandate to spearhead the bio-based economic development. The Bioeconomy Corporation continues to provide technical guidance, facilitation, and support to life science companies under MOSTI's leadership [2, 13].

Malaysia has also made deliberate efforts to become a regional hub for medical tourism, aiming to capitalize on cost-effective and high-quality healthcare. The establishment of the Malaysia Healthcare Travel Council (the "MHTC") in 2009 reflects this commitment [14]. The Council was created to establish Malaysia as a leading healthcare destination by coordinating public–private partnerships, promoting the country's medical offerings internationally, and streamlining the patient experience for inbound medical travelers, thus reflecting a national commitment to growing medical tourism as an economic sector. Substantial investments have been made to upgrade medical facilities in accordance with international standards, particularly in specialty areas like RM, cardiology, orthopedics, and oncology, to attract patients from neighboring and international markets. By integrating medical treatments such as stem cell therapies with tourism experiences, Malaysia not only boosts its RM sector and healthcare delivery into broader development plans but also supports economic diversification. These concentrated investments and institutional reforms have positioned Malaysia as a key contributor to biotechnology and RM in Asia [2, 3].

Innovative Regenerative Medicine Research and Development (R&D) and Products in Malaysia

Malaysian research institutions and universities, such as University Malaya, the National University of Malaysia, and University Putra Malaysia, have all been at the forefront of RM research in the country. These institutions have established dedicated centers for stem cell research and tissue engineering, facilitating cutting-edge studies and clinical trials. Investigators at CTERM have published over 500 articles since 1995 on various stem cells and tissue engineering in an effort to understand their scientific mechanism, and ultimately to use stem cells and tissue engineering to heal wounds and regenerate injured tendons and ligaments [15]. The Genetics & Regenerative Medicine Research Centre (the "GRMRC"), affiliated with Universiti Putra Malaysia, also focuses on genetic and RM research. The GRMRC engages in extensive research aimed at understanding the genetic basis of diseases and the therapeutic potential of induced pluripotent stem cells ("iPSCs"). Their efforts include studying the application of gene editing technologies like CRISPR, exploring novel stem cell therapies, and working on tissue engineering projects to create bioengineered tissues and organs [6].

Malaysia is actively collaborating with international research bodies, which has further enhanced the country's research capabilities and global standing in the field of RM. For example, Malaysia's own cellular therapy biopharmaceutical company, CELLAAX, has established joint ventures with the United Kingdom's SRAM & MRAM group to build a conglomerate to revolutionize stem cell treatment in India and regions worldwide. The CELLAAX approach offers a wide range of therapies, including stem cell secretome therapy, to address conditions such as hair loss, joint pain, and even for anti-aging purposes. CELLAX also offers natural killer cell ("NK cell") therapy, which targets and destroy cancer cells by recognizing stress signals and abnormal protein expression on malignant cells, minimizing harm to surrounding healthy tissues. NK cell therapy is noted for its rejuvenating and restorative effects [16, 17].

Malaysia has achieved notable milestones in securing intellectual property ("IP") for various RM products. One such product is CelltiMax, developed by Cell Tissue Group, a company recognized for its innovative contributions to tissue engineering and regenerative therapies. CelltiMax is a mesenchymal stem cell (MSC)-based injectable biologic derived from Wharton's jelly, designed to support cartilage repair and tissue regeneration. It is primarily used in the treatment of osteoarthritis, musculoskeletal injuries, and degenerative joint diseases. Additionally, other IP-protected products include stem cell therapies aimed at treating conditions such as osteoarthritis, chronic wounds, and certain types of cancer, demonstrating Malaysia's commitment to advancing healthcare through biotechnology [18].

Other collaborations with international and domestic research bodies have yielded favorable results for gene therapy treatments. In 2024, Zolgensma, a gene therapy for treating spinal muscular atrophy ("SMA") in children under 2 years old, was approved in Malaysia. This therapy was authorized by Malaysia's Drug Control Authority ("DCA") under the National Pharmaceutical Regulatory Agency (the

"NPRA") and marked a significant milestone as the first gene therapy approved in the country. Although it was manufactured by Novartis Gene Therapies Inc. in the United States, Novartis Corporation, a Malaysian company, is the product registration holder. The approval of Zolgensma, which is an adeno-associated virus vector-based gene therapy, highlights the nation's capability and readiness to incorporate advanced gene therapies into its healthcare system [19]. Furthermore, stem cell therapies to treat blood cancers (e.g., leukemia, myeloma, lymphoma, thalassemia, and sickle cell anemia) and other disorders—such as orthopedic applications such as cartilage regeneration—are being provided by several large private healthcare providers. These therapies often use autologous (e.g., patient's own) stem cells in combination with hyaluronic acid injections [20].

The dedication of scientists, private pharmaceutical companies, and private healthcare providers in advancing RM in Malaysia through robust R&D, international collaborations, and the protection of intellectual property has placed Malaysia at the forefront of biotechnology in Southeast Asia. This ongoing commitment not only enhances Malaysia's global standing but also promises improved healthcare outcomes for patients worldwide, solidifying the country's reputation as a leader in innovative healthcare solutions.

Ethical Challenges of Regenerative Medicine in Malaysia

The rapid advancements in RM bring a host of context-specific ethical challenges that must be addressed to ensure responsible and equitable progress. In Malaysia, these challenges are influenced by regulatory fragmentation, disparities in patients' access to care, religious plurality, and emerging private sector activity. This section outlines key ethical issues, highlighting the need for strengthened governance and culturally grounded oversight.

A. *Informed Consent*: Ensuring that patients and research participants fully understand the risks, benefits, and uncertainties associated with innovative regenerative therapies is crucial. These therapies, often marketed as "advanced" or "cutting-edge," may in fact still be investigational or unproven, creating the potential for therapeutic misconception. This is especially problematic among terminally ill patients or those with limited treatment options who may feel pressured to pursue novel interventions. Additionally, children and the elderly are increasingly among those targeted for stem cell interventions, particularly for conditions such as cerebral palsy, autism spectrum disorder, and regenerative joint disease. While ethical concerns around decision-making for these groups, such as reliance on guardians for consent or age-related cognitive decline, are not unique to RM, they become more pronounced in this context due to the experimental nature, high cost, and limited long-term safety data associated with many regenerative therapies. The complexity of scientific information, combined with limited health literacy in vulnerable populations, raises questions about the depth and voluntariness of consent. Greater emphasis on plain language communication, ongoing consent, and patient safeguards during experimental procedures is warranted [21].

B. *Equity and Access*: The urban-rural divide and the predominance of private providers in delivering stem cell therapies raise serious concerns regarding equity. While international medical tourists may access these treatments at lower relative costs, underserved local populations may remain excluded due to financial, geographic, or systemic barriers. This difference in accessibility reflects a dual access problem, where RM becomes both a commercial asset and a symbol of health inequality. Moreover, the absence of a comprehensive regulatory framework governing tourist patients' rights, safety, and informed access contributes to an under-regulated market for experimental interventions [22]. Bridging dual access and regulatory gaps—namely domestic health inequity and cross-border regulatory oversight—requires deliberate policy action that prioritizes local needs while also ensuring safe and ethical movement of transnational patients and providers.
C. *Safety and Efficacy*: Many RM therapies in Malaysia are offered in non-academic, private sector settings, sometimes without clear evidence of efficacy or formal regulatory approval. While these interventions hold therapeutic promise, the absence of rigorous clinical trials and long-term follow-up raises concerns about the trial's safety and scientific validity. Thorough testing and regulatory review are essential before such therapies are introduced into clinical practice, especially given the potential risks of immune reactions, tumorigenesis, or unintended long-term effects [21, 22]. Despite these risks, some providers offer highly experimental treatments without confirmed approval from relevant authorities. A key regulatory loophole is the use of the term "therapy" rather than "transplantation," which may allow providers to bypass the more stringent clinical trial pathway—such as registration with the National Medical Research Register (the "NMRR") and ethics committee approval. This practice creates ambiguity about whether patients are receiving medically validated treatment or whether they are receiving research-stage interventions. The current lack of oversight and enforcement mechanisms increases the risk of unsafe practices affecting patients. To address these risks, Malaysia requires a more coherent and enforceable regulatory framework that ensures only well-substantiated therapies are made available and that patient safety is prioritized throughout the development and delivery process.
D. *Commercialization and Exploitation*: The growing commercialization of RM in Malaysia, particularly within the private sector, raises significant ethical concerns about the protection of consumers—especially those who are vulnerable due to serious illness or limited treatment options. Many clinics advertise stem cell-based treatments directly to consumers, often using misleading language, exaggerated claims, and unsubstantiated success rates [23]. These practices can create false hope and lead to significant financial and emotional harm, particularly when interventions are unproven or offered without medical oversight [24, 25]. In Malaysia, although the Medical Practicing Division of the Ministry of Health oversees private practices, there is currently no formal mechanism

requiring pre-approval or vetting of advertisements for RM. As a result, promotional materials are often disseminated without scrutiny, bypassing safeguards that would otherwise be expected in the regulation of medical products and services. The regulatory gap enables unlicensed or inadequately regulated clinics to market directly to both local and foreign patients—fueling a burgeoning stem cell tourism industry that often operates beyond boundaries of formal oversight.

Barbara von Tigerstrom has argued that robust consumer protection laws, particularly in advertising regulation, are essential to curbing misleading and potentially harmful stem cell interventions [26]. Similarly, Ubaka Ogbogu has examined how truthful advertising laws can serve as effective tools in addressing the spread of unlicensed stem cell therapies, especially in cross-border contexts where regulatory jurisdictions may be fragmented [27]. These analyses highlight the importance of integrating advertisement law, health fraud prevention, and cross-border enforcement mechanisms into biomedical governance frameworks.

To ensure ethical commercialization, Malaysia must adopt stronger regulatory strategies, including:

- Mandating pre-screening of promotional materials.
- Empowering regulatory bodies to issue sanctions against false or misleading claims.
- Establishing consumer redress mechanisms for patients harmed by unlicensed or ineffective treatments.
- Strengthening collaboration with international regulators to address cross-border advertising and tourism-related risks.

Without these protections, the commercialization of RM risks undermining both public trust and patient safety.

Ethical Crossroads

Malaysia stands at an ethical and regulatory crossroads as it advances the field of RM. The country's sociopolitical and cultural fabric is unique—it is a multiethnic, multireligious nation comprising Muslims, Buddhists, Christians, Hindus, Sikhs, and diverse indigenous groups with local practices. This pluralism influences not only healthcare decisions (e.g., Muslim patients may decline organ donation to maintain bodily integrity after death) but also ethical reasoning, policymaking, and scientific research governance (e.g., xenotransplantation is illegal in Malaysia in part due to religious objections from Muslim communities regarding the "non-halal" status of porcine-derived materials). Non-halal refers to substances or practices that are not permissible under Islamic dietary and ethical laws. In the context of regenerative medicine, materials derived from pigs (porcine sources) are considered non-halal, and their use in medical products or treatments may be religiously objectionable to many Muslims, influencing both personal healthcare decisions and broader regulatory policies [28].

Biomedical policies in Malaysia often require careful consideration of religious decree (e.g., fatwas), cultural norms, and communal values which intersect with scientific advancement and regulation. This section explores how Malaysia navigates its ethical crossroads across three dimensions: the balance between innovation and ethics, cultural-religious awareness in a diverse population, and the tension between international standards and local realities.

1. *Balancing Innovation and Ethics*: Promoting scientific advancement while ensuring ethical accountability is a constant challenge in low- and middle-income countries ("LMICs"). Malaysia's emphasis on biotechnology as a national economic strategy has accelerated the growth of RM, but this emphasis sometimes outpaces the development and implementation of corresponding ethical oversight systems. More broadly, in rapidly developing research contexts, the drive for innovation may inadvertently outpace the establishment of adequate protections for patients and research participants. In Malaysia, both public research institutions and private clinics are actively advancing stem cell and gene therapy research and applications. However, as with many LMICs, ethical safeguards can be unevenly applied, especially in commercial clinical settings, which may not undergo the same level of scrutiny as government-sponsored trials. To address this uneven application, multi-stakeholder ethical review boards, independent post-market surveillance, and clearer definitions of "innovation treatment" vs. "research" are urgently needed. These measures are essential for aligning scientific progress with ethical integrity and public trust.
2. *Cultural and Religious Considerations*: Malaysia's multicultural and multireligious society presents unique ethical challenges in the governance of RM, as it encompasses a wide range of beliefs and practices across ethnic and religious groups. According to the Department of Statistics Malaysia [29], approximately 63.5% of the population identified as Muslims, 18.7% as Buddhists, 6.1% as Hindus, and 9.1% as Christians, with smaller percentages identifying as Sikhs, other faiths, or as part of indigenous belief systems, particularly among communities in Sabah and Sarawak. In Malaysia, the term indigenous population refers collectively to the Orang Asli of Peninsular Malaysia and the natives of Sabah and Sarawak, such as the Iban, Bidayuh, Kadazan-Dusun, and Murut. While these groups differ in culture, language, and legal recognition across regions, they share distinct ancestral ties to the land and are recognized under national and international definitions of indigeneity. Their worldviews, health beliefs, and traditional practices play a crucial role in shaping culturally sensitive approaches to healthcare and biomedical ethics [30].

   These diverse belief systems influence attitudes toward the use of human tissues, embryonic stem cells, and genetic interventions—necessitating culturally informed and religiously sensitive policymaking. For example, Hindu beliefs emphasize the sanctity and inviolability of the body, both in life and after death, which may affect decisions around organ donation or the use of human biological materials in research and therapy. Islamic bioethics, which significantly influences health policy in Malaysia, permits the use of human embryonic tissues and stem cells under specific conditions.

A key consideration surrounding Islamic ethics specifically is the concept of ensoulment, which Islamic scholars in Malaysia traditionally place at 120 days of gestation. This has informed fatwas allowing research on embryos up to 14 days old, aligning with the international guidelines such as the International Society for Stem Cell Research (ISSCR) while remaining religiously compliant [31]. The Buddhist perspective often emphasizes karmic consequences of medical interventions and the importance of non-harm (*ahimsa*). As such, practices like gene therapy are approached with caution, as they may be seen as interfering with the natural course of life and spiritual development [32]. In addition, Christian and Catholic teachings—particularly regarding the moral status of the embryo—may object to embryonic stem cell research on the grounds that it involves the destruction of potential life. These ethical views are not monolithic but reflect varied interpretations within and across religious communities in Malaysia [33].

An inclusive approach to biomedical policy is essential to adequately address the needs of Malaysia's culturally and religiously diverse population. Engaging with religious authorities, cultural representatives, and community leaders in the policymaking process ensures that scientific advancements proceed in a manner that is ethically sound, socially acceptable, and sensitive to Malaysia's plural values. Such engagement not only strengthens public trust but also promotes responsible innovation grounded in cultural legitimacy.

3. *Global Standards Vs. Local Context*: Adapting global ethical standards to fit local contexts is crucial for the responsible development of RM. International guidelines such as the World Medical Association Declaration of Helsinki and the International Society for Stem Cell Research (ISSCR) Guidelines for Stem Cell Research and Clinical Translation provide robust frameworks to safeguard human subjects in biomedical research and to ensure ethical conduct in emerging therapies [34, 35]. The Declaration of Helsinki outlines core principles such as informed consent, scientific validity, independent ethics committee review, risk-benefit assessment, and transparency in results. These are particularly pertinent to RM clinical trials, where the line between therapeutic care and research can be blurred. The ISSCR Guidelines offer more domain-specific guidance, including requirements for robust preclinical data, stem cell research oversight (or "SCRO") review, restrictions on human embryo research, and a firm stance against unproven therapies offered outside of approved clinical trials, and ethical restrictions on human embryo research—such as the internationally recognized 14-day limit.

   While these frameworks are robust and flexible enough to apply across jurisdictions, directly applying these standards without contextual adaptation risks overlooking Malaysia's socio-cultural, religious, and institutional realities. For instance, ethics committees in Malaysia, where regulatory infrastructure is still evolving, especially outside urban or academic centers—may face resources or expertise constraints in thoroughly reviewing advanced cell or gene therapies. Moreover, ethical concepts such as autonomy may also require reinterpretation within Malaysia's communitarian cultures, where family, religious, or community authorities often play a significant role in medical decision-making. Additionally, Muslim bioethics (guided by the Shafi'i school), fatwas on embryonic research, and traditional indigenous beliefs regarding the body and spirit must be considered in shaping ethical governance. In this context, the Declaration of Helsinki and ISSCR Guidelines do not require reframing, but rather contextual operationalization. Their value lies in offering universal principles that must be translated into culturally sensitive and practically feasible standards at the national level.

To do so effectively, Malaysia could consider the following strategies:

- Develop simplified, localized guidance for the ethics committee on RM-specific protocols.
- Clarify national definitions of clinical innovation versus research, especially in the private sector.
- Adaptation of consent processes to include family and communal decision-making norms while preserving individual autonomy.
- Inclusion of religious authorities, indigenous representatives, and cultural scholars in national ethics policy consultations to ensure moral legitimacy.
- Promote public engagement and dialogue to raise awareness and foster trust in both science and its governance.

By localizing global standards through policy, training, and inclusive consultation, Malaysia can develop an ethical framework for RM that is both scientifically rigorous and culturally responsive. This dual alignment will be critical in maintaining public trust, protecting participants, and ensuring the benefits of RM are responsibly realized.

Regulation and Governance of Regenerative Medicine in Malaysia

Malaysia's regulatory framework for RM remains fragmented, with oversight spread across multiple ministries, advisory bodies, and voluntary guidelines. Unlike some countries that have enacted specific legislation governing regenerative therapies or stem cell research (e.g., the US Food and Drug Administration's Regenerative Medicine Advanced Therapy Designation pathway or Japan's *Act on Safety of Regenerative Medicine*), Malaysia does not yet have a dedicated legal or regulatory instrument tailored to this emerging field. An effective RM regulatory framework must go beyond technical approval; it should also ensure patient safety, uphold ethical research standards, and govern the responsible promotion and commercialization of new therapies and treatments. This section outlines the current instruments and institutional mechanisms relevant to the governance of RM in Malaysia, while highlighting critical regulatory gaps and areas for future policy development.

Regulatory Bodies The MOH and the NPRA play key roles in overseeing the development and approval of biomedical products, including RM treatments for both public and private entities in Malaysia. The MOH is responsible for formulating guidelines, health policies, and legislation and ensuring that RM R&D is regulated. Meanwhile, the NPRA is tasked with licensing biomedical products, including cell- and gene-based therapies, and ensuring their safety, efficacy, and quality prior to market entry. As the field of RM continues to evolve rapidly, both agencies will require enhanced institutional capacity, particularly in terms of technical expertise, scientific evaluation framework, and cross-sector coordination.

Despite MOH's central role, Malaysia currently lacks a specific, consolidated legislation or guidelines dedicated to governing RM. Instead, oversight remains fragmented, guided by a patchwork of non-binding guidelines that address certain aspects—such as stem cell transplantation, blood banking, or cell-based product registration—but fall short of covering the full continuum of RM innovation, clinical application, and commercialization. Several factors contribute to this regulatory gap. First, RM is a cross-cutting, technically complex domain involving cellular manipulation, genetic engineering, and novel delivery platforms, making it difficult for traditional regulatory systems to keep pace. Second, many MOH-issued guidelines, including those from 2008 and 2009, are not legally enforceable, limiting their reach, especially in Malaysia's growing private sector, where unproven therapies may be offered with minimal oversight. Finally, the experimental and rapidly evolving nature of RM often demands regulatory flexibility to avoid stifling innovation—creating a tension between oversight and advancement. Table 9.1 outlines the key stem cell-related guidelines currently in place. While foundational, they are insufficient to govern newer technologies such as induced pluripotent stem cells (iPSCs), gene-edited cell products, or advanced manufacturing platforms.

Notably, a new edition of the Guidelines on Stem Cell and Cell-Based Research and Therapy was recently published by MOH [40]. This updated document expands its scope to include cell-based therapies beyond stem cells, reflecting the diversification of RM in Malaysia. However, it still functions as guidance rather than a binding regulation and does not fully address clinical research standards, direct-to-consumer advertising, or enforcement mechanisms. To better accommodate newer technologies, the NPRA released the Guidance Document and Guidelines for Registration of Cell and Gene Therapy Products (CGTPs) in Malaysia in 2015 [41]. This provides a regulatory pathway for product registration, including safety, efficacy evaluation, manufacturing standards, and post-market surveillance. However, this document is primarily product-focused and does not yet comprehensively govern clinical research, advertising, or the delivery of unproven interventions.

Table 9.1 Stem cell guidelines by the Ministry of Health Malaysia

National Standard for Cord Blood Banking and Transplantation [36]	Establishes ethical and operational standards for cord blood collection, processing, banking, and transplantation. The updated edition, released in October 2019, emphasizes updated donor screening criteria, traceability, adverse event reporting, and alignment with international biobanking standards.
National Guidelines for Hematopoietic Stem Cell Therapy 2nd Edition [37]	Defines criteria for clinical facilities performing hematopoietic stem cell transplants. Addresses personnel proficiency, evaluation of donors and patients, and quality management.
National Standards for Stem Cell Transplantation: Collection, Processing, Storage and Infusion of Haemopoietic Stem Cells and Therapeutic Cells [38, 39]	Outlines protocols for transplantations of stem cells from cord blood, peripheral/placental blood, and bone marrow. Emphasizes cryopreservation and transplantation standards.
Guidelines on Stem Cell and Cell-Based Research and Therapy 3rd Edition [40]	Covers ethical and scientific requirements both for hematopoietic and non-hematopoietic stem cells, including consent, biological material use, and ethical review.

Ethics Committees Research ethics committees ("RECs") and institutional review boards ("IRBs") play a central role in ensuring the ethical conduct of research involving human participants. They are responsible for assessing scientific validity, ensuring informed consent, evaluating the risk-benefit ratio, and safeguarding the rights and welfare of participants. The MOH guidelines mandate that all RM research, including stem cell and gene therapy studies, undergo prior review and approval by REC or IRBs [40]. In addition to institutional review, all stem cell research must be registered with the NMRR and submitted to the National Stem Cell Research and Ethics Sub-Committee ("NSCERT"), a sub-body under the national Medical Research and Ethics Committee. NSCERT holds the authority to review, advise, or reject any proposal in alignment with the National Guideline for Stem Cell Research and Therapy. These guidelines provide detailed requirements on ethical review processes, including the use of human biological materials, consent procedures, and participant protections. The guidelines also outline prohibited activities under the National Standards for Stem Cell Transplantation [38, 39], which are in line with international standards such as those of the ISSCR. These standards include the prohibition of:

- Culturing any intact human embryo in vitro beyond 14 days or until the formation of the primitive streak, whichever comes first, conforming to the international guidelines, such as the ISSCR Guidelines
- Introducing human embryonic stem cells into non-human primate blastocysts or any embryonic stem cells into human blastocysts
- Breeding animals that have received human embryonic stem cells at any stage of development

- Fusing human stem cells or other pluripotent cells with cells of non-human that develop beyond 14 days or the primitive streak, whichever occurs first

RM research as a whole is subject to the same ethical oversight process—including research involving stem cells, gene therapy, and other cellular or genetic interventions. Studies involving clinical interventions (e.g., gene therapy or cellular transplantation) are typically classified as clinical trials and must comply with both ethical and regulatory standards specific to human research.

Legislation and Regulatory Policies Government-funded agencies and privately owned companies researching RM (including stem cells) are required to obtain ethics approval before proceeding with their research or transplantations. Government agencies working on human tissue samples must follow the *Human Tissue Act* [42] and adhere to national guidelines and regular circulars from the Minister and Director General of the MOH. Private companies and healthcare providers face additional obligations, including compliance with:

- The *Private Healthcare Facilities and Services Act* of 1998 [43].
- Applicable business and biotechnology regulations.
- World Health Organization Good Manufacturing Practices [44].
- The joint US Department of Health and Human Services, Centers for Disease Control and Prevention, and National Institutes of Health Biosafety Level 3 (BSL-3) standards for gene therapy involving viral vectors [45].

The diversity in legislation—across sectors and types of research—often results in administrative delays and regulatory ambiguity. A harmonized, centralized framework is needed to streamline approval pathways, clarify legal obligations, and facilitate compliance without stifling innovation.

Clinical Trial Regulations Establishing clear and stringent regulations for clinical trials is essential to safeguard participants and validate the efficacy and safety of new treatments. This includes protocols for long-term follow-up and reporting of adverse events. The Clinical Research Centre (the "CRC") under the Institute for Clinical Research (the "ICR") facilitates clinical research within various public hospitals, ensuring that trials adhere to Good Clinical Practice ("GCP") standards and that they are built on robust scientific foundations [46]. Research involving human adult stem cells, fetal tissues from legally performed terminations of pregnancy, non-human stem cells, and embryonic stem cell lines from surplus IVF embryos are permitted if said research is performed under ethical oversight and in alignment with national religious guidance such as the 2005 decision of the Fatwa Committee of the National Council for Malaysian Islamic Religious Affairs on cloning and assisted reproductive technologies [47].

Public–Private Partnerships Collaboration between public agencies and private entities is vital for advancing RM. These partnerships can facilitate the sharing of resources, expertise, and funding and ultimately drive innovation while ensuring that ethical and regulatory standards are maintained. The Malaysian Bioeconomy Development Corporation ("Bioeconomy Corp.") plays a pivotal role in fostering these partnerships, supporting the commercialization of research, and promoting Malaysia as a hub for biotechnology [2]. However, these partnerships must be supported by clearly delineated regulatory responsibilities to prevent non-compliance, particularly when multiple agencies are involved in research, funding, and implementation. Transparency, coordinated oversight, and ethical alignment are vital to sustaining public trust in the field of RM.

Strategic Pathways for Ethical and Sustainable Advancement of Regenerative Medicine in Malaysia

This section outlines key strategic directions that Malaysia can adopt to ensure that the advancement of RM is both ethically grounded and sustainable, while also aligning with national values, public expectations, and international standards. To ensure that RM progresses responsibly and equitably in Malaysia, a comprehensive strategy must be adopted—one that integrates ethical considerations, cultural sensitivity, regulatory coherence, and scientific innovation. The following strategic pathways are essential to ensure the advancement of RM in Malaysia is ethical and sustainable.

Establishing a Unified and Responsive Regulatory Framework Malaysia currently lacks a single, overarching law or government agency dedicated to the oversight of RM. Establishing comprehensive national guidelines for both research and clinical applications is a critical step toward ensuring responsible innovation. These guidelines should be developed through a consultative and inclusive process involving stakeholders from academia, industry, government, religious authorities, and civil society. A unified framework would provide much-needed clarity for researchers, clinicians, and regulators while also ensuring consistency and adherence to high ethical standards. Presently, existing MOH-issued guidelines cover aspects of stem cell transplantation, cord blood banking, and hematopoietic stem cell research. However, the existing guidelines do not adequately address more innovative developments such as iPSCs, genomic research, gene therapy, the use of viral vectors, and other novel interventions. These areas currently fall outside the scope of existing regulatory instruments, creating uncertainty for researchers and potential risks for patients.

In 2021, the Malaysian MOH and the National Institute of Health released the Guidelines for Conducting Research in Ministry of Health (MOH) Institutions and Facilities, which covers conduct involving stem cell, genomics, and other biomedical domains such as biobanking, clinical trials, and the use of human biological materials [48]. However, there remains an urgent need for both a broader and dedicated national guideline that comprehensively addresses the ethical, procedural, and safety considerations surrounding cutting-edge technologies. Such a document should include clearly articulated allowances and prohibitions, ensure alignment with international standards, and extend applicability beyond public institutions to include the growing private and commercial sectors. A forward-looking regulatory framework will be essential to support innovation while protecting public health and maintaining ethical integrity across all sectors within Malaysia.

Enhancing Institutional Capacity and Ethical Oversight Investing in the training and development of researchers, clinicians, members of ethics committees, and regulatory personnel (including policymakers) is essential. Strengthening capacity in bioethics, clinical research, and regulatory science will provide the necessary growth and ethical oversight in this complex field. Academic programs like the Master of Health Research Ethics ("MOHRE") at Universiti Malaya play a vital role in cultivating the ethical competence of professionals engaged in RM [49]. In addition to formal education, opportunities for continuous professional development—such as participation in regional and international conferences—help researchers and policymakers remain informed of scientific and regulatory advancements.

However, intermittent exposure to professional development is not sufficient. Regular multi-stakeholder roundtable discussions involving researchers, bioethicists, regulators, legal experts, and religious authorities are equally important. These platforms allow for real-time dialogue on emerging ethical dilemmas, evolving research practices, and regulatory challenges. Such engagements foster mutual understanding across disciplines and promote shared responsibility in balancing scientific innovation with ethical or governance imperatives.

Public Engagement and Education Engaging the public in discussions about RM and its ethical implications can help build trust and support for new technologies. Educational initiatives can raise awareness about the benefits and risks of regenerative treatments, fostering an informed and supportive community. Public consultations and outreach programs can also ensure that diverse perspectives are considered in policymaking. For instance, the NSCERT subcommittee organized a public forum in 2008 to gather community input on ethical concerns surrounding stem cell therapies and emerging biomedical technologies [38, 39]. By involving community leaders, patient advocacy groups, and other stakeholders in these dialogues, policymakers and researchers can gain valuable insights into public concerns and values.

International Collaboration Leveraging international partnerships can provide Malaysia and other LMICs with access to global expertise and resources. Collaborating with leading research institutions and regulatory bodies can help Malaysia develop robust and internationally aligned frameworks for RM. For example, cooperation with the United Kingdom's Human Fertilization and Embryology

Authority (HFEA), known for its transparent, inclusive approach to sensitive biomedical issues, and Singapore's Health Sciences Authority (HSA)—a regional leader in regulating advanced therapies in a multicultural context—could offer valuable models for regulatory development. These collaborations facilitate knowledge exchange and the adoption of best practices, ensuring that Malaysian researchers and clinicians are at the forefront of technological and methodological advancements.

International collaborations can also provide opportunities for joint research projects, shared funding mechanisms, and participation in international clinical trials. These partnerships can enhance the capacity of Malaysian institutions to conduct high-quality research and offer innovative treatments. Additionally, being part of global networks can help Malaysia stay updated with the latest ethical standards and regulatory requirements, fostering a more cohesive and globally integrated approach to RM.

Harmonizing Policies and Regulations Establishing common operational policies and regulatory mechanisms across the Association of Southeast Asian Nations (or "ASEAN") region can prevent policy fragmentation and promote easier collaboration and movement of technologies and expertise. Malaysia can play a leading role in regional initiatives to harmonize regulations, streamline clinical trial processes, and facilitate the approval of new treatments.

Harmonizing policies and regulations would involve aligning national guidelines with international. It could also lower industry costs related to protocol amendments, repeated ethics reviews, or region-specific product reformulation, thereby making clinical trials and therapeutic innovation more efficient. This harmonization would reduce administrative costs associated with multi-country approvals and delays caused by varying national compliance requirements and expedite the development and implementation of new therapies. By fostering a unified regulatory environment, ASEAN countries can enhance their collective ability to attract international investments and partnerships, ultimately benefiting patients through faster access to safe and effective RM treatments.

Conclusion

Malaysia stands at a pivotal juncture in navigating the ethical, regulatory, and governance challenges of RM. Strategic investments in biotechnology, coupled with proactive government initiatives and a growing base of research expertise, clinical infrastructure, and private sector engagement, have propelled the field forward. However, sustained progress depends on the ability to balance scientific innovation with ethical integrity and public accountability. Developing comprehensive national guidelines, strengthening institutional capacity, and investing in targeted training for researchers, clinicians, and regulators are critical to ensuring that RM advances in a responsible and sustainable manner. Equally important to the development of

RM is meaningful public engagement through educational initiatives, community dialogue, and transparent communication. Such engagement is essential to build trust, dispel misconceptions, and ensure that diverse perspectives—particularly those rooted in Malaysia's multicultural and multireligious context—are informing or integrated into policymaking.

Malaysia's innovation capacity in RM will also depend on its ability to collaborate and contribute at the international level. By incorporating globally recognized best practices and harmonizing ethical safeguards, Malaysia can further enhance its capacity for scientific innovation, regulatory alignment, and clinical implementation. At the same time, ethical challenges such as informed consent, equity, safety, and the risks of commercialization and exploitation must be addressed with urgency and sensitivity. By establishing a clear, inclusive, and forward-looking regulatory and ethical framework, Malaysia will be well positioned to lead at the global level in the responsible development of RM. This balanced approach will not only promote scientific excellence and economic opportunity but also ensure that the benefits of innovation are equitably shared across society, ultimately contributing to improved healthcare outcomes and public trust in biomedical progress.

References

1. Plummer S, El Miedany Y (2022) Regenerative medicine: challenges and opportunities. In: El Miedany Y (ed) Musculoskeletal ultrasound-guided regenerative medicine. Springer, Cham, pp 539–548
2. Gilbert J (2024) Biotech, bio-based sectors on track to become main drivers of Malaysia's economy. The Sun Daily. https://thesun.my/business-news/biotech-bio-based-sectors-on-track-to-become-main-drivers-of-malaysia-s-economy-HH12737473. Accessed 12 May 2025
3. Siddiqui A (2025) Malaysia turns key med tourism hub. BioSpectrum Asia. https://www.biospectrumasia.com/analysis/30/25518/malaysia-turns-key-med-tourism-hub.html. Accessed 12 May 2025
4. Vermonte P (2024) The global south and the need for economic growth. The Star. https://www.thestar.com.my/business/business-news/2024/05/07/the-global-south-and-the-need-for-economic-growth. Accessed 12 May 2025
5. Fadilah SAW, Leong CF, Cheong SK (2008) Stem cell transplantation in Malaysia and future directions. Med J Malaysia 63:279–280. https://pubmed.ncbi.nlm.nih.gov/19385484/
6. Gan GG, Teh A, Chan LL, Cheong SK, Chang KM, Ibrahim HM (2008) Bone marrow and stem cell transplantation: Malaysian experience. Bone Marrow Transplant 42(S1):S103–S105. https://doi.org/10.1038/bmt.2008.129
7. Saidon N, Anuar NA, Meng CK, Chuan OT, Mui TS, Gin GG, Saw F, Ariffin H, Peng LH, Lee CL, Teh A, Chong BP (2020) Evolution of hematopoietic stem cell transplant programs in Malaysia. Blood Cell Ther 3(3):44–47. https://doi.org/10.31547/bct-2019-017
8. Tissue Engineering and Regenerative Medicine Society of Malaysia (TESMA) (2024) About TESMA. https://tesma.org.my/about-tesma/. Accessed 12 May 2025
9. Amin L, Hashim H, Mahadi Z, Ismail K (2018) Determinants of the willingness to participate in biobanking among Malaysian stakeholders in the Klang Valley. BMC Med Res Methodol 18:163. https://doi.org/10.1186/s12874-018-0619-2
10. Azahar A, Mohd Yusof AN, Azhar ZI (2023) A preliminary study to explore the informed consent approach and the ethical challenges in the Malaysian biobanking for research. Asian Bioeth Rev 15:141–154. https://doi.org/10.1007/s41649-022-00229-y

11. Bioeconomy Corporation (n.d.-a) National Biotechnology Policy 2.0 (DBN 2.0). https://www.bioeconomycorporation.my/corporate/nbp/. Accessed 12 May 2025
12. Ministry of Science, Technology and Innovation (MOSTI) (2022) National biotechnology policy 2.0: towards a Bio-Innovation Society (2022–2030) (English translation of title; original document in Malay). Ministry of Science, Technology and Innovation, Putrajaya. https://www.fao.org/faolex/results/details/en/c/LEX-FAOC224662/. Accessed 12 May 2025
13. Bioeconomy Corporation (n.d.-b) About Us. https://www.bioeconomycorporation.my/corporate/about-us/. Accessed 12 May 2025
14. Malaysia Healthcare Travel Council (n.d.) Malaysia healthcare travel council. https://www.mhtc.org.my/. Accessed 12 May 2025
15. Centre for Tissue Engineering and Regenerative Medicine (n.d.) Google Scholar page. https://scholar.google.com.my/citations?user=7aaFeaoAAAAJ&hl=en. Accessed 12 May 2025
16. BioSpectrum Asia (2023) UK-Malaysia partnership for advancing stem cell treatment in India. https://www.biospectrumasia.com/news/52/22809/uk-malaysia-partnership-for-advancing-stem-cell-treatment-in-india.html. Accessed 12 May 2025
17. CELLAAX Therapy (n.d.) Reimagining the future with CELLAAX therapy. https://cellaax-therapy.com/regenerative-medicine/. Accessed 12 May 2025
18. Cell Tissue Group (n.d.) CelltiMax: a pathway towards improved health & longevity. https://cell-tissue.com/celltimax/. Accessed 12 May 2025
19. CodeBlue (2024) SMA gene therapy Zolgensma gets orphan medicine status in Malaysia. https://codeblue.galencentre.org/2024/02/13/sma-gene-therapy-zolgensma-gets-orphan-medicine-status-in-malaysia/. Accessed 12 May 2025
20. National Pharmaceutical Regulatory Agency (NPRA) (2024) Media Statement by the Minister of Health—Outcome of the 393rd Meeting of the Drug Control Authority: Zolgensma & QDenga (English translation of title; original document in Malay). https://www.npra.gov.my/index.php/en/component/content/article/455-english/press-release/press-release-2024/1527571-keputusan-mesyuarat-pihak-berkuasa-kawalan-dadah-kali-ke-393.html?Itemid=1391. Accessed 12 May 2025
21. Imran SAM, Hamizul MMHA, Khairul Bariah AAN, Wan Kamarul Zaman WS, Nordin F (2022) Regenerative medicine therapy in Malaysia: an update. Front Bioeng Biotechnol 10:789644. https://doi.org/10.3389/fbioe.2022.789644
22. Gopalan N, Noor SNM, Mohamed MS (2021) The pro-medical tourism stance of Malaysia and how it affects stem cell tourism industry. SAGE Open 11:21582440211016837. https://doi.org/10.1177/21582440211016837
23. Sipp D (2017) Direct-to-consumer stem cell marketing and regulatory responses. Cytotherapy 19:721–723. https://doi.org/10.5966/sctm.2013-0040
24. Berger I, Ahmad A, Bansal A, Kapoor T (2016) Global distribution of businesses marketing stem cell-based interventions. Cell Stem Cell 19:158–162. https://doi.org/10.1016/j.stem.2016.07.015
25. Turner L, Knoepfler P (2016) Selling stem cells in the USA: assessing the direct-to-consumer industry. Cell Stem Cell 19:154–157. https://doi.org/10.1016/j.stem.2016.06.007
26. von Tigerstrom B (2017) Regulating the market for unproven stem cell therapies. Regen Med. 12(7):815–826. https://doi.org/10.2217/rme-2017-0057
27. Ogbogu U (2016) Combatting unlicensed stem cell interventions through truthful advertising law: a survey of regulatory trends. McGill J Law Health 9:311–328
28. Wong KL, Faris WF (2024) Islamic bioethics, porcine therapeutics and porcine xenotransplantation: Islamic perspectives. Al-Afkar J Islam Stud 7(2):114–134. https://www.al-afkar.com/index.php/Afkar_Journal/article/view/1198
29. Department of Statistics Malaysia (2024) Demographic Statistics, Fourth Quarter 2024. https://www.dosm.gov.my/portal-main/release-content/demographic-statistics-fourth-quarter-2024. Accessed 12 May 2025
30. International Work Group for Indigenous Affairs (IWGIA) (2025) The Indigenous World 2025: Malaysia. https://iwgia.org/en/malaysia/5668-iw-2025-malaysia.html. Accessed 12 May 2025

31. Gopalan N, Nor SNM, Mohamed MS (2020) Regulation of stem cell technology in Malaysia: current status and recommendations. Sci Eng Ethics 26:1–25. https://doi.org/10.1007/s11948-019-00111-5
32. Keown D (1992) The Nature of Buddhist Ethics. Palgrave Macmillan
33. Sivaraman MA, Noor SN (2016) Human embryonic stem cell research: ethical views of Buddhist, Hindu and Catholic leaders in Malaysia. Sci Eng Ethics 22:467–485. https://doi.org/10.1007/s11948-015-9666-9
34. International Society for Stem Cell Research (ISSCR) (2021) ISSCR guidelines for stem cell research and clinical translation: 2021 update. https://www.isscr.org/guidelines. Accessed 12 May 2025
35. World Medical Association (2013) Declaration of Helsinki: ethical principles for medical research involving human subjects. https://www.wma.net/policies-post/wma-declaration-of-helsinki/. Accessed 12 May 2025
36. Ministry of Health Malaysia (2019) National standards for cord blood banking and transplantation. https://www.moh.gov.my/moh/resources/Penerbitan/Perkhidmatan%20OnG%20&%20Ped/STEM%20CELL/4._The_National_Standards_For_Cord_Blood_Banking_And_Transplantation_.pdf. Accessed 12 May 2025
37. Ministry of Health Malaysia (2023) National guidelines for haemopoietic stem cell therapy 2nd edition. https://www.moh.gov.my/moh/resources/Penerbitan/Perkhidmatan%20OnG%20%26%20Ped/STEM%20CELL/National_Guideline_for_HSCT_2022_%28250224%29.pdf. Accessed 12 May 2025
38. Ministry of Health Malaysia (2009a) Guidelines for stem cell research and therapy. https://www.ummc.edu.my/files/ethic/Guideline%20for%20Stem%20Cell%20Research%20and%20Therapy.pdf. Accessed 12 May 2025
39. Ministry of Health Malaysia (2009b) National standards for stem cell transplantation: collection, processing, storage and infusion of haemopoietic stem cells and therapeutic cells. https://www.moh.gov.my/moh/resources/auto%20download%20images/589d78eb72cbf.pdf. Accessed 12 May 2025
40. Ministry of Health Malaysia (2024) Guidelines on stem cell and cell-based research and therapy 3rd edition. https://www.moh.gov.my/moh/resources/Penerbitan/Perkhidmatan%20OnG%20&%20Ped/STEM%20CELL/5._Guidelines_on_Stem_Cell_3rd_Edition_PDF_.pdf. Accessed 12 May 2025
41. National Pharmaceutical Control Bureau (2015) Guidance Document and Guidelines for Registration of Cell and Gene Therapy Products (CGTPs) in Malaysia. https://www.npra.gov.my/index.php/en/guideline-bio/1527314-guidance-document-and-guidelines-for-registration-of-cell-and-gene-therapy-products-cgtps-in-malaysia-2.html. Accessed 12 May 2025
42. Government of Malaysia (2006) Human Tissues Act 1974: incorporating all amendments up to 1 January 2006. https://www.ummc.edu.my/files/ethic/MCHRS/7%20Human%20Biological%20Materials/Act%20130%20Human%20Tissues%20Act%20(1974).pdf. Accessed 12 May 2025
43. Government of Malaysia (1998) Private Healthcare Facilities and Services Act 1998. https://www.moh.gov.my/index.php/database_stores/attach_download/317/27. Accessed 12 May 2025
44. World Health Organization (WHO) (2014) WHO good manufacturing practices for pharmaceutical products: main principles. In: WHO technical report series, no. 986, annex 2. World Health Organization, Geneva. https://www.who.int/publications/m/item/trs986-annex2. Accessed 12 May 2025
45. U.S. Department of Health and Human Services, Centers for Disease Control and Prevention, National Institutes of Health (2020) Biosafety in microbiological and biomedical laboratories (BMBL), 6th edition. https://www.cdc.gov/labs/pdf/SF__19_308133-A_BMBL6_00-BOOK-WEB-final-3.pdf. Accessed 12 May 2025
46. Goh PP (2018) Establish clinical research centre as national focal point for clinical research in MOH. Ministry of Health Malaysia. Retrieved from https://www.scribd.com/document/398207256/3-Teori-teori-perubahan-perilaku. Accessed 12 May 2025

47. National Council for Islamic Religious Affairs Malaysia (2005) Decision of the fatwa committee of the National Council for Islamic Religious Affairs Malaysia on cloning and assisted Reproductive Technology (ART). Department of Islamic Development Malaysia (JAKIM), Prime Minister's Department. https://www.islam.gov.my/images/ePenerbitan/KOMPILASI_MUZAKARAH_MKI_2016.pdf. Accessed 12 May 2025
48. National Institutes of Health Malaysia (2021) Guidelines for conducting research in Ministry of Health (MOH) institutions and facilities. https://nih.gov.my/images/media/publication/guidelines/NIH_Guideline_2021.pdf. Accessed 12 May 2025
49. Master of Health Research Ethics (MOHRE) (2023) Capacity building in bioethics and clinical research. https://mohre.um.edu.my/course-details. Accessed 12 May 2025

Regenerative Medicine in China: Current Regulatory Challenges and Global Advancements

10

Meng Wang and Li Du

Abstract

The field of regenerative medicine (RM) stands at the precipice of significant breakthroughs in medical science, including developing innovative treatments for degenerative diseases, genetic disorders, and organ failure. However, research and development in the field raises pressing ethical and legal concerns surrounding patient safety, obtaining informed consent, and ensuring equitable access to innovative treatments. China is an important player in the stem cell research and RM space and has implemented various policies to promote biomedical advances. These policies have included the establishment of strong governance measures and related laws to address contentious ethical and legal issues associated with the development of novel therapies. This chapter explores policies and regulatory frameworks that serve to advance RM in China. This chapter will examine key challenges, including issues related to bioethical governance, regulatory gaps, and legal barriers to international cooperation within the field of RM. We will then assess China's proactive measures and current challenges for addressing disparities in access to RM technologies in low- and middle-income countries (LMICs) and outline directions for the country's future efforts to promote a balanced global approach, one that fosters international collaboration, including fair genomic data sharing, to enhance the global availability and affordability of RM products.

Keywords

Regenerative medicine · Bioethical governance · Regulations · China · Low- and middle-income countries

M. Wang · Li Du (✉)
Faculty of Law, University of Macau, Taipa, Macau, China
e-mail: yc07212@um.edu.mo; stephendu@um.edu.mo

U. Ogbogu et al. (eds.), *Regenerative Medicine and the Bioethics of Globalization, Justice and Hegemony*, Collaborative Bioethics 5,
https://doi.org/10.1007/978-3-032-07289-4_10

Introduction

Regenerative medicine (RM) represents a cutting-edge discipline in biomedical science. By leveraging principles from both biology and engineering, RM therapies possess the remarkable ability to repair, replace, or regenerate damaged human tissues and organs. This innovative field holds immense promise in treatment modalities, offering unique therapeutic alternatives for a wide array of serious diseases that were previously thought to be untreatable. For example, novel stem cell therapies constitute a significant technological advancement in RM. By manipulating the pluripotent differentiation potential of stem cells, researchers can guide their transformation into various cell types, including cardiac and neural cells, facilitating the repair or replacement of damaged tissues. Moreover, tissue engineering involves constructing three-dimensional biological structures, such as artificial organs or tissues, to replace damaged structures within the patient's body. Additionally, the use of gene therapy in RM allows direct interventions into a patient's genome to treat diseases, offering precise and personalized treatments [1, 2].

Over the past few decades, many RM therapies, including various cell and gene therapies and products, have obtained market approval in the United States and other countries and regions [3]. China is also recognized as a hub for developing cell and gene therapies and ranked second globally in the number of RM clinical trials conducted within a country [4]. In recent years, several RM-related therapies have been approved or are awaiting market authorization in China. Notably, the country approved its first chimeric antigen receptor T-cell (CAR-T) product for treating adult relapsed or refractory B-cell acute lymphoblastic leukemia in 2023, marking a significant milestone in both regulatory capacity and the clinical translation of advanced RM therapies [5].

In the global market, technology companies, biotech startups, and other pharmaceutical industry players are investing significant resources in RM research and development (R&D). According to a 2024 survey by Custom Market Insights, the global RM market is expected to grow at a compound annual growth rate (CAGR) of 19.2% from 2025 to 2034. By 2034, RM is projected to reach a market size of $248.73 billion USD [6]. The rapid development of artificial intelligence (AI) technology has significantly accelerated progress in the field of RM. In tissue engineering, for instance, AI is able to facilitate the optimization of biomaterial design and scaffold architecture, saving time and ensuring greater accuracy than traditional methods [7, 8]. Moreover, the use of AI tools in image analysis can monitor stem cell differentiation and tissue regeneration processes more effectively, improving the efficiency and accuracy of data interpretation during these stages [9, 10]. These applications not only streamline RM-related R&D but also contribute to more precise and effective therapeutic outcomes. Market analyses specifically attribute the anticipated global market growth in RM from 2022 to 2030 to the application of AI, forecasting a projected CAGR of 21.94% during this period [11].

However, the advancement of RM continues to raise significant ethical and legal concerns. In particular, the surge of unregulated and unproven RM therapies has consistently undermined public trust in this emerging field [12–15]. For instance, it was reported that, by 2021, nearly 1500 businesses in the United States were directly marketing unproven stem cell therapies to consumers [16]. These treatments carry substantial risks, including adverse events of uncontrolled tissue growth, severe infections, and even blindness [17]. According to the Pew Charitable Trusts, at least 360 adverse events related to such therapies, including 20 deaths, were reported by 2020, though underreporting suggests the actual numbers are likely higher [17]. In addition to direct medical harm, these interventions can delay access to proven treatments and impose significant financial burdens on patients.

Similar issues have emerged in China. A 2021 report by Xinhua News Agency exposed hospitals and clinics offering unapproved stem cell injections for as much as ¥98,000 RMB (approximately $14,000 USD) per treatment [18]. These practices blur the line between legitimate and unauthorized treatments, making it difficult for patients to identify evidence-based care and ultimately eroding public trust in RM. In parallel, the use of genetic data and AI analytic tools has also raised ethical and legal concerns regarding the protection of human genetic information and personal health data.

At the international level, there are instruments that provide guidance and regulation for RM in an effort to address and mitigate these concerns. For example, organizations such as the United Nations Educational, Scientific and Cultural Organization (UNESCO) and the World Medical Association have issued declarations that establish a foundation for safeguarding the rights of patients and other human research subjects in biomedical research, medical interventions, and genomics research [19, 20]. International organizations, such as the World Health Organization (WHO) and the International Society for Stem Cell Research (ISSCR), have also engaged in developing rules and guidelines aimed at promoting responsible development of RM-related technologies, as well as in promoting equitable access to RM treatments. These organizations have developed fundamental ethical principles, including the right of patients to give informed consent for RM interventions and the need to protect patients' privacy, including their genetic information and other sensitive health data. Moreover, both the WHO and the ISSCR have devised strategies aimed at reducing disparities in health and technology development, as well as closing gaps in regulatory standards among different countries and regions [21–23].

Notwithstanding these regulatory developments, there are currently no binding international instruments that specifically address or seek to regulate inappropriate practices, nor are there binding instruments that promote the equitable distribution of research outcomes in RM R&D. There is also no global consensus regarding the approval for, and oversight of, RM research and applications. This lack of unified governance and regulation creates challenges for local regulatory agencies, particularly in low- and middle-income countries (LMICs) seeking to mitigate the risks associated with novel RM technologies. LMICs may often lack the institutional capacity, such as regulatory infrastructure and clinical oversight mechanisms, to evaluate the safety and efficacy of advanced therapies, making them vulnerable to becoming "testing grounds" for unproven or experimental interventions. In the long term, this not only compromises patient safety but also exacerbates global disparities in access to trustworthy and evidence-based RM treatments.

Regulatory and ethical review regimes vary widely across countries, with some LMICs facing either insufficient oversight or overly stringent requirements [24–27]. For instance, research indicates that some Asian countries, such as Malaysia, lack comprehensive legislation in this domain [25]. This lack of oversight may present challenges for the responsible development and commercialization of RM therapies in LMICs [28, 29]. A key concern in this regard is the widespread offering of unproven experimental therapies by clinics in LMICs, which exploit regulatory loopholes, put patients' health at risk, and contribute to negative public perceptions of the RM field overall [30, 31]. This phenomenon has also contributed to the rise of "stem-cell tourism," where patients (often from countries with stricter regulations) travel to LMICs seeking access to unproven or experimental RM interventions [32]. Such cross-border medical travel raises further ethical concerns, particularly regarding informed consent, patient safety, and the exploitation of regulatory gaps in host countries [33, 34].

In contrast, several high-income countries have developed robust and expedited frameworks for RM oversight. For instance, the United States Food and Drug Administration (FDA) has issued guidelines such as Fast Track and Breakthrough Therapy to expedite the development of RM products [35]. Similarly, the European Union (EU) has enacted *Regulation (EC) No. 1394/2007*, defining and classifying advanced therapy medicinal products, including RM therapies, with specific regulatory standards for research and market approval [36]. Japan has also implemented an *Act on the Safety of Regenerative Medicine* to regulate RM-related technologies based on risk levels [37].

Over the past decade, China, which is considered a middle-income country, has implemented a series of policies and regulations that aim to promote RM R&D while simultaneously strengthening oversight of cell and gene therapies. For instance, the *Action Plan for Drug Regulatory Science*, initiated under the directive of the State Council of China in 2021, states that developing a comprehensive regulatory framework for cell and gene therapies is a national priority [38]. In addition, China's health and drug regulatory authorities have also issued guidelines for RM-related research and applications to better align domestic practices with global scientific and ethical standards [39, 40]. These international frameworks serve as influential references for national policy development and the advancement of the RM industry, especially in the absence of unified international regulation.

A growing body of literature has outlined the ethical and legal issues associated with the development of stem cell therapies and RM in China. Some authors contend that the governance of RM products, such as cell therapy products, should not predominantly rely on technical guidelines (such as those found in non-binding instruments) that outline recommended standards or best practices for product quality, safety, efficacy, and clinical trial protocols [41, 42]. They contend that the current RM governance model, which relies primarily on technical guidelines, has failed to establish robust and legally enforceable mechanisms for review, regulatory supervision, and risk control across the entire R&D lifecycle of RM in China [41,

42]. Moreover, the absence of binding domestic legal rules also limits the capacity of the relevant regulatory authorities to enforce ethical oversight and to sanction misconduct, while rapid advancements of RM-related technologies, such as gene editing, present new ethical challenges [43, 44]. Scholars have called for the National Science and Technology Ethics Committee to play a more active role in both shaping ethical governance frameworks and participating in regulatory consultations to address the uncertainties and risks associated with biotechnology generally [44]. In addition to advocating for more effective regulations, scholars have also examined strategies aimed toward improving equitable access to innovative RM products. For example, a review of regulatory procedures specific to RM (including approvals for clinical research and trials and marketing authorization for products or therapies with established safety and efficacy parameters) recommended the introduction of fast-track approval channels to increase accessibility to, and support the timely availability of, safe and effective RM treatments [45].

This chapter aims to build on the existing literature by presenting a more comprehensive assessment of the governance frameworks for RM in China. We will begin by examining Chinese policies that support this emerging field, in order to illustrate the government's roadmap for promoting innovation while managing legal and ethical risks. We then examine the laws and regulations that govern various stages of the R&D and commercialization process, including clinical trials, pre-market approval of new pharmaceutical products, and market oversight of final products. Because current RM R&D often involves human genomics and personal health data, we also analyze laws related to the management of human genetic resources, personal data protection, biosecurity measures, and the qualifications and oversight of medical professionals and related institutions. We also discuss challenges within the existing regulatory framework for RM, such as gaps in bioethical governance, insufficient regulation, and legal barriers to international collaboration. In addition, we explore China's role in advancing RM globally. We propose that China take an active role in improving the accessibility and affordability of advanced biomedical technologies and medicines, and in promoting the creation of an international genomic data alliance that ensures equitable participation from LMICs, as no such framework currently exists.

China's Current Policy and Regulatory Regimes Pertaining to RM

Policy Instruments and Initiatives

Both central and local governments in China have demonstrated a strong commitment to advancing research, innovation, and clinical applications of RM technologies. This is reflected in a series of national and ministerial policy instruments introduced since 2013 (see Table 10.1). As summarized in Table 10.1, these policy measures aim to increase capital investment, foster clinical application and commercialization, promote talent development, design new regulatory frameworks,

Table 10.1 Summary of key Chinese policies related to regenerative medicine research and development

Names of policies	Document release date	Issuing authority or authorities	Summary of content
Notice on Issuing the Medium and Long-Term Plan for Key National Technology Infrastructure Construction (2012–2030)	February 2013	The State Council	This plan supports the development of translational medicine research facilities and emphasizes government backing to swiftly translate biomedical research into clinical diagnostic and treatment technologies. The implementation strategy includes establishing clinical resource databases, medical information systems, personalized medicine technology systems, clinical technology development systems, and cell, tissue, and RM technology systems in line with international standards. The overarching goal of the plan is to provide advanced technological platforms and support for cutting-edge medical research and clinical applications.
Notice on the Corporate Income Tax Preferential Policies and Preferential Directory in Guangdong Hengqin New Area, Fujian Pingtan Comprehensive Experimental Zone, Shenzhen Qianhai Shenzhen-Hong Kong Modern Service Industry Cooperation Zone	March 2014	The Ministry of Finance & The State Taxation Administration	This policy explicitly classifies enterprises involved in researching and producing RM technologies and products, including those focused on tissue and organ replacement, repair, enhancement, or regeneration, as encouraged industries. These enterprises are eligible for a reduced corporate income tax rate of 15%.
Notice on Issuing the "Made in China (2025)"	May 2015	The State Council	This policy aims to actively promote the breakthrough development of new technologies and products in biopharmaceuticals, with a strong emphasis on advancing RM fields such as 3D bioprinting and iPSCs.
Opinions on Reform of the System of Evaluation, Review and Approval of Drugs and Medical Devices	August 2015	The State Council	This policy explicitly promotes the R&D of new drugs, emphasizing the optimization of review and approval procedures for innovative drugs and medical devices. It aims to expedite the review of innovative drugs urgently needed in clinical settings, particularly those targeting diseases such as HIV/AIDS, malignant tumors, and rare diseases. Notably, one of the primary objectives of the policy is to optimize the review and approval procedures for these innovative drugs.

Guiding Opinions on Promoting the Sound Development of the Medical Industry	March 2016	The State Council	This policy explicitly supports the industrialization of innovative drugs and the transformation and upgrading of medical devices, including high-throughput gene sequencing systems and artificial joints and spinal devices, which may be closely related to the development and application of RM. It also emphasizes integrating clinical medication with market demand, focusing particularly on developing innovative therapeutic drugs in oncology, cardiovascular and cerebrovascular diseases, neurodegenerative diseases, high-incidence immune diseases, and rare diseases (defined in China as diseases with a neonatal incidence of less than 1 in 10,000, a prevalence of less than 1 in 10,000, or a total patient population of fewer than 140,000 nationwide).
The 13th Five-Year Plan for the National Economic and Social Development of the People's Republic of China (2016–2020)	March 2016	Deliberated and adopted by the National People's Congress	This national plan underscores China's support for accelerating the development of synthetic biology and RM-related technologies, identifying them as strategic industries. It also emphasizes promoting innovative breakthroughs in cutting-edge fields such as biomedicine.
The National Science and Technology Innovation Plan During the "13th Five-Year Plan" Period	July 2016	The State Council	This policy focuses on supporting R&D in novel biopharmaceuticals, advanced biomedical materials, and the utilization of biological resources. It emphasizes accelerating innovation and application development in key technologies like synthetic biology, biological big data, RM, and 3D bioprinting. The policy also aims to enhance the originality of biotechnology, promote technology transfer, and improve the international competitiveness of the bioeconomy. Additionally, it proposes establishing a platform for sharing biological resources and experimental materials, and strengthening the collection and preservation of research reagents, special human genetic resources, genes, cells, microbial strains, etc. to enhance resource quality, availability, and service standards.

(continued)

Table 10.1 (continued)

Names of policies	Document release date	Issuing authority or authorities	Summary of content
Healthy China 2030 Blueprint	October 2016	CPC Central Committee & The State Council	This blueprint proposes that China increase support within health technology innovation, emphasizing advancements in omics technologies, stem cells, and RM, alongside biological therapies as leading medical technologies. Additionally, the blueprint suggests enhancing key technological breakthroughs in innovative drug development, precision medicine, and smart healthcare.
Guidelines for the Development of the Pharmaceutical Industry	October 2016	The National Development and Reform Commission; The Ministry of Science and Technology; The National Health and Family Planning Commission; The State Food and Drug Administration; etc.	These guidelines emphasize China's priority to advance the creation and use of RNA interference drugs, gene therapy, and related medications, as well as cell-based products, including stem cells and immune cells such as CAR-T therapy. The guidelines further advocate for supporting the R&D of novel medical technologies such as gene sequencing, tumor immunotherapy, and stem cell therapy. Emphasis is placed on improving industry access policies and enhancing clinical application management to facilitate the integration of emerging medical technologies with current clinical requirements.

The 13th Five-Year Plan for the Development of Strategic Emerging Industries (2016–2020)	November 2016	The State Council	This plan delineates China's commitment to advancing the biopharmaceutical industry through accelerated research and clinical application in various technologies, including gene sequencing, large-scale cell cultivation, and research dedicated to targeted and sustained drug release specifically. It also supports the development of biological products such as gene therapy and cell therapy. Additionally, the plan advocates for establishing gene banks, stem cell banks, high-level biosafety laboratories, and protein component libraries. China aims to pioneer new medical models based on stem cells and RM technologies, enhance production technology platforms for cells, tissues, and organs, standardize legal regulations and standards in the stem cell and RM fields, and improve intellectual property assessments and conversion mechanisms. Furthermore, the plan seeks to deepen the clinical application of stem cells and RM technologies, including tumor immunotherapy. Moreover, China aims to establish an independent gene-editing technology system with proprietary intellectual property rights and develop new gene therapy technologies for major genetic diseases, infectious diseases, and malignant tumors.

(continued)

Table 10.1 (continued)

Names of policies	Document release date	Issuing authority or authorities	Summary of content
The 13th Five-Year Plan for the Development of the Biotechnology Industry (2016–2020)	December 2016	The National Development and Reform Commission	This plan details measures aimed at accelerating the research and industrialization of biopharmaceutical products and advanced technologies in RM to meet the country's clinical demands and propel medical advancements. Specifically, the plan supports the innovation, research, and industrialization of biopharmaceuticals such as stem cells and CAR-T immunotherapy. Additionally, the plan promotes integrating biotechnology with material technology and advocates for the application of 3D printing in biomedical, RM, and tissue engineering fields to expedite technological development. Moreover, the plan proposes establishing specialized service platforms aligned with international standards for gene therapy, cell therapy, immunotherapy, and other therapies to facilitate the application and transformation of novel therapeutic technologies. It also supports the development of high-end technical service institutions for gene synthesis, gene editing, and related domains. Furthermore, the plan emphasizes establishing an individualized immune cell therapy technology application demonstration center to standardize and promote individualized immune cell therapy, providing medical institutions with high-quality cell therapy products. The center aims to collaborate with partners to achieve technological breakthroughs and establish platforms focusing on new cellular therapy technologies and viral vector production processes. Additionally, the plan recommends constructing a biotechnology industry standard substance repository to enhance traceability and standard research on protein, nucleic acid, stem cell, and microbial quality control.
The 13th Five-Year Plan for Scientific and Technological Innovation in Medical Devices (2016–2020)	May 2017	The Ministry of Science and Technology	This special plan prioritizes and supports the R&D of new medical devices across various fields, including tissue engineering, RM, molecular imaging, precision diagnosis, and precision therapy.

The 13th Five-Year Plan for Scientific and Technological Innovation in the Health Industry (2016–2020)	June 2017	The Ministry of Science and Technology; The National Health and Family Planning Commission; The State Food and Drug Administration; etc.	This special plan outlines China's focus on key medical technological breakthroughs, targeting malignant tumors, cardiovascular and cerebrovascular diseases, metabolic disorders, and rare diseases. Key technologies for advancement include next-generation gene sequencing, tumor immunotherapy, stem cells and RM, and biomedical big data analysis. The plan also supports establishing precision application solutions and decision support systems, such as early screening and personalized treatment, to enhance the precision and efficiency of disease diagnosis, treatment, and patient management.
The 13th Five-Year Plan for Health and Health Technology Innovation (2016–2020)	June 2017	The Ministry of Science and Technology; The National Health and Family Planning Commission; The State Food and Drug Administration; etc.	This special plan proposes that the Chinese national government supports strengthening basic research and commercialization in cutting-edge RM technologies. It emphasizes advancing research in stem cells and RM (including the scaled cultivation of embryonic, induced pluripotent, and adult stem cells) to facilitate tissue and organ regeneration. The plan also supports the commercialization of technologies such as stem cells, RM, immunotherapy, gene therapy, cell therapy, and genetic manipulation. The plan further aims to explore applications in xenotransplantation, enhancing cellular immunity against viruses, and developing anti-cancer cell capabilities, accelerating the clinical application of these biotechnologies to improve clinical treatment and care.

(continued)

Table 10.1 (continued)

Names of policies	Document release date	Issuing authority or authorities	Summary of content
The 13th Five-Year Plan for National Basic Research (2016–2020)	June 2017	The Ministry of Science and Technology; The Ministry of Education; The Chinese Academy of Sciences; etc.	This specialized plan explicitly supports research on strategic and forward-looking major scientific issues, including stem cells and translational research. It highlights the importance of research on directed differentiation and cellular reprogramming of stem cells, tissues, and organ reconstruction based on stem cells, including the establishment of stem cell repositories. Additionally, the plan emphasizes the clinical translation of stem cells in recognition of both the strategic goal of strengthening China's core competitiveness in this field and the urgent need to address ongoing health needs. The plan proposes strengthening national laboratories and major scientific infrastructure by providing targeted funding (e.g., through national key R&D programs), improving institutional coordination (such as the establishment of inter-agency working mechanisms), and optimizing technology resource sharing and service platforms. It also emphasizes strengthening the training of basic research talents through more national talent development initiatives and reinforcing significant international scientific and technological cooperation via funding schemes and major collaborative projects.
Strengthening the "from 0 to 1" Basic Research Work Program	January 2020	The Ministry of Science and Technology; The National Development and Reform Commission; etc.	This policy explicitly states that the national science and technology plan will continue to support frontier research in nanoscience, stem cells, synthetic biology, protein machinery, and related areas. Additionally, the policy outlines long-term projects aimed at supporting young scientists in their R&D efforts in frontier fields such as life sciences.
Notice on Preferential Enterprise Income Tax Policies for the Guangdong-Macao In-Depth Cooperation Zone in Hengqin	January 2021	The Ministry of Finance & The State Taxation Administration	This notice states that enterprises in cooperation zones that meet the proposed criteria and conditions can benefit from a reduced corporate income tax rate of 15%. Entities which are involved in the research and production of emerging scientific technologies and products—Such as gene sequencing, stem cells, functional proteins, and life health—As well as innovative anticancer drugs, novel cardiovascular and cerebrovascular drugs, new neurologic drugs, antibody drugs, gene therapy drugs, cell therapy drugs, recombinant protein drugs, and nucleic acid drugs—May qualify for this benefit.

The Recommendations on Comprehensively Strengthening the Capacity Building of Drug Supervision	May 2021	The State Council	These policy recommendations propose that China should implement the *Action Plan for Drug Regulatory Science*, which includes enhancing policy research on drug regulation, establishing research bases dedicated to drug regulation science, and advancing the research and application of new tools and standards in drug regulation. According to this initiative, research on drug regulation is encouraged to be integrated into relevant national scientific and technological plans, with a specific focus on supporting regulatory scientific research in biologics, gene therapy drugs, cell therapy drugs, and other related fields. Additionally, the policy expresses China's commitment to enhancing the internationalization of drug regulation through participation in international coordination mechanisms, fostering international exchanges and cooperation, and promoting mutual recognition of drug regulatory standards. In practice, this refers to efforts to align China's regulatory procedures with international norms for certain key products and sectors, as well as to pursue reciprocal agreements or recognition arrangements with major trading partners or regions to facilitate market access and improve regulatory efficiency. While this initiative is not legally binding and does not set specific timelines, it reflects the Chinese government's policy direction for improving its drug regulatory capacity.
The Promotion of High-Quality Development in Public Hospitals (2021–2025)	September 2021	The National Health Commission & The National Administration of Traditional Chinese Medicine	This policy underscores the imperative for public hospitals to prioritize the clinical application of advanced technologies in life sciences and biopharmaceuticals, encompassing precision medicine, RM, artificial intelligence, and 3D printing, among others, to align with clinical needs. It advocates enhancing incentive mechanisms for medical innovation and outcomes evaluation tailored to practical applications. Additionally, the policy encourages public hospitals to take the lead or participate in establishing institutions and centers dedicated to technology transfer and transformation.

(continued)

Table 10.1 (continued)

Names of policies	Document release date	Issuing authority or authorities	Summary of content
The 14th Five-Year Plan for the Development of Pharmaceutical Industry (2021–2025)	December 2021	The National Development and Reform Commission; The Ministry of Science and Technology; The National Health Commission; The National Medical Products Administration; etc.	This plan aims to drive innovation and industrialization in biopharmaceutical products by prioritizing the development of novel and improved drugs targeting recombinant protein therapy. The plan also emphasizes R&D in immunotherapy, stem cell therapy, gene therapy products, and specific immunoglobulins. Additionally, it supports advancements in technologies such as large-scale cell culture and industrial preparation techniques for novel biopharmaceuticals. Another key focus of the plan is to strengthen the talent pool in the biopharmaceutical sector by targeting efforts directly at improving talent training mechanisms and cultivating professionals in drug discovery, clinical trial design, biopharmaceutical manufacturing, and multidisciplinary expertise.
The 14th Five-Year Plan for Bioeconomic Development (2021–2025)	December 2021	The National Development and Reform Commission	This plan strongly supports innovative frontier biotechnologies, including gene diagnosis and therapy, stem cell therapy, and immunotherapy. It promotes translating these innovations into products and clinical applications to advance models in RM and precision medical treatment. The plan also encourages developing new technologies such as biocomputing and DNA storage, alongside advancing biopharmaceuticals like recombinant proteins, peptides, cell-based therapies, and gene therapies. The plan underscores the necessity of continuously updating and formulating technical standards within the state by specifically focusing on clinical applications such as stem cell therapy, cellular immunotherapy, and gene therapy products. It highlights the need to establish both research bases for drug regulation science and platforms for quality and safety assessments to scientifically conduct clinical evaluations. However, the plan does not specify exact locations; such research bases may potentially be piloted in regions with well-developed biomedical translation capacities or initiated within leading universities and public research institutions, subject to future implementation details.

Opinions on Several Special Measures for Relaxing Market Access in Shenzhen's Building of a Pioneering Demonstration Zone for Socialism with Chinese Characteristics	January 2022	The National Development and Reform Commission & The Ministry of Commerce	This policy proposes Shenzhen as a pilot city to explore relaxed approval processes for human genetic resources, including establishing a management platform to approve human genetic resources and supporting R&D in novel medical products like stem cell therapy, immunotherapy, and gene therapy. Additionally, the policy recommends streamlining approval procedures for international collaborations involving human genetic resources, suggesting risk-based classification and management for research samples such as human tissues and blood at entry and exit points. Expedited clearance procedures for low-risk items would be implemented while ensuring biosafety. Furthermore, the policy encourages Shenzhen to establish regional ethics centers meeting international standards to guide ethical reviews at clinical trial institutions. It supports these centers in accepting ethical review commissions from institutions lacking such capabilities. Additionally, the policy promotes collaborations between Shenzhen's medical institutions and Contract Research Organizations to enhance the clinical trial and quality management capabilities of medical technologies.
The 14th Five-Year Plan for the Development of Health and Medical Talents (2021–2025)	August 2022	The National Health Commission	This plan details China's strategy to leverage national major projects and significant talent programs to cultivate and identify individuals possessing strategic acumen, profound scientific expertise, and forward-thinking judgment in critical biomedical fields. These fields include areas such as stem cells and RM, novel vaccines, and biopharmaceuticals.

(continued)

Table 10.1 (continued)

Names of policies	Document release date	Issuing authority or authorities	Summary of content
Guiding Opinions on Enhancing Clinical Specialty Capacity Building	July 2023	The National Health Commission	These guiding opinions state that all Chinese medical institutions should prioritize addressing clinical practical issues and enhance innovation, applied research, and the transformation of clinical diagnosis and treatment technologies. Specifically, eligible medical institutions are urged to innovate and achieve breakthroughs in cutting-edge areas of biomedicine, such as RM and precision medicine. This innovation is encouraged through multiple mechanisms, including establishing institutional development plans, improving performance-based funding distribution, promoting interdisciplinary collaboration, and optimizing internal governance structures. Moreover, the guiding opinions underscore the importance of achieving efficient clinical translation and independently developing new technologies, products, drugs, and therapies.
The Implementation Plan for the Comprehensive Pilot Program of Further Opening Up the Service Industry in Wuhan City	September 2023	Wuhan Municipal Government	This local policy explicitly supports the construction of a national-level human genetic resource sample bank in the province of Hubei's capital, Wuhan, focusing on umbilical cord blood hematopoietic stem cells, mesenchymal stem cells, and immune cells. Additionally, the policy encourages Wuhan to explore the establishment of a big data center and collaborate with biopharmaceutical companies to share data, thereby supporting new drug development and disease research efforts.

The Work Plan to Support Beijing in Pursuing the Initiative for the National Comprehensive Demonstration Zone for Further Opening-Up of the Service Sector	November 2023	The State Council	This policy supports Beijing in exploring equity incentive mechanisms for foreign, Hong Kong, Macao, and Taiwan personnel in pharmaceutical R&D enterprises focused on stem cells and genetics. It also encourages qualified medical institutions to conduct clinical trials involving stem cells and related treatments. Moreover, the policy promotes international cooperation in stem cell and genetic R&D, facilitating management services for human genetic resources by enterprises from Hong Kong and Macao in Beijing. Additionally, the policy aims to develop guidelines and important data directories for bioindustry data classification and categorization to clarify identification standards for critical data (including medical data, biological and genetic information, and data related to autonomous driving) and to ensure robust data security protection. This initiative was launched in response to a formal request jointly submitted by the Beijing Municipal Government and the Ministry of Commerce. Beijing was selected not only because it is China's capital but also because of its strategic positioning and administrative capacity to pilot comprehensive institutional reforms. The central government approved the plan to enable Beijing to lead in service sector liberalization and to generate replicable experiences.

(continued)

Table 10.1 (continued)

Names of policies	Document release date	Issuing authority or authorities	Summary of content
Opinions on Special Measures for Supporting the Guangdong-Macao In-Depth Cooperation Zone in Hengqin in Relaxing Market Access	December 2023	The National Development and Reform Commission & The Ministry of Commerce	This policy explicitly supports the research and application of new medical technologies involving stem cells. It includes supporting relevant institutions and laboratories within the cooperation zone (such as those located in the Hengqin area of Zhuhai City, adjacent to Macao) to establish qualification standards for the quality review and inspection of stem cell preparations. The policy also encourages the introduction and construction of third-party drug inspection institutions and laboratories specializing in cell therapy product testing. Furthermore, it supports pharmaceutical enterprises within the cooperation zone to conduct international collaborative clinical trials and encourages medical institutions to utilize advanced testing methods for health analysis and diagnosis. Additionally, the policy proposes streamlining approval procedures for activities involving human genetic resources in international collaborative clinical trials. Specifically, for activities not involving the export of human genetic resource materials, the authority for filing can be delegated from the national-level competent authority to the provincial-level competent authority.

Other policy documents supporting the research and translational application of RM-related technologies such as immune cells and stem cells in pilot areas approved by the state include: the Framework Plan for China (Hainan) Pilot Free Trade Zone (released by the State Council in October 2018); the General Plan for Six Newly Established Pilot Free Trade Zones (released by the State Council in August 2019); the Implementation Plan for Supporting the Construction of Hainan Bo'ao Lecheng International Medical Tourism Pilot Zone (released by the National Development and Reform Commission, the National Health Commission, the National Medical Products Administration, and other ministries in September 2019); and, the Framework Plan for the Comprehensive Pilot Program of Further Opening up the Service Sector in Shanghai Municipality (released by the Ministry of Commerce in April 2021). These documents represent key national-level initiatives, but additional regional policies continue to emerge as China places increasing emphasis on advancing RM innovation and translational applications.

introduce tax incentives for relevant enterprises, and establish special pilot zones with relaxed industry accessibility standards.

China has consistently designated RM-related technologies as crucial to related key research areas and has funded these technologies accordingly through state-backed programs. For instance, the *Medium and Long-Term Plan for Key National Technology Infrastructure Construction (2012–2030)* released by the State Council in 2013, along with initiatives such as the National Key Basic Research and Development Program, the National High-Tech Research and Development Program, and the National Science and Technology Support Program issued by the Ministry of Science and Technology (MOST) in 2014, have all explicitly supported the research and application of cutting-edge technologies, including RM. These plans and programs support R&D primarily through dedicated funding allocations. Furthermore, the MOST has issued guidelines for special projects, such as the National Key Research and Development Program for Stem Cell Research and Organ Repair, which provides funding for young scientists and foreign scientists employed by mainland institutions [46, 47]. Our review of China's policy landscape reveals that projects focusing on stem cell and organ repair, biomaterials, tissue and organ replacement, as well as gene and cell therapy, are regularly included in national funding initiatives (see Table 10.1). These funding programs support various actors along the innovation chain, including both public institutions (such as universities and public research institutions) and private enterprises, so long as the institutions or enterprises meet eligibility criteria regarding research capacity and legal status [48, 49].

With the support of national funding, especially funding allocated under the National Key Research and Development Program for Stem Cell and Translational Research during the 13th Five-Year Plan (2016–2020), multiple national key laboratories and stem cell germplasm resource banks have been established, built to match the quality standards of leading countries in the field of RM [50]. Research teams in China's public sector have reported groundbreaking advancements in stem cell research. These advancements include studies on guiding human pluripotent stem cells toward liver cell lineages and improving immune tolerance and immune system reconstruction following hematopoietic stem cell transplantation [51, 52].

Evidence also suggests that research led by private enterprises has likewise benefited from national funding programs [49, 53]. Notable examples include the establishment of an induced pluripotent stem cell (iPSC) "super donor" bank by a private biotechnology firm, and the development of iPSC-derived cell therapy products that have since entered the drug registration process [54]. These therapies are expected to offer new treatment options for tissue and organ repair, as well as for a range of challenging diseases, including Parkinson's disease, diabetes, and malignant tumors. In addition, stem cell transplantation and cell storage projects have been launched within nationally funded industrial parks—specialized innovation zones established to advance the research, clinical translation, and commercialization of cutting-edge medical technologies—with active participation from private enterprises [55].

Furthermore, Chinese policies prioritize clinical translation and real-world applications of stem cell and RM technologies. National policies such as the *Guiding*

Opinions on Promoting the Sound Development of the Medical Industry and *the 13th Five-Year Plan for Scientific and Technological Innovation in the Health Industry (2016–2020)* emphasize the development of novel RM treatments for rare diseases, cardiovascular problems, tumors, and other medical conditions. In China, rare diseases are typically defined as conditions that meet one or more of the following criteria: an incidence rate of less than one in 10,000 among newborns, a prevalence rate of less than one in 10,000 in the general population, or a total patient population of fewer than 140,000 nationwide [56]. These policies encourage both public institutions and private enterprises to participate in the clinical application of RM technologies by providing targeted funding programs, promoting public–private and international partnerships, and streamlining procedures for clinical trials and product registration.

Some more recent policies focus specifically on strengthening the capacities of public-sector healthcare institutions. For example, the issuance of the *Guiding Opinions on Enhancing Clinical Specialty Capacity Building* by the National Health Commission (NHC) in July 2023 emphasizes public medical institutions' roles in RM-related clinical research and translation. This policy aims to align research with clinical needs by promoting demand-oriented research planning, facilitating collaboration between clinical and research teams, and supporting translational activities within public medical institutions, thereby improving patients' access to innovative therapies. Moreover, policy initiatives such as the *Promotion of High-Quality Development in Public Hospitals (2021–2025)* have introduced incentive mechanisms and evaluation systems for the translation of RM research achievements.

Policies at both the central and local levels of government prioritize the commercialization and industrial development of RM. Together, the policies outline a strategic approach to commercialization involving expedited approval procedures for eligible innovative RM drugs (and related medical devices) and enhanced support for products like CAR-T cell therapies to enter the market [57–59]. Recent policies also focus on establishing gene banks and stem cell banks and promoting data sharing among research institutions and compliant biopharmaceutical companies [60, 61]. Such proactive policy initiatives have accelerated the introduction of cutting-edge RM technologies and medications. For example, in early 2025, China approved its first stem cell-based drug, Amimestrocel Injection, for the treatment of graft-versus-host disease (GVHD), following an expedited review process supported by recent national regulatory reforms [62].

Several Chinese national policies, including the *Guidelines for the Development of the Pharmaceutical Industry* and the *13th Five-Year Plan for the Development of Strategic Emerging Industries (2016–2020)*, emphasize the enhancement of entry standards and conditions for the RM industry. These policies also call for improving the management of clinical applications of RM technologies, refining evaluation and transfer mechanisms to strengthen intellectual property rights protection, and modifying national regulatory systems to better accommodate innovative medical technologies [58, 60]. Additionally, the *Action Plan for Drug Regulatory Science* issued by the State Council of China has proposed measures to strengthen research, international cooperation, and mutual recognition of regulatory mechanisms for gene and cell therapies. Taken together, these initiatives aim to align China's regulatory practices with international standards developed by organizations such as the WHO and the ISSCR, and to ensure the safety of RM technologies and products. They are also expected to expand development opportunities for RM and enhance China's ability to compete at the international level in the biotech sector.

China's latest national strategic plan, namely, the *14th Five-Year Plan for the Development of Health and Medical Talents (2021–2025)* prioritizes talent development in the field of RM to build capacity for R&D. This talent strategy supports the cultivation of new health industry talent through university and postgraduate programs, while strengthening the capacities of existing researchers via targeted funding, competitive salary schemes, and clear promotion pathways. China continues to provide tax incentives for the biotech industry and support commercialization through expedited and priority approval processes for certain innovative drugs and new therapies—particularly those targeting life-threatening diseases, rare conditions, or urgent public health needs—as well as international cooperation [63–67]. These policies and measures collectively advance China's RM industry through enhanced support for discovery and innovation and improved access to related products and technologies within the country.

Regulatory Regimes

So far, no dedicated national law has been established to specifically regulate clinical research or the translation of RM in China. Instead, RM research and clinical translation are primarily governed by administrative regulations, issued by government agencies such as the NHC or the MOST, to guide specific areas of biomedical R&D and clinical oversight. As early as 1993, the State Scientific and Technological Commission of China (later reorganized as the MOST in 1998) issued the *Measures on the Safe Management of Genetic Engineering*, which remain in effect today. These measures established requirements for the classified declaration and safety grade assessment of genetically related research, experiments, and products.

RM research in China has been, and continues to be, subject to rigorous safety assessments and regulatory approval before proceeding. At the same time, the former Ministry of Health (MoH) and the former State Food and Drug Administration (SFDA) issued regulations on somatic cell therapy and gene therapy, within the scope of drug regulations [68, 69]. These regulations remain valid, and health administrative departments and drug regulatory authorities are designated as the entities responsible for relevant reviews and supervision. The SFDA, a former regulatory body in China, was dissolved in 2018, and its responsibilities were transferred to the State Administration for Market Regulation (SAMR). The National Medical Products Administration (NMPA), which operates under the jurisdiction of the SAMR, is responsible for ensuring the safety and quality of drugs, medical devices, and cosmetics. Its mandates include developing regulatory policies, drafting laws and regulations, and overseeing their enforcement. Additionally, the NMPA conducts research and formulates policies to enhance the management and service of new technologies and products in the pharmaceutical, medical device, and cosmetics sectors. Furthermore, it regulates the qualifications of practicing pharmacists, establishes entry requirements for pharmacists, and guides and supervises the pharmacist registration process.

Early regulations regarding RM and related fields in China also targeted the use of, and research surrounding, umbilical cord blood hematopoietic stem cells. These regulations establish the framework for qualification requirements for personnel and practitioners, licensing standards for institutional practice, procedures for umbilical cord blood storage, and an explicit ban on profit-seeking activities in this field [70, 71]. While subsequent rules issued by the NHC and other agencies have introduced updated provisions in this field, the original regulatory framework remains in force. Additional regulations and ethical guidelines have also been issued to strengthen quality control in assisted reproductive technology and human embryonic stem cell research [72, 73]. These regulations establish principles of informed consent and requirements for ethical review and explicitly prohibit practices such as the buying and selling of gametes, fertilized eggs, and embryos [72, 73]. Collectively, these measures form a foundational framework for safety assessment and bioethical governance in early-stage R&D involving high-risk RM materials, such as umbilical cord blood hematopoietic stem cells, gametes, fertilized eggs, and embryos. While not comprehensive, these regulations demonstrate the core principles that guide oversight and risk management throughout the RM sector.

Regulations on Research and Clinical Translation of RM-Related Technologies

For the clinical translation of RM-related technologies and products, the former MoH issued a series of management measures and guidelines. Based on these regulations, medical institutions and healthcare personnel must refrain from using RM technologies that have not been approved or clinically verified by the relevant health authorities or drug regulatory agencies [72, 74]. RM-related research projects conducted in hospitals must undergo ethical review and approval by the hospital's ethics committee, adhere to medical ethical principles such as those outlined in national and international ethical standards, and fully respect patients' rights to informed consent and autonomy [72, 74]. These measures remain in effect and continue to provide the regulatory foundation for current clinical and ethical oversight of RM research in China.

The *Measures for the Administration of the Clinical Application of Medical Technologies*, enacted in 2009, provided an early framework for regulating the clinical application of RM-related medical technologies [75]. During that period, RM therapies could be classified as medical technologies subject to administrative approval. The measures categorized cloning therapy technology, autologous stem cell and immune cell therapy technology, gene therapy technology, and other technologies related to reproductive organs as Class III medical technologies, which must undergo additional safety assessment, efficacy evaluation, and an ethical review process conducted by the former MoH before they can be used in clinical practice for the first time [75]. Under that regulatory framework, technologies such as xenogeneic stem cell therapy, xenogeneic gene therapy, and human somatic cell cloning technology were prohibited from clinical application. Although no explicit announcement was made to lift this ban, the 2009 *Measures for the Administration of the Clinical Application of Medical Technologies* were formally declared invalid by the NHC in 2024 [76]. The 2009 measures have since been replaced by the updated 2018 version of the *Measures for the Administration of the Clinical Application of Medical Technologies*, which no longer includes specific bans on technologies such as xenogeneic stem cell therapy, xenogeneic gene therapy, or human somatic cell cloning. Instead, the 2018 version prohibits clinical use under four general conditions: when safety or efficacy is uncertain, when significant ethical concerns exist, when the technology has been clinically phased out, or when it has not undergone appropriate clinical validation [77].

In May 2009, the former MoH issued the *Notice on the Release of the First Batch of Catalogue of Class III Medical Technologies Approved for Clinical Application*, which permitted various RM treatment methods, such as autologous immune cell therapies, hematopoietic stem cell applications (including cord blood), and tissue-engineered transplants, to be approved as Class III medical technologies. This catalogue encompasses various RM-related technologies, including autologous immune cell therapy (T cell, NK cell), cell transplantation therapy (excluding stem cells), umbilical cord blood hematopoietic stem cell therapy, hematopoietic stem cell therapy (excluding umbilical cord blood stem cell therapy), tissue-engineered tissue transplantation therapy, and others [78]. However, the catalogue did not include any RM technologies involving reproductive organs, such as ovarian stem cell therapy, autologous testicular cell therapy, or reproductive tissue engineering. The former MoH also developed several management standards and solicited public opinions to enhance the oversight of clinical applications of RM-related medical technologies. Examples of these management standards include the *Management Specifications for Autologous Immune Cell (T Cell, NK Cell) Therapy (Draft for Comment)*, the *Management Specifications for Cell Transplantation Therapy (Draft for Comment)*, the *Management Specifications for Umbilical Cord Blood Hematopoietic Stem Cell Therapy (Draft for Comment)*, and the *Management Specifications for Tissue-Engineered Tissue Transplantation Therapy (Draft for Comment)* [79].

Despite these efforts by the former MoH, a comprehensive regulatory framework on the clinical application of RM had not been established at the national level. The lack of clear procedures and standards for approval, supervision, and evaluation of RM therapies resulted in the extensive adoption of unverified and unregulated RM therapies in Chinese medical institutions [39].

To address this regulatory gap, in 2011, the former MoH mandated that all medical institutions and research entities engaged in RM research and applications (such as stem cell therapies) must conduct self-inspections and implement self-correction measures in accordance with the (jointly issued by the former SFDA and the former MoH) and the *Measures for the Administration of the Clinical Application of Medical Technologies* (issued by the former MoH) [80]. A national task force was jointly established by the former MoH and the former SFDA, supported by provincial working groups, to oversee local implementation. Participating institutions were required to complete standardized self-assessment forms, summarize past and ongoing research, identify compliance issues, and propose rectifications. These reports were submitted to provincial authorities and ultimately compiled at the national level. Medical institutions were prohibited from making unauthorized changes to clinical trial protocols, and any stem cell clinical research or application projects lacking approval from the former MoH and the former SFDA were to be suspended [80].

Subsequently, health and drug administration departments in China have issued multiple administrative measures and drafts, introducing strict quality control standards and detailed normative principles to further regulate clinical trials and preclinical studies involving RM [81–83]. According to these regulations, medical institutions conducting RM-related clinical research are required to set up special clinical research management committees and ethics committees and must keep written review records and opinions. In instances where significant safety risks or breaches of ethical principles occur, the research projects are to be suspended or discontinued and promptly reported to the local health authority [82]. If institutions violate these regulations, they are subject to penalties for all such violations. For example, violations related to clinical trials and RM research, such as unauthorized collection of fees from subjects, will result in strict disciplinary measures. Specifically, the clinical trial research base's qualification will be revoked, and legal actions will be taken against responsible parties [84]. The relevant local health and drug regulatory authorities determine and impose penalties for violations in accordance with national laws and regulatory guidelines. Assessment criteria for penalties included, but are not limited to: failure to correct non-compliance, violations of research integrity or ethical principles, inflicting harm on donors or participants, unauthorized collection of research-related fees, and the commercial promotion of unapproved stem cell therapies [82, 83].

In 2015, following a State Council directive on administrative approval reform, the National Health and Family Planning Commission (NHFPC)—the successor to the former MoH—formally discontinued the Class III medical technology pathway [85]. The accompanying technology permission catalogue was revoked. Following this regulatory change, RM-related therapies such as stem cell-based interventions have been reclassified into two distinct pathways: being regulated as a medicine subject to China's pharmaceutical approval system or being managed as clinical research requiring prior scientific and ethics reviews and official filing [39]. Accompanying this reclassification, the NHFPC and the former SFDA jointly issued the *Management Measures for Stem Cell Clinical Research (Provisional)*, providing a clearer and more standardized framework for compliance and ethical review in stem cell R&D in China. This jointly issued regulation requires institutions planning to conduct clinical stem cell research to carry out in-house scientific and ethics committee reviews before starting the project [83]. The results of the in-house reviews must then be submitted to the respective provincial health and drug regulatory authorities for assessment, after which these authorities will file the materials with national-level regulatory bodies [83, 86].

In addition, institutions conducting stem cell research must register their stem cell research projects in the national medical research registration system. Researchers are required to submit annual reports through this platform to maintain compliance. In the event of premature termination, the institution must immediately notify regulatory authorities, providing both the reasons for discontinuation and details of any remedial measures taken. The accompanying guidelines also stipulate that the quality control standards for stem cell preparations must comply with China's *Good Manufacturing Practices for Pharmaceutical Products*, as defined by the former MoH [81]. In 2015, the MOST also issued a service guide regarding the use of human genetic resources in research, clinical trials, and international cooperation involving RM. This service guide specifies that, where any activities, such as RM research, fall under the above uses, administrative permission is required from the MOST before the projects can begin [87].

The NHFPC, the former SFDA, and the MOST have continued to standardize the regulatory approach for RM. These institutions have introduced detailed technical management regulations, refined administrative approval processes, and enhanced ethical review methods for biomedical research involving humans and biotechnology. For instance, the *Measures for the Ethical Review of Biomedical Research Involving Humans*, issued in 2016 and still in force, require that ethics committees established within medical and healthcare institutions play a central role in overseeing biomedical research involving human subjects. Although these measures do not specifically address RM, RM-related research and clinical trials that involve human participants are subject to the same ethical oversight requirements. Medical institutions that do not have ethics committees are not permitted to conduct RM-related research and clinical trials [88]. Ethics committees are responsible for reviewing proposed research projects to ensure ethical integrity, including evaluating whether they meet principles such as obtaining informed consent, protecting participant privacy, selecting participants fairly, refraining from charging fees, providing appropriate compensation for reasonable expenses, offering special safeguards for vulnerable populations, and assessing the risk-benefit ratio. For research projects that have received approval, designated committee members are responsible for conducting continuing review to monitor ethical compliance during implementation [88]. Ethics committees are also granted the authority to suspend or terminate projects in cases where significant safety risks or ethical violations are identified [88].

Lastly, the *Regulation on the Prevention and Handling of Medical Disputes*, established in 2018 by the State Council of China, states that any medical institutions and personnel involved in the use of new medical technologies must undergo technical assessments and ethical review processes. Violations of these regulations or ethical standards may result in penalties ranging from fines and temporary practice suspensions to more severe consequences such as revocation of practice licenses and potential criminal liability [89].

Regulations for Genetic Resources Management

Genetic resources have played a crucial role in innovation within the field of RM. In China, both the *Regulation on the Administration of Human Genetic Resources* and its *Detailed Rules for Implementation* emphasize the legal and ethical management of human genetic resources [90, 91]. These regulations and rules establish provisions for international collaborative research and strictly prohibit the buying and selling of human genetic resources. They further stipulate that the collection, utilization, and provision of Chinese human genetic resources to foreign entities must adhere to ethical principles and undergo ethical review processes [90, 91]. Where RM-related activities involve such uses, these requirements would likewise apply.

Foreign entities and individuals are prohibited from collecting or preserving human genetic resources within China or exporting Chinese genetic resources without proper authorization [90, 91]. However, foreign collaborations with Chinese research institutions or enterprises are permitted under the regulation. RM R&D projects involving Chinese genetic resources may fall within the scope of such permissible collaborations. Notably, the Chinese Ministry of Commerce has issued a list that restricts foreign access and investment in the field of RM. The list includes areas such as human stem cells, as well as tools for genetic diagnoses and related treatment technologies. For instance, in 2020, the Ministry of Commerce of China updated the *Agreement Concerning Amendment to the Agreement on Trade in Services of the Mainland and Hong Kong Closer Economic Partnership Arrangement* [92]. This revision included provisions prohibiting Hong Kong service providers from engaging in the development and application of human stem cells, gene diagnosis, and treatment technologies. Furthermore, in 2021, the *Special Administrative Measures (Negative List) for the Access of Foreign Investment in Pilot Free Trade Zones (2021)* further strengthened these restrictions. The *Negative List* explicitly prohibits foreign investors from investing in the fields of human stem cells and gene diagnosis and treatment technologies.

National Laws Relevant to RM

In addition to the policies and regulations discussed above, Chinese national legislation passed by the Standing Committee of the National People's Congress also applies to RM-related research and application. For instance, *Amendment (XI) to the Criminal Law of the People's Republic of China* (*Criminal Law*), enacted in 2020, categorizes the implantation of gene-edited and cloned embryos as a criminal offense [93]. The *Civil Code of the People's Republic of China* (*Civil Code*), which came into effect in its entirety in January 2021 as the country's first unified civil code, also imposes legal constraints on the field of RM, specifically the ethical and lawful use of human biological resources [94]. The code explicitly prohibits transactions involving the buying and selling of human cells, tissues, and organs. Furthermore, pursuant to Article 1008 of the *Civil Code*, regarding the development of new drugs, medical devices, and clinical trials for new treatment methods, approval from relevant regulatory authorities and ethics committees is mandatory [94]. Participants (or their guardians, where applicable) must be fully informed about the purposes, procedures, and potential risks of the trial. Participants, or their guardians, are also required to provide voluntary written consent, and no fees may be charged for participation in clinical trials [94].

The *Biosecurity Law of the People's Republic of China (Biosecurity Law)*, implemented in April 2021, introduces a risk-based classification system for biotechnology. The *Biosecurity Law* marks a significant step toward systematically integrating formerly fragmented bioethical and biosafety provisions into a comprehensive legal framework for biosecurity governance. The law reaffirms fundamental requirements that have long been in effect, including strict adherence to ethical principles, mandatory ethical review procedures, and professional qualification standards for engaging in RM research and clinical applications. For example, according to its requirements, RM projects are expected to comply with ethical principles and undergo ethics review within accredited medical institutions. Clinical procedures involving human subjects in RM should be performed by qualified healthcare professionals. Additionally, institutions engaged in RM R&D must establish internal biosafety training, monitoring, and reporting systems. Penalties will be enforced in cases of illegal research or non-compliance with applicable biosafety regulations [95].

Furthermore, the *Biosecurity Law* reiterates restrictions on providing Chinese human genetic resources to foreign entities (or individuals) as well as restrictions on the collection, preservation, or use of these resources by foreign entities (or individuals). International collaboration involving Chinese genetic resources in RM requires approval from national regulatory authorities to ensure a standardized process for all RM-related genetic resource projects [95]. This standardized process also provides a more transparent framework, promoting international cooperation in RM clinical studies and trials. The framework also mitigates concerns surrounding the provision of Chinese human genetic resources to foreign entities or individuals by providing express regulations and procedures concerning interaction with these groups.

The enactment of China's *Personal Information Protection Law of the People's Republic of China (PIPL)* in November 2021 has significantly strengthened the framework for safeguarding personal information for patients, subjects, or trial participants involved in RM-related R&D [96]. Institutions and practitioners involved in RM-related clinical research, trials, or therapies must strictly adhere to the *PIPL* when handling sensitive personal information such as biometric and medical health data. Specifically, before conducting RM-related treatments or clinical studies and trials, patients, subjects, or trial participants must be informed in advance about the purpose, use, potential impact, privacy protection measures, as well as their rights to voluntarily participate and withdraw from the collection and processing of personal information. Additionally, the *PIPL* states that the processing of personal data can only proceed after ascertaining explicit consent of the patient, subject, or trial participant.

The *Law on Doctors of the People's Republic of China (Law on Doctors of China)*, adopted on August 20, 2021, and effective as of March 1, 2022, is national legislation that comprehensively regulates physicians' rights, duties, and ethical obligations [97]. It replaced the *Law on Licensed Medical Practitioners of the People's Republic of China* [97]. In addition, the *Regulation on the Administration of Medical Institutions*, issued by the State Council, explicitly mandates that physicians disclose both medical risks and alternative treatment options to patients and obtain informed consent from them [98]. Both regulatory instruments require physicians and medical institutions to obtain valid medical licenses before conducting RM-related clinical treatments, and they prohibit non-healthcare technical personnel from participating in such treatments within medical institutions. Medical institutions must also obtain the appropriate licenses in accordance and are required to conduct diagnoses and treatments strictly in line with their approved registrations or filed treatment protocols [97, 98]. An ethics review by an ethics committee established within the medical institution is mandatory for all RM-related clinical studies [97].

Violations, such as failing to properly notify patients of medical risks or alternative treatment options, failing to obtain informed consent, or engaging in the clinical application of prohibited medical technologies, will result in administrative penalties imposed on the responsible individual or institution by the applicable health administrative department. These penalties may include official warnings, fines, or even revocation of medical licenses [97, 98].

Other laws in China, such as the *Basic Healthcare and Health Promotion Law of the People's Republic of China* and the *Law of the People's Republic of China on Scientific and Technological Progress (Law on Scientific and Technological Progress)*, emphasize the crucial role of national legislative support surrounding innovative technologies [99, 100]. These laws actively promote the R&D of new technologies and products, including RM, in alignment with the principles of inclusivity and caution, which reflect the careful balance between promoting technological advancement and maintaining ethical and safety standards. For instance, Article 5 of the *Law on Scientific and Technological Progress* provides for technological risk prevention and management. Furthermore, Article 103 requires institutions to establish science/technology and ethics review mechanisms, and Article 107 prohibits any research activities that endanger human health or violate scientific ethics. It also includes various incentive measures relevant to the development of RM, such as tax exemptions. Generally, these national laws provide support to institutions, universities, social organizations, and researchers in advancing cutting-edge technological developments. These supports also promote international cooperation and participation in the formulation of relevant international standards in the field of RM [99, 100].

Pursuant to Articles 32, 33, and 44 of the *Basic Healthcare and Health Promotion Law*, researchers and institutions conducting RM clinical trials are subject to stringent legal requirements, including obtaining informed consent from patients, ensuring privacy protection, and securing ethical review approvals [99]. The *Law on Scientific and Technological Progress* specifically emphasizes China's commitment to progressively enhancing the protection of intellectual property rights and safety review mechanisms in international scientific cooperation. These laws not only provide comprehensive national legislative support for RM alongside existing policies but also serve to highlight the imperative of integrating ethical and safety considerations in technological innovation within the country.

Local Legislation for RM

At the local legislative level, numerous provinces and cities have established rules for governing RM research and its application. For example, in 2023, the city of Shenzhen enacted the *Regulations on the Promotion of the Cellular and Genetic Industry in the Shenzhen Special Economic Zone*, which comprehensively outline key processes in the city's cell and gene technologies sector [101]. These regulations cover critical aspects, including cell collection and storage, R&D of cell and gene products, market approval procedures, and product manufacturing. A significant feature of this local legislation is its provision for expanded access to clinical trials for cell and gene products, specifically targeting patients who may not qualify for conventional trials, thus expanding patient access to RM therapies. It also emphasizes that fulfilling ethical review requirements, obtaining informed consent, and meeting relevant approval conditions are prerequisites for expanded access to clinical trials for cell and gene products. The regulations also support the use of safety data obtained from expanded access to investigational cell and gene therapies as crucial references for subsequent drug approval processes.

The Hainan Province offers another notable example of regional legislative progress within the field of RM. The province has designated the Boao Lecheng Pilot Zone (located in Qionghai City) as a hub for biomedical development. As the only officially designated "special medical zone" in China, the Boao Lecheng Pilot Zone serves as a national testbed for healthcare system reforms, including expedited patient access to cutting-edge biomedical technologies such as stem cell therapies. It also facilitates and supports international collaboration in both biomedical research and clinical translation.

In November 2024, Hainan enacted the *Regulations on Promoting Biomedical New Technologies in the Boao Lecheng International Medical Tourism Pilot Zone of the Hainan Free Trade Port* [102]. As a key component of China's national strategy to promote trade liberalization and innovation, the Hainan Free Trade Port initiative is designed to cultivate a globally competitive, business-friendly ecosystem. This includes implementing streamlined regulatory frameworks specifically for priority sectors like healthcare and tourism, while fostering greater international connectivity and openness [103]. The regulations govern the conduct and oversight of clinical research as well as the translational application of innovative biomedical technologies within the Boao Lecheng Pilot Zone, including cell therapies, gene therapies, and tissue engineering. As part of its institutional support framework, the regulations promote the establishment of biomedical research facilities in the pilot zone and encourage local biomedical R&D and manufacturing entities to build collaborative partnerships with medical institutions, universities, and industry stakeholders at both the domestic and international levels. Notably, to safeguard patient rights, the regulations encourage the development of insurance products that cover the costs of new biomedical treatments and provide financial compensation for subjects or patients in cases of injury or death resulting from biomedical clinical research or translational applications.

Building on this regulatory framework, the Hainan Provincial Health Commission, the Hainan Medical Products Administration, and the Boao Lecheng International Medical Tourism Pilot Zone Administration jointly issued the *Interim Measures for the Implementation of Translational Applications of Biomedical New Technologies in the Boao Lecheng International Medical Tourism Pilot Zone of the Hainan Free Trade Port* in January 2025 [104]. These measures serve as the implementing mechanism for the 2024 Regulations by detailing the application, review, and supervisory procedures necessary for the translational use of advanced biomedical therapies. Importantly, the measures introduce a risk-based classification and management system to ensure that oversight is proportionate to the potential risks associated with each RM translational project.

In February 2025, the Hainan Provincial Health Commission and the Hainan Medical Products Administration further issued the *Administrative Provisions on the Clinical Research of Biomedical New Technologies in the Boao Lecheng International Medical Tourism Pilot Zone* [105]. These provisions specify the conditions required to initiate clinical research involving biomedical technologies within the pilot zone, outline detailed procedures for project registration and approval, and define the qualification requirements for both sponsoring institutions (typically hospitals or research centers responsible for overseeing the conduct of the research) and principal investigators (the lead researchers who design and manage the clinical studies). The provisions emphasize that all clinical research must be guided by a dual focus on innovation and quality assurance, be scientifically justified and ethically sound, and must not endanger human health, biological safety, or the public interest.

While these provisions apply only within the Boao Lecheng Pilot Zone and lack nationwide legal force, they represent a meaningful progress in institutionalizing ethical and scientific norms for RM in China. Their implementation will likely help to establish approval pathways for RM products, increase transparency across both research and clinical translation processes, and define clearer accountability mechanisms for both sponsoring institutions and principal investigators.

In summary, although China has not enacted a comprehensive national law specifically targeting RM research and its application, the Chinese government has established a regulatory framework that consists of administrative regulations and applicable national laws (see Tables 10.2 and 10.3). This framework currently serves as the regulatory foundation and offers specific guidance for RM-related research, clinical trials, and clinical translation. It establishes an ethical oversight system and mandates ethical review as a legal prerequisite for drug trials, biomedical innovation, and all clinical research involving human subjects. The framework also promotes advances in RM in a sustainable and responsible manner,

Table 10.2 Summary of current laws and regulations applicable to regulating regenerative medicine research and development in China

Names of laws and regulations	Document release date	Issuing authority or authorities	Summary of content
Regulation on the Prevention and Handling of Medical Disputes	July 2018	The State Council	This regulation mandates that medical institutions provide medical technology services commensurate with their capabilities to ensure clinical safety and manage medical risks. Introducing new technologies requires technical assessments and ethical reviews to ensure safety, efficacy, and adherence to ethical standards. Medical personnel must comprehensively inform patients about their conditions and procedures during diagnosis and treatment. For surgeries or high-risk procedures, timely disclosure of risks and alternatives, along with obtaining written consent, is mandatory. Institutions performing high-risk activities must develop contingency plans to address any unforeseen risks. Non-compliance with disclosure requirements or conducting high-risk activities without contingency planning may lead to penalties, including fines, demotions, or criminal liability as per applicable laws.
Regulations for the Implementation of the Drug Administration Law (2019 Amendment)	The current version was revised and released in March 2019 (the original version was first released in August 2002).	The State Council	These regulations expand upon and provide detailed explanations of the principles outlined in the *Drug Administration Law of the People's Republic of China*. The regulations cover various aspects of drug production, distribution, packaging, and utilization, aiming to enhance regulatory oversight of the pharmaceutical market and promote scientific and standardized management of drugs.
Regulation on the Administration of Human Genetic Resources	May 2019	The State Council	This regulation aims to standardize and safeguard the management and utilization of human genetic resources in China. It covers the collection, preservation, utilization, and provision of genetic resources to foreign individuals and entities. Responsibility for review and supervision lies with the science and technology administrative departments. Compliance with principles like legality, fairness, privacy protection, voluntary participation, and informed consent is mandatory for all involved parties. Ethical principles must guide related R&D activities, which must undergo ethical review. Penalties for violations, such as failure to undergo ethical review, include fines, industry bans, and potential criminal liability.

(continued)

Table 10.2 (continued)

Names of laws and regulations	Document release date	Issuing authority or authorities	Summary of content
Civil Code of China	May 2020	The National People's Congress of China	The *Civil Code* mandates that clinical trials must obtain approval from relevant authorities and ethics committees before proceeding. Participants or their guardians must be fully informed about the trial's purpose, procedures, and risks by providing written consent. Charging fees to participants is strictly prohibited. All activities involving human genes, embryos, and other research must comply with all applicable laws to protect human health, uphold ethical standards, and safeguard public interests. Furthermore, the *Civil Code* explicitly forbids the sale or purchase of human cells, tissues, organs, or corpses in any form, rendering such transactions void. It details medical liability, covering compensation for expenses like medical care, transportation, nutrition, hospital meals, lost income, and rehabilitation costs.
Biosecurity Law of China	October 2020	The Standing Committee of the National People's Congress of China	This legislation aims to mitigate risks associated with hazardous biological agents, ensuring the safe progression of biotechnology while protecting public health and ecological systems. It establishes a framework for biosafety risk prevention and control, with principles and leadership responsibilities outlined for biosafety management. Local governments are mandated to oversee biosafety within their jurisdictions. The legislation also sets stringent requirements for the safe management of biotechnology research, development, and application, prohibiting activities that threaten biosafety. Ethical principles and national biosafety regulations must guide biotechnology R&D, with institutions required to implement biosafety protocols, including training, monitoring, and reporting. Furthermore, the legislation introduces a classification system for biotechnology R&D, applying varying levels of review and supervision based on risk. It emphasizes responsible use and conservation of human genetic and biological resources, regulating their collection, preservation, utilization, and provision. Penalties for violations include fines, permit revocation, and industry bans to uphold compliance and safety standards.

Amendment (XI) to the Criminal Law of China	December 2020	The Standing Committee of the National People's Congress of China	This amendment to the *Criminal Law* introduces Article 336-1, which was added after Article 336. The new Article 336-1 explicitly prohibits the implantation of genetically edited or cloned human embryos into humans or animals, or the implantation of genetically edited or cloned animal embryos into humans. Offenders found guilty of these acts may face imprisonment or detention for up to three years in addition to fines. For more severe cases, where circumstances are particularly grave, the penalty may range from three to seven years of imprisonment, along with fines.
Personal Information Protection Law of China	August 2021	The Standing Committee of the National People's Congress of China	This legislation aims to regulate the handling of personal information, including medical and health-related data, to ensure individuals' data security. It outlines specific rules governing the collection, use, processing, and protection of personal data, emphasizing principles of legality, legitimacy, and necessity. The legislation stipulates that data processors bear primary responsibility for data protection and must implement robust management and protective measures. These measures include obtaining informed consent for data collection, limiting data collection to necessary purposes only, and prohibiting excessive data gathering—Such as data collection that is outside the scope of a particular entity's focus or jurisdiction. The legislation also identifies that sensitive personal information, such as biometric and medical information, requires enhanced protective measures. Violations of these regulations may result in penalties such as service suspension, public disclosure in credit records, fines, and other sanctions.

(continued)

Table 10.2 (continued)

Names of laws and regulations	Document release date	Issuing authority or authorities	Summary of content
Law on Doctors of China	August 2021	The Standing Committee of the National People's Congress of China	This legislation comprehensively regulates the registration, practice, assessment, and other procedures pertaining to physicians, while outlining their core responsibilities and ethical obligations. It defines the ethical principles that physicians must adhere to, including protecting patient rights and maintaining medical ethics, such as respecting patient autonomy, ensuring informed consent, practicing with honesty and integrity, avoiding conflicts of interest, and providing care without discrimination. The legislation mandates that physicians provide patients with thorough information about their condition, proposed treatments, and other relevant matters during diagnosis and treatment. For procedures involving surgery, specialized examinations, or treatments with elevated risks, physicians must promptly disclose all medical risks, alternative treatment options, and obtain explicit consent from patients before proceeding with any and all treatment. The legislation imposes strict penalties on physicians who violate professional ethics or engage in criminal activities. Offenses such as disclosing patient privacy, using prohibited medical technologies in a clinical setting, or severely breaching medical ethics may lead to fines, suspension or revocation of medical licenses, or permanent expulsion from the profession.

Law on Scientific and Technological Progress of China	The current version was revised and released in December 2021 (the original version was first released in July 1993).	The Standing Committee of the National People's Congress of China	This legislation comprehensively addresses various aspects of technology R&D, covering research, commercialization, enterprise innovation, policy support, outcome assessment, international cooperation, and intellectual property protection. It advocates for developing technologies, products, and services in a manner that is inclusive—Ensuring that scientific progress benefits the broader public, including vulnerable or underserved populations—And cautious, with due consideration to ethical and safety concerns. The legislation outlines China's intention to strengthen intellectual property protection, technological ethics, and safety review mechanisms within international technology R&D collaborations. This includes establishing technological ethics committees, enhancing education and research on technological ethics, and reinforcing review, assessment, and regulatory frameworks. Moreover, the legislation promotes integrating technology R&D with the establishment of product and service standards. It encourages collaborative advancements in research, development, and lawful adoption of significant national technological innovations and standards for products and services. The legislation also highlights China's aims to actively participate in international standard-setting processes and offers tax incentives to enterprises involved in high-tech R&D and production. Additionally, the legislation mandates that research institutions, universities, enterprises, and related entities fulfill their responsibilities in managing technological ethics, as stipulated in articles 103 and 107, which require the establishment of robust technological ethics review mechanisms and prohibit activities that may threaten national security, undermine the public interest, endanger human health, or violate research integrity and technological ethics in any way.

(continued)

Table 10.2 (continued)

Names of laws and regulations	Document release date	Issuing authority or authorities	Summary of content
Regulation on the Administration of Medical Institutions	The current version was revised and released in March 2022 (the original version was first released in February 1994).	The State Council	This regulation provides detailed procedures for medical institutions to obtain licenses to conduct clinical diagnosis and treatment, and outlines the role of health administrative departments in supervising and managing the entire licensing process. Specifically, the health administrative departments at or above the county level are responsible for approving the establishment of medical institutions, handling practice registration, conducting institutional evaluations, and ensuring compliance with laws, regulations, and medical technology standards. This regulation also specifies the requirements for managing medical personnel, including physicians, nurses, and other staff, regarding their professional qualifications and practice standards. Only entities or individuals with proper licenses or registrations are authorized to conduct diagnosis and treatment activities, and these entities or individuals must also ensure they are adhering strictly to any applicable laws, regulations, and medical technology standards. Additionally, medical institutions are prohibited from allowing non-health technical personnel to engage in technical medical work. During diagnosis and treatment procedures, medical personnel must disclose to the patient their condition and any proposed medical interventions. For procedures involving surgery, specialized examinations, or other specific treatments, medical personnel must inform patients or their next of kin about potential medical risks and alternative treatment options and ensure that they secure explicit consent before performing any procedures. Violations, such as exceeding authorized diagnostic and treatment activities or employing non-health technical personnel for medical tasks, may lead to penalties (for both individuals and institutions), including fines, revocation of practice licenses, or orders to cease practice.

Regulations on the Promotion of the Cellular and Genetic Industry in the Shenzhen Special Economic Zone	January 2023	The Standing Committee of Shenzhen Municipal People's Congress	This local legislation mandates that all activities concerning the research, development, production, operation, use, and safeguarding of cellular and genetic products for disease diagnosis and treatment within the designated zone must comply with strict regulations. These regulations encompass the collection and storage of cells, the development of cell and genetic products, the expansion of drug clinical trials, the application of gene technology, and the processes for marketing authorization and product production. For instance, the legislation strictly prohibits any cellular and genetic research, development, or application activities that pose risks to public health, national security, or societal interests. It stipulates that all R&D regarding cells and genes must adhere to national ethical review regulations. Ethical compliance would be enforced by institutional ethics committees established within qualified medical institutions, under the supervision of the Shenzhen Municipal Health Commission and, where applicable, the city's administration for market regulation. The collection of cells must be conducted exclusively by qualified medical institutions with the necessary qualifications and facilities. Enterprises or research institutions must engage qualified medical institutions to perform these activities. Medical institutions are required to adhere to established medical technology standards for cell collection and must implement rigorous quality management systems and standard operating procedures, as explicitly required under Article 14 of the legislation. Furthermore, cellular and genetic drugs undergoing clinical trials for the treatment of severe diseases must undergo thorough medical analysis to ensure that their potential benefits outweigh the risks. The legislation also emphasizes the protection of subjects' rights throughout extended clinical trials, including dignity, informed consent, and privacy. It also mandates compliance with pharmaceutical production quality management standards to ensure the safety and efficacy of cellular and genetic drugs.

(continued)

Table 10.2 (continued)

Names of laws and regulations	Document release date	Issuing authority or authorities	Summary of content
Regulations on Promoting Biomedical New Technologies in the Boao Lecheng International Medical Tourism Pilot Zone of the Hainan Free Trade Port	November 2024	The Standing Committee of Hainan Provincial People's Congress	This regulation establishes a structured legal framework governing the development and application of cutting-edge biomedical technologies, such as cell therapy, gene therapy, and tissue engineering, within the pilot zone. It applies to all related clinical research, translational uses, and oversight activities. It outlines governance responsibilities to be shared among provincial health and drug authorities, including the Boao Lecheng Administration, and other relevant governmental bodies. The regulation also supports the establishment of biomedical research institutions, encourages international and domestic cooperation, and promotes the development of public platforms for cell preparation, storage, and testing. Foreign-invested enterprises are permitted to engage in research and product development within the zone, provided they comply with all applicable regulatory standards. It also encourages the development of insurance products covering injuries or deaths associated with biomedical research and application. Commercial insurance claims are encouraged to be processed through simplified procedures and direct settlement mechanisms, as stated in the regulation, although detailed implementation rules are not specified.

				The regulation requires that medical institutions conducting translational projects meet specific qualification standards, including tertiary-level capabilities, established academic and ethics committees, and comprehensive internal systems for quality management. The regulation also stipulates standards regarding medical institutions' risk control and adverse event response. These institutions are designated as the primary entities responsible for ensuring the safety, efficacy, and legal compliance of biomedical innovation activities. Informed consent, ethical review, and compliance with national clinical trial and data protection standards are mandatory. The regulation also introduces a credit scoring system to evaluate institutional compliance with legal, ethical, and procedural requirements governing biomedical research and translational application activities. While the regulation does not clearly define all scoring criteria, key evaluation domains may include: Quality control protocols, risk management systems, ethical review processes, data reporting practices, and adverse event handling mechanisms. Severe violations, such as conducting unauthorized biomedical applications, are subject to fines, license suspension, and other legal penalties.

Table 10.3 Summary of other normative documents applicable to regulating regenerative medicine research and development in China

Names of normative documents	Document release date	Issuing authority or authorities	Summary of content
Key Points for Quality Control in Clinical Research of Human Somatic Cell Therapy and Gene Therapy	May 1993	The Ministry of Health	This normative document explicitly mandates that clinical research involving human somatic cell therapy and gene therapy must adhere to regulations stipulated under national drug administration laws. Entities interested in conducting related R&D are required to submit applications to the health administrative department within the MoH. These applications undergo thorough review processes, including qualification assessments for research units and personnel, among others. Approval from the health administrative department is mandatory before commencing either clinical trials or clinical validation studies.
Measures on the Safe Management of Genetic Engineering	December 1993	State Scientific and Technological Commission (which has been restructured)	This normative document sets forth standardized protocols for research, experimentation, and industrial production involving genetic engineering. It highlights the imperative of conducting comprehensive safety assessments for genetic engineering technologies and outlines detailed procedures for their application and approval. Moreover, the document delineates the legal responsibilities and liabilities associated with violations of these procedures.
Cord Blood Hematopoietic Stem Cell Bank Management Measures (Trial)	May 1999	The Ministry of Health	This normative document explicitly mandates that approval from the health administrative department of the State Council is required to establish umbilical cord blood hematopoietic stem cell banks. Furthermore, the collection of umbilical cord blood must adhere strictly to principles of voluntariness and informed consent. Both donors and recipients are required to sign informed consent forms and to abide by relevant medical ethics guidelines. Importantly, umbilical cord blood collection activities are strictly prohibited from being conducted for profit. The document also outlines penalties for unauthorized activities, illegal collection practices, and instances of medical accidents caused by preventable errors—Such as quality control failures or pathogen contamination—For which cord blood banks bear legal responsibility.

Cord Blood Hematopoietic Stem Cell Bank Establishment and Management Standards (Trial)	January 2001	The Ministry of Health	This normative document specifies personnel qualification requirements, facility and equipment standards, and necessary management systems required for umbilical cord blood hematopoietic stem cell banks.
Measures on the Management of Assisted Reproductive Technologies in Humans	February 2001	The Ministry of Health	This normative document regulates medical techniques and methods used to manipulate gametes, zygotes, and embryos for achieving pregnancy. These medical techniques include artificial insemination, in vitro fertilization-embryo transfer ("IVF-ET"), and related technologies, including cell and gene-editing therapies. It mandates that human-assisted reproductive technologies adhere to the principle of informed consent, requiring participants to sign an informed consent form before the onset of any of the stipulated medical techniques. Ethical considerations must be reviewed by a medical ethics committee. Moreover, the document explicitly prohibits the commercial buying and selling of gametes, fertilized eggs, and embryos.
Guiding Principles for Research and Quality Control Techniques of Human Cell Therapy Preparations	March 2003	The State Food and Drug Administration	These guiding principles require that all procedures and final products related to somatic cell therapy research and preparation strictly adhere to standard operating procedures. Research institutions must comply with GMP standards—Formulated by China's national drug regulatory authority—For producing clinical products to ensure standardized production processes. Furthermore, adherence to GCP is mandatory to uphold ethical integrity in both research and clinical practice.

(continued)

Table 10.3 (continued)

Names of normative documents	Document release date	Issuing authority or authorities	Summary of content
Ethical Guiding Principles for the Research of Human Embryonic Stem Cell	December 2003	The Ministry of Science and Technology & The Ministry of Health	These guiding principles explicitly outline that the research of human embryonic stem cells conducted in China must adhere to both ethical standards and internationally recognized bioethical guidelines, though no particular international declaration is explicitly cited. Some prohibited practices are explicitly listed in the Chinese guiding principles, including human reproductive cloning and the mixing of human reproductive cells with those of other species. These restrictions are also consistent with widely held international bioethical standards, although the guiding principles themselves do not cite any specific international instruments in this regard. Additionally, the guiding principles specify that the in vitro culture period for blastocysts derived from either in vitro fertilization, somatic cell nuclear transfer, parthenogenesis, or genetic modification must not exceed 14 days. This 14-day limit is widely recognized in international bioethics as the point before the emergence of the primitive streak, which marks the beginning of neural development and individuation. Culturing embryos beyond this stage raises profound ethical concerns regarding the moral status of the embryo and potential sentience. To safeguard patient and subject rights, the principles strictly require practitioners to obtain informed consent and decision-making from patients and subjects to ensure patient and subject privacy. Institutions involved in human embryonic stem cell research must also establish an internal ethics committee to oversee the ethical and scientific aspects of the research. Finally, the buying and selling of gametes, fertilized eggs, and embryos are explicitly prohibited under these guidelines.

Guidelines for Hospital Management Evaluation	May 2008	The Ministry of Health	These guidelines consolidate and operationalize several regulatory principles that were previously dispersed across instruments such as the *Law on Doctors of the People's Republic of China*, the *Regulation on the Administration of Medical Institutions*, and various national ethical codes. These guidelines mandate that medical technologies used in hospital-based research projects must comply with applicable legal requirements, adhere to established medical ethical principles, and also undergo formal approval procedures before implementation. Medical personnel are required to prioritize patient safety throughout the research process and are expressly prohibited from using unapproved or clinically unverified medical technologies. Additionally, medical personnel must prioritize and respect their patients' right to give informed consent and patients' right to choose their individual treatment paths. Rather than introducing entirely new regulatory obligations, these guidelines serve to harmonize and standardize existing compliance requirements within a unified national hospital evaluation framework. They enhance institutional accountability by linking legal, ethical, and procedural compliance with hospital performance assessments, while also making relevant obligations more detailed and practically enforceable.
Notice on the Release of theFirst Batch of Catalogue of Class III Medical Technologies Approved for Clinical Application	May 2009	The Ministry of Health	According to this catalog, several advanced biomedical technologies related to RM were officially approved for clinical use as part of the first batch of high-risk (Class III) medical technologies. These technologies include homologous organ transplantation, autologous immune cell (T-cell and natural killer ("NK") cell) therapy, gene chip diagnostic technology, cell transplantation therapy (excluding stem cells), cord blood hematopoietic stem cell therapy technology, hematopoietic stem cell therapy (excluding cord blood hematopoietic stem cells), and tissue engineering tissue transplantation therapy. The approval for clinical application of these technologies is supervised by the MoH or provincial-level health administrative departments.

(continued)

Table 10.3 (continued)

Names of normative documents	Document release date	Issuing authority or authorities	Summary of content
Good Manufacturing Practices for Pharmaceutical Products	Updated in January 2011 (originally 1988)	The Ministry of Health	This normative document provides detailed and enforceable requirements for pharmaceutical production management and quality control. The document prescribes the essential conditions for pharmaceutical manufacturing, including appropriate facilities, specialized equipment, calibrated instruments, qualified personnel, and standardized operating procedures to be used in pharmaceutical production. Furthermore, it specifies comprehensive operational standards related to institutional responsibilities, staff training and qualifications, production environments, raw materials and finished products, as well as protocols for validation and verification. In addition, it mandates the implementation of corrective and preventive measures specifically designed to address quality-related risks such as contamination, cross-contamination, mix-ups, and manufacturing deviations. These measures are further supported by regulatory requirements for adverse reaction reporting, complaint investigation, and product recall systems to ensure prompt mitigation of residual risks and to uphold post-market quality surveillance.
Notice on Conducting Self-Inspection and Self-Correction Work on Stem Cell Clinical Research and Application	December 2011	The Ministry of Health & The State Food and Drug Administration	This notice mandates that medical institutions involved in stem cell clinical research and applications must adhere to the GCP and the *Measures for the Administration of the Clinical Application of Medical Technologies*. It requires these institutions to conduct self-inspections in accordance with GCP and related measures and to make corrections accordingly. Additionally, the notice stipulates that unapproved stem cell clinical research and application projects are to be suspended. Clinical trials for stem cell products are required to follow the relevant domestic regulatory requirements for drug clinical trial registration. Approved stem cell product clinical trials must strictly follow the approval document issued by the State Food and Drug Administration and related clinical trial regulations. The notice also clarifies that medical institutions involved in stem cell clinical research and applications are prohibited from making arbitrary changes to the research protocol or from charging participants fees.

Management Measures for Clinical Research Projects Conducted by Medical and Health Institutions	October 2014	The National Health and Family Planning Commission; The State Food and Drug Administration; The National Administration of Traditional Chinese Medicine	These management measures establish jurisdiction over all clinical research involving human drugs and all medical devices utilized within medical and health institutions, including investigational drugs and in vitro diagnostic reagents. According to these measures, medical and health institutions conducting clinical research must establish an internal Clinical Research Management Committee and an Ethics Committee. The measures also specify the responsibilities of the ethics committee, the application procedures for the establishment of research projects, and the conditions for the appointment of research project leaders. Additionally, the measures detail procedures for suspending or terminating research projects in cases of significant safety risks or ethical violations, with requirements to report any such risks or violations to the provincial-level health administrative authorities.
Notice on the Cancellation of Approval for the Clinical Application of Third-Class Medical Technologies	June 2015	The National Health and Family Planning Commission	This notice eliminates the approval procedure for the clinical application of third-class medical technologies. It mandates that medical institutions must refrain from clinically applying technologies with significant safety or efficacy concerns, as well as from those posing significant ethical issues (such as cloning technology or surrogate pregnancy technology), or those technologies that are explicitly prohibited by domestic health administrative departments at or above the provincial level. Medical technologies that have met the required thresholds for safety and efficacy but may involve high technical difficulty and risks, or significant ethical risks, must undergo strict internal management by the medical institutions themselves, under the oversight and filing system of provincial health administrative departments, and their clinical application must be limited under specified conditions. Medical institutions that fail to register their use of restricted medical technologies with the administrative department that issued their medical institution practice license, fail to file such registration with the provincial health administrative authorities, or that conduct prohibited clinical applications of medical technologies, will face penalties imposed by local or provincial health administrative departments.

(continued)

Table 10.3 (continued)

Names of normative documents	Document release date	Issuing authority or authorities	Summary of content
Management Measures for Stem Cell Clinical Research (Trial)	July 2015	The National Health and Family Planning Commission & The State Food and Drug Administration	These management measures establish comprehensive standards for conducting clinical research on stem cells in medical institutions, emphasizing principles of standardization, transparency, professional ethics, and protection of research subjects' rights. The management measures outline specific requirements and responsibilities that medical institutions must fulfill during clinical stem cell research. Provisions include prohibitions against charging subjects any fees related to research and against publishing or indirectly promoting advertisements for stem cell clinical trials. Additionally, the measures detail procedures for project initiation and registration, informed consent protocols, and ethical guidelines for sourcing and acquisition of stem cells. In cases of research misconduct, punitive actions are specified, including public criticism, temporary project suspension, mandatory rectification within a designated timeframe, and recording the relevant individuals and institutions in scientific misconduct records.
Guiding Principles for Quality Control and Preclinical Studies of Stem Cell Preparations (Trial)	July 2015	The National Health and Family Planning Commission & The State Food and Drug Administration	These guiding principles delineate fundamental standards applicable to the preparation and pre-clinical research phases of various types of stem cells intended for clinical applications (excluding hematopoietic stem cell transplantation, which is separately regulated). They mandate compliance with GMP and establish rigorous standard operating procedures aimed at ensuring the quality control of stem cell preparations and the safety and effectiveness of treatments.
Service Guide for Administrative Permits on the Collection, Acquisition, Trade, Export, and Exit of Human Genetic Resources	September 2015	The Ministry of Science and Technology	This service guide regulates and manages activities involving the collection, acquisition, sale, export, and transfer of human genetic resources within China. According to the guidelines, applicants intending to engage in activities related to the collection or acquisition of human genetic resources must meet specific requirements. These requirements include undergoing ethical reviews, obtaining informed consent from providers of human genetic resources, and developing feasible collection or acquisition plans. Participation of foreign entities in these activities, which is categorized under international cooperation, necessitates compliance with additional conditions such as intellectual property rights and data-sharing agreements. The Ministry of Science and Technology serves as the competent authority responsible for receiving and adjudicating these activities.

Measures for the Ethical Review of Biomedical Research Involving Humans	October 2016	The National Health and Family Planning Commission	This normative document explicitly mandates that medical institutions conducting biomedical research involving human subjects must establish an independent ethics committee. This committee is then entrusted with conducting ethical reviews autonomously. Institutions lacking an ethics committee are prohibited from conducting biomedical research involving human subjects. Furthermore, the document provides comprehensive guidance on the responsibilities of the ethics committee, including its composition and the necessary documentation for registration. It emphasizes principles such as informed consent, risk management, fair compensation, privacy protection, legal liability, and special protection measures within the ethical review process.
Measures for the Safety Management of Biotechnology Research and Development	July 2017	The Ministry of Science and Technology	This normative document specifies that biotechnology R&D must strictly comply with applicable laws and regulations, uphold ethical standards, and ensure the protection of national security, public interests, and the legitimate rights and interests of affected individuals. The document advocates for the implementation of a hierarchical management system for biotechnology R&D, categorizing activities into high, moderate, and general risk levels based on their potential risks. Moreover, comprehensive assessments are required for public disclosure, transfer, promotion, industrialization, and commercial application of biotechnology R&D outcomes. These assessments aim to more effectively mitigate significant biosafety risks.
Guiding Principles for Research and Evaluation of Cell Therapy Products (Trial)	December 2017	The State Food and Drug Administration	These guiding principles mandate that the sourcing, handling, and clinical trials of cell therapy products adhere strictly to ethical standards and comply with pharmaceutical management regulations governing R&D registration and approval. Additionally, these principles outline specific regulations for risk control, quality control, and safety assessment procedures to ensure the integrity and safety of cell therapy products throughout their development, including for clinical trials.

(continued)

Table 10.3 (continued)

Names of normative documents	Document release date	Issuing authority or authorities	Summary of content
Measures for the Administration of the Clinical Application of Medical Technologies	Updated in August 2018 (originally 2009)	The National Health Commission	This normative document (the 2009 version) restricts several RM-related medical technologies, including xenogeneic stem cell therapy, xenogeneic gene therapy, and human somatic cell cloning technology, from clinical application. Cloning therapy, autologous stem cell and immune cell therapy, gene therapy, and xenogeneic stem cell transplantation technology are classified as involving significant ethical issues, based on the document and its annexed directory of third-class medical technologies. These therapies require further safety and efficacy verification through standardized clinical trials and are categorized as third-class medical technologies. Before initial clinical application, these technologies must also undergo safety and efficacy demonstrations and ethical reviews organized by the MoH. In the updated 2018 version of the document, new medical technologies characterized by uncertain safety and efficacy in clinical application, significant ethical concerns, clinical obsolescence, or technologies lacking sufficient clinical research evidence, remain prohibited.
Rules for Investigation and Handling of Scientific Research Integrity Cases (Trial)	September 2019	The Ministry of Science and Technology; The Supreme People's Court; The National Health Commission; etc.	This normative document provides detailed constraints and provisions concerning actions that violate scientific research principles and standards in scientific research and related activities. The document clearly defines the scope of research misconduct, stipulates the responsibilities of various supervisory departments in investigating and handling research misconduct, and outlines the procedures for investigation and handling. Additionally, the document includes detailed punitive measures to ensure misconduct is handled seriously.
The Agreement Concerning Amendment to the Agreement on Trade in Services of the Mainland and Hong Kong Closer Economic Partnership Arrangement	November 2019	The Ministry of Commerce	These amendments prohibit Hong Kong service providers from participating in the development and application of human stem cell and gene diagnosis and treatment technologies.

Good Clinical Practice for Drugs Clinical Trials	Updated April 2020 (originally 2003)	The National Medical Products Administration & The National Health Commission	These specifications outline the quality standards that must be maintained throughout the entire process of drug clinical trials, covering program design, organization and implementation, monitoring, recording, summary, and reporting. The specifications underscore the importance of ethical review and informed consent as critical measures to protect the rights of trial participants. Clinical trials are required to strictly adhere to ethical requirements, prioritizing the safety and rights of participants above scientific and societal benefits. Furthermore, these specifications emphasize that clinical trials must be scientifically rigorous, taking into account the anticipated risks and benefits to both participants and society. Trials can only commence or continue if the anticipated benefits outweigh the risks, with well-defined protocols that are subject to approval by the ethics committee before implementation.
The Procedures for the Evaluation of Breakthrough Therapy Drugs (Trial)	July 2020	The National Medical Products Administration	This document specifies that applicants conducting drug clinical trials may apply for inclusion in the breakthrough therapy program under specific conditions. This program is intended for innovative or improved new drugs designed to prevent or treat life-threatening diseases, diseases that significantly impair quality of life without effective treatments, or drugs that have substantial clinical evidence demonstrating significant advantages over existing treatments.
The Procedures for the Evaluation and Approval of Conditional Drug Marketing Applications (Trial)	July 2020	The National Medical Products Administration	This document outlines that applicants conducting drug clinical trials have the option to submit a conditional approval application to the Center for Drug Evaluation of the National Medical Products Administration. This option is available for innovative drugs that meet the conditions specified in the technical guidelines for conditional approval of drugs. Such conditions typically include drugs that demonstrate significant clinical value and that are needed urgently in clinical settings.
The Measures for the Priority Review and Approval of Drug Marketing Authorizations (Trial)	July 2020	The National Medical Products Administration	This normative document specifies that drugs of significant clinical value, including innovative drugs and those that seek to improve existing medications addressing urgent clinical needs such as prevention and treatment of major infectious diseases, rare diseases, or those diseases included in the breakthrough therapy program, are eligible to apply for priority review and approval procedures during the drug marketing authorization application process.

(continued)

Table 10.3 (continued)

Names of normative documents	Document release date	Issuing authority or authorities	Summary of content
Interim Provisions on the Handling of Misconduct in Scientific and Technological Activities	July 2020	The Ministry of Science and Technology	These interim provisions dictate procedures for handling violations and related scientific misconduct by institutions and individuals in scientific and technological R&D. The provisions delineate the scope of such violations as well as the measures and procedures for addressing scientific research misconduct. Furthermore, the provisions designate the science and technology administrative departments at all levels as the competent authorities responsible for dealing with violations in scientific and technological R&D.
Code of Conduct for Medical Research Integrity and Related Behaviors	January 2021	The National Health Commission; The Ministry of Science and Technology; The State Administration of Traditional Chinese Medicine	This normative document applies comprehensively to all activities conducted by medical research institutions and their personnel across various stages, including project application, research implementation, result reporting, process management, publication of findings, and evaluation and review of research proposals, project outcomes, and academic performance. The document mandates that medical researchers strictly adhere to research ethics guidelines, actively seek ethical review, and accept both internal and external ethical supervision—Such as that conducted by institutional ethics committees and regulatory authorities—To effectively protect the legitimate rights and interests of research subjects. After the conclusion of research, the storage, sharing, and disposal of human or animal samples, hazardous substances, data, or other materials must meticulously comply with relevant biosafety and research management regulations to ensure the legality and safety of all research activities.
Guiding Principles for Non-Clinical Research and Evaluation of Gene Therapy Products (Trial)	November 2021	The National Medical Products Administration	These guiding principles recommend that non-clinical research involving gene therapy products should be designed following a comprehensive risk analysis. The risk analysis proposed would include a thorough assessment of both potential risks and benefits to clinical trial subjects. The principles assert that safety studies for gene therapy products should be conducted in institutions certified according to GMP standards; moreover, compliance with relevant drug quality management specifications must be strictly ensured throughout the process.

Guiding Principles for Non-Clinical Research on Gene-Modified Cell Therapy Products (Trial)	November 2021	The National Medical Products Administration	These guiding principles emphasize the consideration of emerging safety risks from genetic modification of cells, including off-target effects from gene editing, mutations resulting from vector insertions, and other risks associated with the expression of transgenic products. The principles assert that research plans should integrate factors such as intended indications, patient demographics, administration routes, and dosing protocols. Furthermore, during risk-benefit assessments, careful attention is needed to address challenges in transitioning from non-clinical to clinical studies and uncertainties in predicting risks.
Special Administrative Measures (Negative List) for the Access of Foreign Investment in Pilot Free Trade Zones (2021)	December 2021	The National Development and Reform Commission & The Ministry of Commerce	This negative list expressly forbids foreign investment (including individuals and enterprises) in the development and application of human stem cells, gene diagnosis, and treatment technologies. However, in practice, certain pilot free trade zones (e.g., Hainan) are planning to launch programs aimed at gradually opening this sector under strict regulatory oversight.
Standards for Accelerating the Review of Innovative Drug Marketing Authorization Applications by the Center for Drug Evaluation (Trial)	March 2023	The National Medical Products Administration	This normative document aims to establish a transparent, standardized, and scientifically grounded framework for reviewing innovative drug applications. The document prioritizes innovative drugs for children, drugs for treating rare diseases, and drugs included in the breakthrough therapy designation program—a regulatory mechanism designed to expedite the development and review of drugs that target serious or life-threatening conditions and demonstrate substantial clinical advantages over existing therapies. The document also standardizes aspects such as review timelines and workflow to ensure the scientific rigor and efficiency of the drug review process.

(continued)

Table 10.3 (continued)

Names of normative documents	Document release date	Issuing authority or authorities	Summary of content
Detailed Rules for the Implementation of the Regulation on the Administration of Human Genetic Resources	May 2023	The Ministry of Science and Technology	These detailed rules further elaborate on the provisions and requirements outlined in the *Regulation on the Administration of Human Genetic Resources*. The rules elaborate on the detailed procedures and conditions that must be met for the collection, preservation, utilization, and external provision of human genetic resources. Moreover, the rules reaffirm that the collection and preservation of Chinese human genetic resources within China, as well as the provision of these resources abroad, must be conducted by either Chinese scientific research institutions, medical institutions, or enterprises. Furthermore, the rules outline the administrative licensing and filing procedures for international cooperation involving human genetic resources. The rules also mandate a prior reporting procedure for the provision and open use of genetic resources to foreign countries. Additionally, the rules specify the responsibilities of entities involved in managing human genetic resources. The rules also establish penalties and supervision mechanisms for violations, aiming to deter unauthorized activities and uphold ethical standards in the management and utilization of genetic resources.
Guiding Principles for Clinical Trials of Human-Source Stem Cells and Associated Derivative Cell Therapy Products (Trial)	June 2023	The National Medical Products Administration	These guiding principles require that when human-source stem cell-related products enter the clinical trial phase, they should strictly follow the requirements of GCP. The document stresses that clinical trial projects involving human-source stem cell-related products must undergo adequate ethical review. Moreover, the safety and efficacy of the products must be appropriately evaluated in accordance with applicable laws and customs. Effective risk control measures should also be implemented, and long-term follow-up of clinical subjects should be conducted to maximize subjects' safety and rights. In addition, after obtaining the marketing authorization for the applicable product, the applicant should consider collecting real-world effectiveness and safety information through the implementation of post-marketing observational studies or key monitoring. The applicant should also regularly submit drug safety update reports or drug re-registration procedures to the regulatory authorities to ensure effective communication with the regulatory authorities.

The Interim Measures for the Implementation of Translational Applications of Biomedical New Technologies in the Boao Lecheng International Medical Tourism Pilot Zone of the Hainan Free Trade Port	January 2025	The Hainan provincial Health commission; the Hainan medical products Administration; the Boao Lecheng International Medical Tourism Pilot Zone Administration	These interim measures establish a normative framework to regulate the translational application of biomedical technologies, such as cell therapy, gene therapy, and tissue engineering within the Boao Lecheng Pilot Zone specifically. The measures aim to ensure medical safety, ethical compliance, and technical controllability of these technologies in the pilot zone. The instrument covers the entire translational application process, including project application, technical review, clinical implementation, quality tracking, information management, risk prevention, and regulatory oversight. The measures clarify the respective responsibilities of governmental authorities, review bodies, and medical institutions, while emphasizing requirements such as informed consent, privacy protection, adverse event monitoring, quality control of preparations, institutional qualifications, and accountability mechanisms. The overall goal of the measures is to provide institutional safeguards for the safe and responsible advancement of innovative medical technologies.
The Administrative Provisions on the Clinical Research of Biomedical New Technologies in the Boao Lecheng International Medical Tourism Pilot Zone	February 2025	The Hainan Provincial Health Commission & The Hainan Medical Products Administration	This normative document establishes a comprehensive regulatory framework for conducting clinical research on innovative biomedical technologies, such as cell therapy, gene therapy, and tissue engineering, within the Boao Lecheng Pilot Zone. The document aims to ensure scientific rigor, patient safety, and ethical compliance throughout the entirety of the research process. To meet these aims, the document outlines requirements for project registration, ethical and academic review, institutional qualifications, informed consent, quality management, data retention, and adverse event reporting. The document also introduces risk-based oversight mechanisms, including dynamic evaluations and conditions for the suspension or termination of research. The document further emphasizes government supervision while clearly delineating the responsibilities of relevant stakeholders.

emphasizing ethical principles, stringent safety standards, and prioritizing patient welfare throughout the entire process of research and clinical translation.

Still, in the absence of a dedicated national law governing RM, regulatory authority remains fragmented across multiple agencies, resulting in inconsistent oversight and enforcement. Therefore, it is imperative to establish comprehensive legislation that provides unambiguous legal authority, harmonizes existing regulatory requirements, and ensures consistent implementation of ethical and scientific standards for RM research and clinical applications across China.

Challenges Concerning the Governance of RM

While China has implemented supportive policies and regulatory measures to guide and promote advances in RM, the country also faces several challenges in this evolving field. These challenges include issues surrounding bioethical governance, regulatory loopholes, and imperfect mechanisms for international cooperation.

Bioethical Governance Issues

In China, existing ethical guidelines and regulations concerning the use of embryos in research lack detailed provisions regarding their sources, the management of surplus embryos, and criteria for ethical reviews. For example, the *Ethical Guiding Principles for the Research of Human Embryonic Stem Cell* only outline permissible sources of human embryonic stem cells but do not address ethical risks such as commodification or improper use of surplus embryos [73]. Similarly, the *Guidelines for the Establishment of Ethics Review Committees for Clinical Research Involving Humans (2020 Edition)*, issued by the National Health Commission Medical Ethics Expert Committee, stress the legal review of embryo and stem cell sources but lack clear criteria and practice guidelines for ethical reviews [106].

In the absence of clear, legally binding national protocols for operational standards, ethics committees at Chinese institutions currently rely on a hybrid system of ethical guidelines. These typically combine domestic normative frameworks, particularly ministerial-level guidance documents, with internationally recognized standards such as the *Declaration of Helsinki* and Council for International Organizations of Medical Sciences guidelines [20, 107]. However, as these guidelines are typically formulated in broad terms, they leave room for varied interpretations [108, 109]. Ethics committees, including those based at research institutions, universities, corporations, or other entities, may develop their own review standards based on different understandings of key ethical principles. This often leads to disparity in the application of key principles including informed consent, risk-benefit assessment, and privacy protection, all of which increase the potential for inconsistent applications of review criteria applied to RM-related R&D. Some ethics committees (likely in an effort to standardize their procedures) may refer to international instruments, such as the *International Ethical Guidelines for Health-related*

Research Involving Humans (2016) issued by the Council for International Organizations of Medical Sciences, or the *ICH E6(R2): Good Clinical Practice (GCP) Guidelines* developed by the International Council for Harmonisation of Technical Requirements for Pharmaceuticals for Human Use (ICH) to guide their review of RM R&D [107, 110]. However, there are currently no national guidelines on whether and how such international standards should be applied. It is thus unclear which instruments are appropriate for reference, and no process exists to determine which instruments should then be integrated into China's domestic ethical review procedures.

Moreover, since the ethical review processes and documents are typically not publicly accessible, it is difficult to determine how ethics guidelines are being applied in practice—making it nearly impossible to assess whether the various ethical committees are both referring to and interpreting the same international frameworks consistently. This lack of transparency and coordination significantly increases the risk of varied review standards across institutions, particularly in ethically complicated RM R&D involving stem cells. Ethical committees may then have highly varied requirements for documentation, informed consent procedures, and privacy protection measures. They may also vary in their ethical justifications for embryo use, especially regarding the necessity of using embryos, the availability of alternatives, and the ethical management of surplus embryos [108, 111]. This lack of standardization and inconsistent application of ethical principles may also lead to inadequate ethical review of RM R&D or even lead to the approval of research that might otherwise fail to meet stricter standards applied elsewhere. Such inconsistencies may create loopholes in ethical oversight, leading to significant risks for participant, subject, and patient protection, particularly regarding potential physical safety, informed consent, or even personal information security [112, 113]. Ethical ambiguity and inconsistency are particularly concerning in cases involving public consultation or cases involving ethically sensitive technologies, as varied review standards may trigger public skepticism about the fairness and reliability of the ethical review process generally.

Additionally, utilizing RM technologies for personalized treatments necessitates extensive collection and utilization of genetic data from individuals, raising further ethical and legal considerations related to data protection, such as informed consent, privacy protection, data security, and transparency of data processing [114, 115]. While China's current regulatory frameworks, including the *Civil Code* and *PIPL*, provide legal grounds for safeguarding personal genetic information, the rapid development of new RM-related technologies presents novel challenges that may strain existing regulatory mechanisms. For example, the use of AI in biomedical research and clinical applications has raised concerns about potential data misuse, unauthorized access, and issues related to genetic discrimination [116–118]. These concerns stem primarily from AI systems' extensive collection, processing, and storage of sensitive genetic and health information. Such large-scale data handling creates significant privacy risks—especially when datasets are repurposed without proper safeguards, or when cross-referencing multiple sources enables individual re-identification. The inherent opacity of many AI algorithms further complicates these issues, as their automated decision-making processes often lack transparency in data handling practices and challenge traditional informed consent frameworks. Compounding these risks, AI models trained on biased or incomplete datasets may generate unreliable predictions or systematically discriminatory results, potentially compromising both patient safety and the equity of healthcare delivery. Moreover, while China's current ethical review procedures and regulations specify important issues such as institutional qualifications, the responsibilities of healthcare providers and researchers, and procedural aspects of ethics committees, there remains a significant gap in effectively addressing and providing robust remedies for breaches of patient or subject genetic privacy and improper data use [119].

The Challenge of Insufficient Regulatory Frameworks

In rapidly advancing fields like biology and medicine, regulatory frameworks and oversight mechanisms often fail to keep up with the speed of scientific innovation [39, 40]. This discrepancy can result in new technologies being commercialized and marketed before undergoing comprehensive scientific, ethical, or legal review, posing risks to patients' health and safety and eroding public trust when it comes to innovative technologies. In addition to the concerns posed by the premature commercialization of innovative technologies, inadequate regulatory and governance frameworks also raise several critical concerns. These concerns include a lack of meaningful and standardized procedures for obtaining informed consent, insufficient safeguards for sensitive genetic and other health data, and uncertainty for patients regarding how privacy protections are implemented in clinical and research settings [39, 120]. Additionally, the absence of mechanisms to ensure equitable access to innovative RM products and treatments raises questions of fairness. While formal oversight responsibilities are divided between governmental health authorities and institutional ethics committees, enforcement remains inconsistent. This is largely due to the fragmented regulations, lack of a centralized enforcement agency, and the undefined procedures for investigating violations [121, 122]. This deficient framework allows misconduct to go unchecked while leaving affected participants without recourse. Moreover, the dual role of the government as both a promoter and regulator of biotechnology may give rise to conflicts of interest that compromise independent oversight.

Concerns about the lack of centralized and authoritative procedures governing RM and related fields are not merely theoretical. Investigations have revealed instances of Chinese clinics offering unproven stem cell therapies as a direct consequence of this inadequate oversight [123]. This situation and similar instances jeopardize the protection of patients' rights and ultimately jeopardize patients' well-being. Tragically, there have even been cases of patients who have lost their lives after receiving unregulated biological immunotherapy treatments. For instance, the Wei Zexi incident in 2016 underscored the risks encountered by patients seeking novel medical treatments. Wei Zexi, a 21-year-old Chinese university student diagnosed with advanced synovial sarcoma, exhausted traditional chemotherapy and radiotherapy options without significant improvement. Seeking alternative treatments, Wei Zexi chose tumor biological immunotherapy (DC-CIK therapy) offered by a medical institution. Despite substantial investment of resources and time, Wei Zexi's condition did not improve, leading to his death [124]. Such incidents highlight the real-world consequences of regulatory failure and emphasize China's need for a swift and comprehensive regulatory response.

Following the He Jiankui incident—where a Chinese scientist claimed to have created the world's first genetically edited babies using CRISPR-Cas9 technology, igniting global outrage for breaching scientific and ethical standards [125]—the Chinese government amended the *Criminal Law* to specifically address unethical uses of innovative biomedical treatments. The newly established *Civil Code* also reflects the government's commitment to promoting the ethical and lawful utilization of human biological resources and dedication to establishing legitimate procedures for conducting clinical trials for new drugs, medical devices, or novel treatment methods. Nevertheless, the important issues remain unaddressed by the current regulatory regimes. For instance, the legal liabilities surrounding the misuse and abuse of RM R&D are not clearly defined in existing laws, creating inconsistencies and unpredictability when determining liabilities in legal proceedings. Moreover, the *Amendment (XI) to the Criminal Law* only applies to reproductive cloning and deems certain gene-editing activities illegal, without extending to other RM technologies. The narrow focus of this amendment leaves substantial regulatory gaps, failing to cover emerging and potentially controversial RM technologies such as iPSCs and tissue engineering.

Regulatory Challenges for International Cooperation

International collaboration plays a crucial role in advancing RM; it enables open communication among researchers, bolstering international endeavors to tackle scientific challenges within the field. Furthermore, the ethical and regulatory challenges arising from the development of RM are not restricted to one country but rather are global issues. International cooperation is vital for establishing common standards and regulations to effectively manage these challenges. The Chinese government has actively promoted international research cooperation in RM through various policies and laws, such as supporting foreign investment pilots and facilitating international cooperative clinical trials [66]. However, China's current laws and regulations on personal data protection and human genetic resources management impose restrictions on personal data sharing and cross-border transfer that may hinder Chinese international collaborative research in the RM domain. For instance, the *Regulation on the Administration of Human Genetic Resources* and the *Biosecurity Law* prevent the provision of Chinese human genetic resources to foreign entities and individuals, and impose stringent requirements on the collection, utilization, and export of human genetic resources. Specifically, RM projects involving international collaboration face intricate application, approval, and filing procedures to adhere to regulatory requirements, such as extensive documentation, rigorous ethical reviews, and multiple layers of governmental approval. These procedures are intended to ensure compliance with national security and bioethical standards. However, they may inadvertently impede the efficiency and timeliness of international research collaborations in the RM field. Additionally, national legislation on personal data protection like *PIPL* mandates security assessments and individual consent for transferring personal information abroad. While these measures aim to safeguard biological resources and personal data, they also introduce complexities and procedural hurdles for RM projects involving international cooperation.

Another area that remains insufficiently addressed within China's current laws and regulations is data ownership and intellectual property (IP) rights in the context of international cooperation. Current regulations, such as Articles 22 and 24 of the *Regulation on the Administration of Human Genetic Resources*, require that international research collaborations involving Chinese genetic resources define ownership and establish benefit-sharing arrangements for research outcomes [91]. They also mandate that all data and records generated from such research be fully accessible to, and backed up by, the Chinese partner [91]. However, these provisions lack clear guidelines for actual implementation. For instance, while Article 28 of the *Regulation on the Administration of Human Genetic Resources* permits both parties to use the human genetic resource information generated during international collaborations, it does not specify how such data may be jointly accessed, reused in subsequent research, or shared with third parties. Similarly, Articles 24 and 25 provide that patents must be jointly applied for and co-owned by the two parties, and that the use, transfer, and profit-sharing of other research outcomes should be governed by a contract. In the absence of such an agreement, outcomes should be shared proportionally based on each party's contribution, with joint consent required for transfer to third parties. However, these provisions lack detailed guidance on several critical aspects, including how to assess contribution levels, manage and enforce licensing to third parties, and apply default regulatory mechanisms when contracts do not specify such terms.

In practice, the lack of enforceable operational rules can lead to uncertainty and related disputes, such as who holds the right to publish research results first, how data can be reused or transferred across borders, and how to allocate commercial benefits [126, 127]. This gap may be particularly problematic in collaborations involving private-sector actors, where commercial stakes are often high [128, 129]. Without a clear and enforceable framework governing data ownership and IP attribution, long-term international cooperation may be jeopardized due to legal ambiguity and a lack of mutual trust.

Addressing the challenges surrounding international collaboration requires finding a regulatory balance between safeguarding national interests (including data security, technology sovereignty, and IP protection) and promoting international collaboration. This pursuit involves exploring simplified approval procedures for international collaborative research projects, clarifying rules on data ownership and benefit-sharing, and engaging in developing regional and international rules that promote secure and ethical data sharing in RM. These changes will not only advance scientific research and the development of new therapies but will also strengthen China's position and credibility in the international community.

Challenges Concerning Global Disparities in RM

The uneven development of RM in LMICs poses a significant challenge to patients' ability to obtain equitable access to life-saving treatments. The disparity in accessibility reflects the dominant role of high-income countries (HICs) in biomedical research and innovation. In consideration of global disparities in development, existing research indicates that the terms Global North and Global South have evolved beyond their original geographical meanings to represent political and economic constructs highlighting disparities in development and influence among nations [130–132]. Coined in the 1950s and 1960s, these terms initially categorized the Global South as comprising LMICs, contrasted with the Global North, which referred to HICs [133]. The Global North traditionally holds more advanced economic systems, technological capabilities, and political influence, thus dominating international affairs. According to the United Nations Conference on Trade and Development (UNCTAD), the Global South encompasses all countries in Asia (excluding Japan, South Korea, and Israel), Africa, Latin America, the Caribbean, and Oceania (excluding Australia and New Zealand) [130]. These regions share common characteristics such as lower income levels, dense populations, inadequate infrastructure, and political and cultural marginalization within the global context [131, 132, 134].

Studies have shown that diseases affecting both LMICs and HICs usually attract more research focus and resources for innovative health products and precision treatments tailored for populations in developed economies [135–138]. Research also suggests that there is significant global inequality in the dissemination and adoption of emerging medical technologies and therapies between HICs and LMICs [135–139]. The Global Health and Healthcare Strategic Outlook: A Shared Vision for 2035 (China Edition) (2023) presented at the World Economic Forum identified challenges for LMICs regarding the accessibility and affordability of advanced medical technologies and drugs, such as payment and reimbursement issues, inadequate dissemination of health and technological knowledge, and the lack of standardized frameworks for evaluating clinical outcomes [140]. In addition to the traditional leadership role of the HICs in R&D, challenges in the LMICs also include constrained research resources and infrastructure, scarcity of trained research personnel, lack of technical guidance, limited funding, underdeveloped policy and regulatory frameworks, and limited international cooperation efforts [135].

Interventions by the International Community

The United Nations (UN), WHO, and other international organizations are addressing the aforementioned issues through various measures. For example, the United Nations Industrial Development Organization (UNIDO) consistently provides technical assistance for pharmaceutical and biologics manufacturing in less developed regions by promoting partnerships, investments, and technology transfers, through collaborative platforms between the Global North and South, and within the Global South [141]. Moreover, the WHO has consistently reiterated the need for global access to safe and effective new drugs and medical technologies [142–146]. To improve access to innovative biopharmaceuticals in LMICs, the WHO has launched pilot projects for the prequalification of specific biosimilars [145]. At the 2019 UN High-level Meeting on Universal Health Coverage, initiatives were launched to enhance global accessibility and affordability of cell and gene therapies [147]. Increasing price and cost transparency across the value chain is one of the pathways highlighted by these initiatives. These initiatives prioritize the implementation of innovative incentives, financing mechanisms, and transparent public–private partnerships to drive the R&D of new health products and technologies. Additionally, the initiatives underscore the critical importance of training, building, and retaining a skilled healthcare workforce, particularly in LMICs, to address the shortage of healthcare professionals capable of engaging in biomedical innovation. Furthermore, international organizations advocating for the rights of patients with rare diseases have issued consensus documents stressing the critical need for more efficient sharing of diagnostic expertise and RM-related data on a global scale [148–151].

China's Approaches to Addressing Global Disparities Within RM

China has launched the Global Development Initiative and the Belt and Road Initiative to promote global development and south–south cooperation among countries in the Global South [152–154]. These initiatives aim to foster infrastructure connectivity, deepen economic collaborations, and support greater self-sufficiency and advancement among LMICs. In August 2023, Chinese President Xi Jinping affirmed China's commitment to safeguarding the collective interests of LMICs [154]. In the field of innovative medicines, China has taken measures such as increasing economic support, knowledge sharing, and technological assistance to ensure the equitable distribution of cutting-edge biotechnology and medical innovations worldwide, including addressing regional disparities [155–157]. For instance, through the country's commitment to south–south cooperation, China has allocated over $56 billion USD in foreign aid, supporting nearly 3000 aid projects in sectors such as innovative technology, healthcare, and wellness [158–160]. Within the scope of these measures, several recent initiatives have focused on promoting international health collaborations in the field of RM. For example, the China Stem Cell Group, through its affiliated hospital in Boao, Hainan, has launched a major project focused on stem cell transplantation for thalassemia [161]. This project aims to establish a world-class stem cell transplantation center, with plans to extend its services to other LMICs in Southeast Asia.

Another example of China's efforts toward global development is the Centre Hospitalier Universitaire de Brazzaville (CHU) Sickle Cell Research and Treatment Center in the Republic of the Congo. The center was constructed in 2015 by the Power Construction Corporation of China in an aim to support RM-related infrastructure in LMICs [162]. Although publicly available data on patient outcomes from this specific center remain limited, to date, the center is recognized as the most specialized facility of its kind in Central and Western Africa and has significantly improved the region's capacity to diagnose and treat sickle cell disease by using advanced biomedical technologies [162].

In addition to building infrastructure, China has assisted other LMICs by deploying medical teams and assisting in establishing pharmaceutical factories. Key institutions like the Chinese Academy of Sciences have offered training programs for professionals involved in RM R&D [163]. Moreover, Chinese research institutions have collaborated with multiple countries in the Global South to establish transnational technology transfer platforms and joint laboratories in the health sector. These collaborations promote cross-border medical cooperation, facilitate the exchange of knowledge and innovative medical outcomes, and provide targeted technical support for prevalent diseases in recipient countries [164–166]. These partnerships are essential for enhancing RM's capabilities within the Global South countries.

However, China's cooperation with other LMICs still presents challenges, especially regarding data governance. As discussed earlier, China has instituted strict health data localization laws and approval procedures for international collaboration on research involving Chinese human genetic data. These rigorous data protection regulations could hinder international research collaboration between China and LMICs in the field of RM, especially those involving human genetic data and personal health records. Ideally, a balanced regulatory framework would uphold robust data privacy protections while facilitating secure and efficient cross-border data sharing for scientific research and clinical applications. Such a system would not only protect individual rights but also facilitate collaborative innovation in RM between China and global partners. In this regard, Chinese scholars have been advocating for the establishment of health and medical data flow regulations that reflect China's position, which emphasizes promoting international collaboration while safeguarding data security, protecting individual privacy, and maintaining national data sovereignty, particularly in partnerships with other LMICs [167–169]. Chinese legal scholars have also urged the creation of a data flow framework that serves the common interests of multiple parties, including healthcare and research institutions, patients and the general public, and international collaboration partners, to ensure equitable access to medical research and advancements [170].

Considering China's dedication to protecting the collective interests of LMICs in ensuring the accessibility and affordability of advanced biomedical technologies and medications, the Chinese government should actively promote the establishment of international genomic data-sharing initiatives. In this regard, China actively engages with international organizations like the International Stem Cell Forum (ISCF) and commits to supporting the global sharing of RM-related research resources and data. As a member of the ISCF, China promotes the international circulation and exchange of cell lines and other research resources while advocating for unified standards for international stem cell banks [170]. While some HICs have already developed sophisticated data-sharing infrastructures and participate in global platforms, such as the Global Alliance for Genomics and Health, many LMICs still lack comparable institutional capacity and negotiating power [171,

172]. Establishing a formal international genomic data alliance could enable LMICs in particular to devise data-sharing initiatives, policies, and programs, providing a platform to negotiate fair and equitable data-sharing terms.

Conclusion

In this chapter, we provided an overview of the regulatory environment for innovation in the field of RM in China. Our analysis indicates that consistent financial investments and supportive policies have enabled significant progress in the development of RM therapies. To further technological advancements, the Chinese government has established rigorous regulatory regimes to promote responsible research and innovation. These systems encompass departmental rules for biomedical R&D, administrative regulations for the supervision of medical institutions and personnel, and the management of human genetic resources, as well as other applicable national laws. Our analysis suggests that while China has developed proactive measures to strengthen regulatory oversight and to guarantee the legality and ethical integrity of RM research and clinical translation, several challenges remain to be addressed. These include issues related to bioethical governance, regulatory gaps, and other barriers impacting international cooperation between China and other countries.

Additionally, we scrutinized issues regarding the accessibility and affordability of RM products and applications in LMICs. We explored Chinese measures for promoting biomedical research and improving access to innovative RM therapies in LMICs. We suggest that China is in a position to take a leadership role in creating a formal international genomic data alliance with LMICs, largely by offering a platform for negotiating fair and equitable data-sharing terms. Such international collaboration would not only promote technological advancements but could also enhance accessibility and affordability of RM-related treatments in LMICs specifically.

Acknowledgments This research is funded by the University of Macau under the research grant No. MYRG-GRG2023-00025-FLL.

References

1. Hosseinkhani H, Domb AJ, Sharifzadeh G, Nahum V (2023) Gene therapy for regenerative medicine. Pharmaceutics 15(3):856–895. https://doi.org/10.3390/pharmaceutics15030856
2. Mayo Clinic (2024) Stem cells: what they are and what they do. Mayo Foundation for Medical Education and Research. https://www.mayoclinic.org/tests-procedures/bone-marrow-transplant/in-depth/stem-cells/art-20048117. Accessed 20 Jun 2025
3. U.S. Food and Drug Administration (2025) Approved cellular and gene therapy products. FDA. https://www.fda.gov/vaccines-blood-biologics/cellular-gene-therapy-products/approved-cellular-and-gene-therapy-products. Accessed 18 Jun 2025
4. Exactitude Consultancy (2023) Cell therapy technology market. 2025 © Exactitude Consultancy. https://exactitudeconsultancy.com/zh-CN/reports/14414/cell-therapy-technologies-market/#report-details. Accessed 19 Jun 2025

5. Chen J (2023) China's original CD19 CAR-T product Yuanruida® (Nakiolunsai injection) was approved for marketing. China News Service Shanghai. https://www.sh.chinanews.com.cn/yljk/2023-11-08/118129.shtml. Accessed 18 Jun 2025
6. Custom Market Insights (2024) Global regenerative medicine market 2025–2034. CMI Consulting LLC. https://www.custommarketinsights.com/report/regenerative-medicine-market/. Accessed 18 Jun 2025
7. Barrera MDB, Franco-Martínez F, Díaz-Lantada A (2021) Artificial intelligence aided design of tissue engineering scaffolds employing virtual tomography and 3D convolutional neural networks. Materials (Basel) 14(18):5278–5298. https://doi.org/10.3390/ma14185278
8. Wu Y, Ding X, Wang Y, Ouyang D (2024) Harnessing the power of machine learning into tissue engineering: current progress and future prospects. Burns Trauma 12:tkae053. https://doi.org/10.1093/burnst/tkae053
9. Issa J, Abou Chaar M, Kempisty B, Gasiorowkski L, Olszewski R, Mozdziak P, Dyszkiewicz-Konwińska M (2022) Artificial-intelligence-based imaging analysis of stem cells: a systematic scoping review. Biology (Basel) 11(10):1412–1426. https://doi.org/10.3390/biology11101412
10. Marzec-Schmidt K, Ghosheh N, Stahlschmidt SR, Küppers-Munther B, Synnergren J, Ulfenborg B (2023) Artificial intelligence supports automated characterization of differentiated human pluripotent stem cells. Stem Cells 41(9):850–861. https://doi.org/10.1093/stmcls/sxad049
11. 360iResearch (2023) Artificial intelligence in regenerative medicine market worth $240.63 million by 2030—exclusive report by 360iResearch. EIN Presswire. https://www.einpresswire.com/article/668401736/artificial-intelligence-in-regenerative-medicine-market-worth-240-63-million-by-2030-exclusive-report-by-360iresearch. Accessed 18 Jun 2025
12. Harris AR, Walker MJ, Gilbert F (2022) Ethical and regulatory issues of stem cell-derived 3-dimensional organoid and tissue therapy for personalised regenerative medicine. BMC Med 20:499–509. https://doi.org/10.1186/s12916-022-02710-9
13. Master Z, Matthews KRW, Abou-El-Enein M (2021) Unproven stem cell interventions: a global public health problem requiring global deliberation. Stem Cell Reports 16(6):1435–1445. https://doi.org/10.1016/j.stemcr.2021.05.004
14. Ogbogu U, Du L, Rachul C, Bélanger L, Caulfield T (2013) Chinese newspaper coverage of (unproven) stem cell therapies and their providers. Stem Cell Rev Rep 9(2):111–118. https://doi.org/10.1007/s12015-012-9425-0
15. Raposo VL, Du L (2020) Stem cell based products in Europe and in China: where are we and where should we go? Eur Pharm Law Rev 4(3):161–171. https://doi.org/10.21552/eplr/2020/3/6
16. Thompson D (2024) Government crackdowns can cripple bogus stem cell 'cures' industry. HealthDay. https://www.usnews.com/news/health-news/articles/2024-10-11/government-crackdowns-can-cripple-bogus-stem-cell-cures-industry. Accessed 18 Jun 2025
17. Wai KM, Leng T, Goldberg J (2022) Putting stem cell-based therapies in context. National Eye Institute Research News. https://www.nei.nih.gov/about/news-and-events/news/putting-stem-cell-based-therapies-context. Accessed 18 Jun 2025
18. Xinhua News Agency (2021) 98,000 yuan per shot: the black market for "stem cell therapy" is growing. Economic Information Daily. http://www.xinhuanet.com/fortune/2021-04/14/c_1127327015.htm. Accessed 18 Jun 2025
19. United Nations Educational, Scientific and Cultural Organization (2006) Universal declaration on bioethics and human rights. https://unesdoc.unesco.org/ark:/48223/pf0000146180. Accessed 28 Jul 2025
20. World Medical Association (2024) Declaration of helsinki. https://www.wma.net/what-we-do/medical-ethics/declaration-of-helsinki/. Accessed 28 Jul 2025
21. International Society for Stem Cell Research (2019) The ISSCR releases informed consent standard for stem cell-based interventions. ISSCR. https://www.isscr.org/isscr-news/the-isscr-releases-informed-consent-standard-for-stem-cell-based-interventions. Accessed 20 Jun 2025

22. World Health Organization (2021) WHO considerations on regulatory convergence of cell and gene therapy products. WHO Press. https://cdn.who.int/media/docs/default-source/biologicals/ecbs/who-public-consultation_cgtp-white-paper_16_dec_2021.pdf?sfvrsn=18f6c549_5. Accessed 18 Jun 2025
23. World Health Organization (2023) Cell, tissue and gene therapy products. https://www.who.int/teams/health-product-policy-and-standards/standards-and-specifications/cell%2D%2Dtissue-and-gene-therapy-products. Accessed 18 Jun 2025
24. European Medicines Agency (2024) Advanced therapy medicinal products: overview. EMA. https://www.ema.europa.eu/en/human-regulatory-overview/advanced-therapy-medicinal-products-overview. Accessed 19 Jun 2025
25. Imran SAM, Hamizul MHAM, Bariah AANK, Zaman WSWK, Nordin F (2022) Regenerative medicine therapy in Malaysia: an update. Front Bioeng Biotechnol 10:789644. https://doi.org/10.3389/fbioe.2022.789644
26. Tobita M, Konomi K, Torashima Y, Kimura K, Taoka M, Kaminota M (2016) Japan's challenges of translational regenerative medicine: act on the safety of regenerative medicine. Regen Ther 4:78–81. https://doi.org/10.1016/j.reth.2016.04.001
27. U.S. Food and Drug Administration (2019) Framework for the regulation of regenerative medicine products. FDA. https://www.fda.gov/vaccines-blood-biologics/cellular-gene-therapy-products/framework-regulation-regenerative-medicine-products. Accessed 18 Jun 2025
28. Alemayehu C, Mitchell G, Nikles J (2018) Barriers for conducting clinical trials in developing countries: a systematic review. Int J Equity Health 17(1):37–47. https://doi.org/10.1186/s12939-018-0748-6
29. World Health Organization (2022) Towards a global guidance framework for the responsible use of life sciences: summary report of consultations on the principles, gaps and challenges of biorisk management, May 2022. WHO. https://www.who.int/publications/i/item/WHO-SCI-RFH-2022.01. Accessed 18 Jun 2025
30. Fears R, Akutsu H, Alentajan-Aleta LT, Caicedo A, de Carvalho ACC, Čolić M, Cornish J, Cossu G, Debré P, Dierckxsens G, El-Badri N, Griffin G, Hsieh PC, Inamdar MS, Kumar P, Abraham CM, Maciulaitis R, Mahtab MA, O'Brien FJ, Pepper MS, ter Meulen V (2021) Inclusivity and diversity: integrating international perspectives on stem cell challenges and potential. Stem Cell Reports 16(8):1847–1852. https://doi.org/10.1016/j.stemcr.2021.07.003
31. Sugarman J, Clark A, Fishkin J, Kato K, McCormack K, Munsie M, Peluso MJ, René N, Solomon SL (2023) Critical considerations for public engagement in stem cell-related research. Stem Cell Reports 18(2):420–426. https://doi.org/10.1016/j.stemcr.2023.01.002
32. Lyons S, Salgaonkar S, Flaherty GT (2022) International stem cell tourism: a critical literature review and evidence-based recommendations. Int Health 14(2):132–141. https://doi.org/10.1093/inthealth/ihab050
33. Jiménez C (2025) Ethical, legal, and social implications of stem cell tourism. MedicalTourism.com. https://www.medicaltourism.com/articles/ethical-legal-and-social-implications-of-stem-cell-tourism. Accessed 20 June 2025
34. Langford L, Foong P (2024) Unproven stem cell therapies: an evaluation of patients' capacity to give informed consent. Griffith Law Review 33(1):58–88. https://doi.org/10.1080/10383441.2024.2422138
35. U.S. Food and Drug Administration (2023) Fast track, breakthrough therapy, accelerated approval, priority review. FDA. https://www.fda.gov/patients/learn-about-drug-and-device-approvals/fast-track-breakthrough-therapy-accelerated-approval-priority-review. Accessed 28 Jul 2025
36. European Parliament and Council of the European Union (2007) Regulation (EC) No 1394/2007 of the European Parliament and of the Council of 13 November 2007 on advanced therapy medicinal products and amending Directive 2001/83/EC and Regulation (EC) No 726/2004 (Text with EEA relevance). EUR-Lex. https://eur-lex.europa.eu/eli/reg/2007/1394/oj/eng. Accessed 25 Jul 2025

37. Yokomizo R, Hosoya S, Umezawa A, Okamoto A (2025) Amendment of the act on the safety of regenerative medicine: a new voyage of regenerative medicine in Japan and the dawn of a new era for young investigators. Regen Ther 29:237–246. https://doi.org/10.1016/j.reth.2025.03.011
38. General Office of the State Council of China (2021) Implementation opinions of the general office of the State Council on comprehensively strengthening the construction of drug regulatory capacity. General Office of the State Council Operation and Maintenance Unit: China Government Network Operation Center. https://www.gov.cn/zhengce/content/2021-05/10/content_5605628.htm. Accessed 18 Jun 2025
39. Du L (2021) Research and development of cutting-edge biotechnology: legal challenges and responses. Wuhan University Press, Wuhan. https://wdp.com.cn/books/2024-05-05/7814.html. Accessed 29 Jul 2025
40. Yuan B (2016) The law-regulation-guidance regulatory system for stem cell-based medicinal products. Chinese Bull Life Sci 8:949–957. https://doi.org/10.13376/j.cbls/2016128
41. Diao Y, Liang Y (2019) Management of cell therapy products in Japan and its implications for China. Prog Pharm Sci 43(12):908–913. https://pps.cpu.edu.cn/en/article/id/578c9402-24d6-43a4-ad05-e3bc2633849a. Accessed 25 Jul 2025
42. Lu J, Liu B, Luo J (2019) Research progress and regulatory perspectives for stem cell-based regenerative medicine products. Scientia Sinica (Vitae) 49(1):18–27. https://doi.org/10.1360/N052018-00138
43. Peng Y, Huang X, Zhou Q (2020) Ethical and policy considerations for human embryo and stem cell research in China. Cell Stem Cell 27(4):511–514. https://doi.org/10.1016/j.stem.2020.09.010
44. Peng Y, Zhou Q (2021) China's strategies for governance of biotechnological changes and new ethical challenges. Bull Chin Acad Sci 36(11):1288–1297. https://doi.org/10.16418/j.issn.1000-3045.20210823001
45. Zhang Y (2023) Opportunities and challenges for development and industrialization of new drugs for cell therapy in China. Prog Pharm Sci 47(1):1–5. https://doi.org/10.20053/j.issn1001-5094.2023.01.001
46. National Natural Science Foundation of China (2023) Announcement on the release of the 2023 project guidelines for the major research program on information decoding and orderly regulation of tissue and organ regeneration and repair. National Natural Science Foundation of China, Department of Medical Sciences. https://www.nsfc.gov.cn/publish/portal0/tab1482/info90381.htm. Accessed 20 Jun 2025
47. University of Science and Technology of China (2023) NSFC notice: 2023 project guidelines for five major research programs. Research Department of University of Science and Technology of China. https://kyb.ustc.edu.cn/2023/0903/c6077a611004/page.htm. Accessed 18 Jun 2025
48. Ministry of Science and Technology of China (2019) Application guidelines for six key projects including "stem cell and translational research" were released. ScienceNet.cn. https://news.sciencenet.cn/htmlnews/2019/1/422475.shtm. Accessed 20 Jun 2025
49. State Council of China (2016a) Notice of the State Council on issuing the 13th five-year national science and technology innovation plan. General Office of the State Council Operation and Maintenance Unit. https://www.gov.cn/zhengce/content/2016-08/08/content_5098072.htm. Accessed 20 Jun 2025
50. National Health Commission of China (2021) Response to recommendation no. 7090 from the fourth session of the 13th National People's Congress. https://www.nhc.gov.cn/wjw/jiany/202112/70ba1d99e75c4acab6215f651badef45.shtml. Accessed 25 Jul 2025
51. Peking University Health Science Center (2022) PKU Medicine-led national key R&D programs receive outstanding evaluations in comprehensive performance assessment. https://www.bjmu.edu.cn/yjcg/e1b71f439d174be7a063db4b16685bae.htm. Accessed 26 Jul 2025
52. Yang Z, Chen Q, Guo W, Wang J, Zhang Y, Lu S, Yu Z, Shen J (2022) Organization, implementation and thinking of the key projects of stem cell and transformation research in China.

China Biotechnology 42(9):116–123. https://www.sinomed.ac.cn/article.do?ui=2022493785. Accessed 28 Jul 2025
53. Zhu P, Liu X (2018) Stem cell and translational research receives 585 million yuan funding with industrialization as key focus. https://m.21jingji.com/article/20180606/08f986e4b856fe1189262d05ac3d3c4a.html. Accessed 18 Jun 2025
54. Nuwacell (2022) Nuwacell undertakes the key project of "stem cells and organ repair" of the national key R&D program. Zhongsheng Suyuan Biotechnology Co., Ltd. https://m.nuwacell.com/index.php/news/23/51.html. Accessed 20 Jun 2025
55. People's Daily (2016) "13th Five-Year Plan" stem cell and translational research projects launched. http://www.xinhuanet.com/politics/2016-09/20/c_129288552.htm. Accessed 26 Jul 2025
56. Public Hygiene and Health Commission of Shenzhen Municipality (2024) International rare disease day: fighting the one-in-ten-thousand probability. Public Hygiene and Health Commission of Shenzhen Municipality. https://wjw.sz.gov.cn/wzx/content/post_11169478.html. Accessed 28 Jul 2025
57. General Office of the State Council of China (2016) Guiding opinions of the General Office of the State Council on promoting the healthy development of the pharmaceutical industry. General Office of the State Council Operation and Maintenance Unit. https://www.gov.cn/zhengce/content/2016-03/11/content_5052267.htm. Accessed 20 Jun 2025
58. Industry Coordination Department (2016) Notice on the issuance of the "Guidelines for the Development of the Pharmaceutical Industry". National Development and Reform Commission of the People's Republic of China. https://www.ndrc.gov.cn/fggz/cyfz/zcyfz/201611/t20161108_1149864.html. Accessed 20 Jun 2025
59. State Council of China (2015) Opinions of the State Council on reforming the review and approval system for drugs and medical devices. General Office of the State Council Operation and Maintenance Unit. https://www.gov.cn/zhengce/content/2015-08/18/content_10101.htm. Accessed 20 Jun 2025
60. State Council of China (2016b) Notice of the State Council on issuing the 13th five-year plan for the development of strategic emerging industries. General Office of the State Council Operation and Maintenance Unit. https://www.gov.cn/zhengce/content/2016-12/19/content_5150090.htm. Accessed 18 Jun 2025
61. Wuhan Municipal Government (2023) The implementation plan for the comprehensive pilot program of further opening up the service industry in Wuhan City. Wuhan Municipal People's Government Office Documents. https://sw.wuhan.gov.cn/zfxxgk/zc/zcfg/202312/P020231225542173801488.pdf Accessed 18 Jun 2025
62. Xinhua News Agency (2025) The National Medical Products Administration conditionally approved the marketing of Amimestrocel injection. Xinhuanet. https://www.news.cn/health/20250103/333eb0c67e0f4a13b47aea9b97f9b071/c.html>. Accessed 18 Jun 2025
63. Department of Commerce of Guangdong Province (2015) Notice of the ministry of finance and the state administration of taxation on the preferential corporate income tax policies and preferential catalogues for Guangdong Hengqin New Area, Fujian Pingtan Comprehensive Experimental Zone, Shenzhen Qianhai Shenzhen-Hong Kong Modern Service Industry Cooperation Zone. Ministry of Finance. https://com.gd.gov.cn/zggdzymysyq/zcfg/gjzc/content/post_4072938.html. Accessed 26 Jun 2025
64. HKTDC Research Business Advisory Team (2022) Hengqin further implements preferential corporate income tax policies. HKTDC Research. https://research.hktdc.com/sc/article/MTA5MzkwMjkyOQ. Accessed 19 Jun 2025
65. National Development and Reform Commission of China (2022) Opinions on several special measures to relax market access in Shenzhen's construction of a pilot demonstration zone for socialism with Chinese characteristics. Ministry of Commerce of the People's Republic of China. https://www.gov.cn/zhengce/zhengceku/2022-01/26/content_5670555.htm. Accessed 20 Jun 2025
66. National Development and Reform Commission of China, Ministry of Commerce of China (2023) The National Development and Reform Commission and the Ministry of Commerce

issued a statement on supporting the Guangdong-Macao cooperation in Hengqin, opinions on relaxing special measures for market access in the cooperation zone. National Development and Reform Commission of China. https://www.ndrc.gov.cn/xxgk/zcfb/tz/202312/t20231226_1362933.html. Accessed 25 Jul 2025
67. State Council of China (2023) The State Council's reply on the "work plan to support Beijing in deepening the construction of a national comprehensive demonstration zone for expanding and opening up the service industry. General Office of the State Council Operation and Maintenance Unit: China Government Network Operation Center. https://www.gov.cn/zhengce/content/202311/content_6916720.htm. Accessed 18 Jun 2025
68. Ministry of Health of China (1993) Key points for quality control in clinical research of human somatic cell therapy and gene therapy. http://www.ssctcell.com:10000/Upload/164068028696.pdf. Accessed 28 Jul 2025
69. State Food and Drug Administration of China (2003) Guiding principles for research and quality control techniques of human cell therapy preparations. https://yiqi-oss.oss-cn-hangzhou.aliyuncs.com/aliyun/technology/275923/120589.pdf. Accessed 27 Jul 2025
70. Ministry of Health of China (1999) Notice on printing and distributing the cord blood hematopoietic stem cell bank management measures (provisional). https://www.nhc.gov.cn/qjjys/s7935/200804/2315cbfb0e8b4bc4b2d75ca43b1d5832.shtml. Accessed 28 Jul 2025
71. National Health and Family Planning Commission of China (2001) Notice of the Ministry of Health on the issuance of the "Cord Blood Hematopoietic Stem Cell Bank Establishment and Management Standards (Trial)". National Health and Family Planning Commission of the People's Republic of China. http://www.nhc.gov.cn/zwgkzt/glgf/201308/b886839e9b3241bdb710b753bfbe962c.shtml. Accessed 20 Jun 2025
72. Ministry of Health of China (2001) Measures on the management of assisted reproductive technologies in humans. General Office of the State Council Operation and Maintenance Unit. https://www.gov.cn/gongbao/content/2002/content_61906.htm. Accessed 28 Jul 2025
73. Ministry of Science and Technology of China, Ministry of Health of China (2003) Ethical guiding principles for the research of human embryonic stem cell. https://www.scrcnet.org/download/eccr_31.pdf. Accessed 28 Jul 2025
74. Ministry of Health of China (2008) Guidelines for hospital management evaluation. ZhengYiZhiKu. https://zk.faxin.cn/fl/zyfl/content.html?gid=A75124&libid=all. Accessed 28 Jul 2025
75. Ministry of Health of China (2009a) Notice of the Ministry of Health on the issuance of the "measures for the administration of the clinical application of medical technologies". General Office of the State Council Operation and Maintenance Unit. https://www.gov.cn/gongbao/content/2009/content_1388686.htm. Accessed 20 Jun 2025
76. National Health Commission of China (2024) Announcement on declaring the fifth batch of commission documents invalid (Guo Wei Ban Fa [2024] No. 3). https://www.nhc.gov.cn/bgt/c100023/202401/8f5e1e8062f6402f8a2e419a42019370.shtml. Accessed 25 Jul 2025
77. National Health Commission of China (2018) Measures for the administration of the clinical application of medical technologies. General Office of the State Council Operation and Maintenance Unit. https://www.gov.cn/gongbao/content/2018/content_5346680.htm. Accessed 20 Jun 2025
78. General Office of the Ministry of Health of China (2009) Notice on the release of the first batch of catalogue of class III medical technologies approved for clinical application. https://www.nhc.gov.cn/yzygj/c100068/200905/7756a86d63d94af4acbc983c13a1990d.shtml. Accessed 28 Jul 2025
79. Ministry of Health of China (2009b) Letter on seeking opinions on the management specifications for third-class medical technologies, including technical specifications for gender reassignment surgery (draft for comment). Department of Medical Administration. https://www.cma.org.cn/attachment/2009/11/27/20091127677c696e62fa47609683657a602a917c.pdf. Accessed 20 Jun 2025
80. Ministry of Health of China (2011) Notice on conducting self-inspection and self-correction work on stem cell clinical research and application. Ministry of Health of the People's

Republic of China. http://www.nhc.gov.cn/zwgkzt/pkjjy1/201201/53890.shtml. Accessed 20 Jun 2025
81. General Office of the National Health and Family Planning Commission of China, General Office of the State Food and Drug Administration of China (2015) Notice on issuing the guidelines for quality control and preclinical research of stem cell preparations (trial implementation). National Medical Products Administration. https://www.nmpa.gov.cn/xxgk/fgwj/gzwj/gzwjyp/20150731120001226.html?type=pc&m=. Accessed 20 Jun 2025
82. National Health and Family Planning Commission of China, State Food and Drug Administration of China, National Administration of Traditional Chinese Medicine of China (2014) Management measures for clinical research projects conducted by medical and health institutions. https://www.nhc.gov.cn/yzygj/c100068/201410/2dd7982842ac4945ab52da7277537f4a.shtml. Accessed 26 Jul 2025
83. National Health and Family Planning Commission of China, State Food and Drug Administration of China (2015) Notice on issuing the management measures for stem cell clinical research (provisional). National Medical Products Administration. https://www.nmpa.gov.cn/directory/web/nmpa//yaopin/ypfgwj/ypfgbmgzh/20150720120001607.html. Accessed 26 Jul 2025
84. Ma T (2013) Ministry of Health: those who conduct stem cell clinical trials in violation of regulations will be disqualified. China News Network. https://www.chinanews.com.cn/gn/2013/03-07/4623841.shtml. Accessed 19 Jun 2025
85. National Health and Family Planning Commission of China (2015) Notice on the cancellation of approval for the clinical application of third-class medical technologies. https://www.nhc.gov.cn/yzygj/c100068/201507/57e5e35a3b484b9986ba41b0d5b2ff96.shtml. Accessed 26 Jul 2025
86. National Medical Products Administration of China (2019) Implementation rules for the management measures of stem cell clinical trial research projects (provisional). National Medical Products Administration. https://www.nmpa.gov.cn/directory/web/nmpa/images/ufrOwLDsv8a9zLqvoblyMDE1obMyMjW6xSC4vbz+MS5kb2M=.doc. Accessed 26 Jul 2025
87. Ministry of Science and Technology of China (2015) Notice of the General Office of the Ministry of Science and Technology on the implementation of administrative licensing for the collection, acquisition, trade, export, and exit of human genetic resources. Ministry of Science and Technology of the People's Republic of China. https://www.most.gov.cn/xxgk/xinxifenlei/fdzdgknr/fgzc/gfxwj/gfxwj2015/201509/t20150930_121850.html. Accessed 20 Jun 2025
88. National Health and Family Planning Commission of China (2016) Measures for the ethical review of biomedical research involving humans. https://www.nhc.gov.cn/wjw/c100221/202201/558653bd3d1e4a078904152a30605a03/files/85629ae9fbe64956aeeb13c380bd313e.pdf. Accessed 26 Jul 2025
89. State Council of China (2018) Regulation on the prevention and handling of medical disputes. General Office of the State Council Operation and Maintenance Unit. https://www.gov.cn/zhengce/content/2018-08/31/content_5318057.htm. Accessed 18 Jun 2025
90. Ministry of Science and Technology of China (2023) Implementation rules of the regulations on the administration of human genetic resources issued by the Ministry of Science and Technology no. 21. Ministry of Science and Technology of the People's Republic of China. https://www.most.gov.cn/xxgk/xinxifenlei/fdzdgknr/fgzc/bmgz/202306/t20230601_186416.html. Accessed 20 Jun 2025
91. State Council of China (2019) Regulations of the People's Republic of China on the administration of human genetic resources. General Office of the State Council Operation and Maintenance Unit. https://www.gov.cn/zhengce/content/2019-06/10/content_5398829.htm. Accessed 18 Jun 2025
92. Ministry of Commerce of China (2020) Agreement concerning amendment to the agreement on trade in services of the Mainland and Hong Kong closer economic partnership arrangement. General Office of the State Council Operation and Maintenance Unit. https://www.gov.cn/gongbao/content/2020/content_5506449.htm. Accessed 28 Jul 2025

93. Standing Committee of the 13th National People's Congress (2020a) Amendment (XI) to the criminal law of the People's Republic of China. General Office of the State Council Operation and Maintenance Unit. https://www.gov.cn/xinwen/2020-12/27/content_5573660.htm. Accessed 28 Jul 2025
94. National People's Congress of China (2020) Civil Code of the People's Republic of China. General Office of the State Council Operation and Maintenance Unit. https://www.gov.cn/xinwen/2020-06/01/content_5516649.htm. Accessed 26 Jul 2025
95. Standing Committee of the 13th National People's Congress (2020b) Biosecurity law of the People's Republic of China. General Office of the State Council Operation and Maintenance Unit. https://www.gov.cn/xinwen/2020-10/18/content_5552108.htm. Accessed 20 Jun 2025
96. Standing Committee of the 13th National People's Congress (2021a) Personal information protection law of the People's Republic of China. General Office of the State Council Operation and Maintenance Unit. https://www.gov.cn/xinwen/2021-08/20/content_5632486.htm. Accessed 27 Jul 2025
97. Standing Committee of the 13th National People's Congress (2021b) Law on doctors of the People's Republic of China. General Office of the State Council Operation and Maintenance Unit. https://www.gov.cn/xinwen/2021-08/20/content_5632496.htm. Accessed 20 Jun 2025
98. State Council of China (2022) Regulation on the administration of medical institutions (2022 revision). Chinalawinfo Co., Ltd. http://www.lawinfochina.com/display.aspx?id=40901&lib=law. Accessed 18 Jun 2025
99. Standing Committee of the 13th National People's Congress (2019) Basic healthcare and health promotion law of the People's Republic of China. General Office of the State Council Operation and Maintenance Unit. https://www.gov.cn/xinwen/2019-12/29/content_5464861.htm. Accessed 20 Jun 2025
100. Standing Committee of the 13th National People's Congress (2021c) Law of the People's Republic of China on scientific and technological progress (2021 Revision). Ministry of Science and Technology of the People's Republic of China. https://www.most.gov.cn/xxgk/xinxifenlei/fdzdgknr/fgzc/flfg/202201/t20220118_179043.html. Accessed 20 Jun 2025
101. Standing Committee of Shenzhen Municipal People's Congress (2023) Bulletin (No. 1, 2023). https://www.szrd.gov.cn/attachment/0/70/70771/1133794.pdf. Accessed 28 Jul 2025
102. Standing Committee of Hainan Provincial People's Congress (2024) Regulations on promoting biomedical new technologies in the Boao Lecheng International Medical Tourism Pilot Zone of the Hainan Free Trade Port. https://www.hnftp.gov.cn/zczdtx/cyzc/202412/t20241205_3780052.html. Accessed 19 Jun 2025
103. Xinhua News Agency (2020) Issued by the CPC Central Committee and the State Council plan for the construction of Hainan Free Trade Port. State Council Gazette. https://www.gov.cn/gongbao/content/2020/content_5519942.htm. Accessed 18 Jun 2025
104. Hainan Provincial Health Commission (2025) Notice on issuing the implementation measures for the transformation and application of new biomedical technologies in the Boao Lecheng International Medical Tourism Pilot Zone of Hainan Free Trade Port (interim). Hainan Provincial Health Commission. https://wst.hainan.gov.cn/swjw/xxgk/0200/0202/202501/t20250124_3808738.html. Accessed 19 Jun 2025
105. Hainan Provincial Health Commission, Hainan Medical Products Administration (2025) Administrative provisions on the clinical research of biomedical new technologies in the Boao Lecheng International Medical Tourism Pilot Zone. https://wst.hainan.gov.cn/swjw/ywdt/tzgg/202502/P020250227387446914633.pdf. Accessed 19 Jun 2025
106. National Health Commission Medical Ethics Expert Committee, Chinese Hospital Association (2020) Guidelines for the establishment of ethics review committees for clinical research involving humans (2020 edition). https://kjb.szu.edu.cn/info/1146/6827.htm. Accessed 25 Jul 2025
107. Council for International Organizations of Medical Sciences (2016) International ethical guidelines for health-related research involving humans. CIOMS, WHO Press. https://cioms.ch/wp-content/uploads/2017/01/WEB-CIOMS-EthicalGuidelines.pdf. Accessed 18 Jun 2025

108. Chen R, Hu Z (2018) Acceptance and adjustment: spread of ethics of human embryonic stem cell research in China. J Eng Stud 10(5):518–526. https://doi.org/10.3724/SP.J.1224.2018.00518
109. Liu Y, Bian L (2024) On various relations in the development of medical ethics in China. Med Philos 45(18):13–18. https://qikan.cqvip.com/Qikan/Article/Detail?id=7113168414. Accessed 28 Jul 2025
110. International Council for Harmonisation of Technical Requirements for Pharmaceuticals for Human Use (ICH) (2016) ICH harmonised guideline integrated addendum to ICH E6(R1): guideline for good clinical practice E6(R2). ICH. https://database.ich.org/sites/default/files/E6_R2_Addendum.pdf. Accessed 19 Jun 2025
111. Li J (2022) How to strengthen the institutional support for scientific and technological ethics governance. China Agricultural University News. https://news.cau.edu.cn/mtndnew/809779.htm. Accessed 18 Jun 2025
112. Cao Y (2023) Subjects, procedures and method optimization in the construction of science and technology ethical review system. Sci Technol Prog Policy 40(6):121–129. https://doi.org/10.6049/kjjbydc.2022090612
113. Li J, Wang T (2022) The historical motivation of the establishment of the ethical review mechanism of scientific research institutions and its problems in operation. Stud Dialect Nat 38(3):51–57. https://cohd.cau.edu.cn/module/download/downfile.jsp?classid=0&filename=1ffbd821a63947029143075fbc9e21c1.pdf. Accessed 28 Jul 2025
114. Clayton EW, Evans BJ, Hazel JW, Rothstein MA (2019) The law of genetic privacy: applications, implications, and limitations. J Law Biosci 6(1):1–36. https://doi.org/10.1093/jlb/lsz007
115. Kacetl J, Marešová P, Maskuriy R, Selamat A (2020) Ethical questions linked to rare diseases and orphan drugs: a systematic review. Risk Management and Healthcare Policy 13:2125–2148. https://doi.org/10.2147/RMHP.S260641
116. Cao S, Wei X, Du Z, Xu Y, Chen H, Xue D (2020) Public perceptions of ethical issues in precision medicine and its use in Shanghai and Lanzhou. Chinese Health Resources 23(5):468–472. https://doi.org/10.13688/j.cnki.chr.2020.20001
117. Ji D (2017) Legal challenges of precision medicine on genetic privacy protection. Chinese Med Ethics 30(8):936-939+944. https://stlaw.pku.edu.cn/hd/kjflt/kjcxyfl/3641.htm. Accessed 29 Jul 2025
118. Liu C, Chen H (2018) Relative security of health care big data protection. J Zhejiang Univ (Med Sci) 6:563–576. https://doi.org/10.3785/j.issn.1008-9292.2018.12.01
119. Zhu B, Wang C (2020) Analysis on privacy governance in scientific data sharing in China. Library Inf Serv 64(22):37–47. https://doi.org/10.13266/j.issn.0252-3116.2020.22.004
120. Lovell-Badge R, Carpenter M, Charo RA, Clark A, Daley GQ, Hyun I, Knoblich J, Rooke H, Rossant J, Sipp D (2021) ISSCR guidelines for stem cell research and clinical translation. International Society for Stem Cell Research. https://static1.squarespace.com/static/611faaa8fee682525ee16489/t/62ed69b184e2ed258e6eb7e4/1659726257773/isscr-guidelines-for-stem-cell-research-and-clinical-translation-2021.pdf. Accessed 19 Jun 2025
121. Tian Y, Li X (2021) The rule of law and logical transformation of the governance mechanism of the scientific ethics. Sci Technol Prog Policy 38(2):121–127. https://doi.org/10.6049/kjjbydc.2020040351
122. Yao W (2022) Improvement of ethical review mechanism for the application of gene editing technology. J Beijing Univ Technol (Soc Sci Ed) 22(6):123–134. https://doi.org/10.12120/bjutskxb202206123
123. Brophy J (2017) Navigating the 'grey areas': Australian medical travellers in China's stem cell bionetwork. Asia Pac Viewpoint 58(2):216–227. https://doi.org/10.1111/apv.12162
124. Hua Z (2016) How much legal responsibility remains to be clarified in the "Wei Zexi incident". Beijing News. http://www.xinhuanet.com/politics/2016-05/03/c_128950543.htm. Accessed 19 Jun 2025
125. Normile D (2019) Chinese scientist who produced genetically altered babies sentenced to 3 years in jail. Science Insider, American Association for the Advancement of Science. https://

www.science.org/content/article/chinese-scientist-who-produced-genetically-altered-babies-sentenced-3-years-jail. Accessed 20 Jun 2025
126. Dong C, Zhu X, Jiang L, Dai N (2023) Application and management of human genetic resources in China. Shanghai J Prev Med 35(5):513–517. https://doi.org/10.19428/j.cnki.sjpm.2023.22602
127. Lei R, Jiao H, Liu H, Shi J, Zhang JY, Zhang H, Yao P, Wu C, Zhang D, Wang Y, Peng Y, Kang H, Li W, Xu S, Chen T, Chen H, Chen H, Su Y, Zhang H, Ou Y, Zhu W, Ma Y (2022) Policy recommendations for China's 'implementing rules for human genetic resources management'. HUST Center for Bioethics. https://kar.kent.ac.uk/94062/. Accessed 29 Jul 2025
128. Deloitte (2021) Focus on China's human genetic resources management to safeguard biopharmaceutical R&D. https://pdf.dfcfw.com/pdf/H3_AP202106171498420547_1.pdf?1623952761000.pdf. Accessed 25 Jul 2025
129. King & Wood Mallesons (2021) Human genetic resources management legislation, new law enforcement trends and compliance challenges for companies. https://www.chinalawinsight.com/2021/02/articles/dispute-resolution/%E4%BA%BA%E7%B1%BB%E9%81%97%E4%BC%A0%E8%B5%84%E6%BA%90%E7%AE%A1%E7%90%86%E7%AB%8B%E6%B3%95%E3%80%81%E6%89%A7%E6%B3%95%E6%96%B0%E5%8A%A8%E6%80%81%E5%8F%8A%E5%B8%A6%E7%BB%99%E4%BC%81%E4%B8%9A%E7%9A%84/. Accessed 18 Jun 2025
130. El-Gamasy H (2024) Does 'Global South' need to be retired? International Policy Digest. https://intpolicydigest.org/does-global-south-need-to-be-retired/. Accessed 29 Jul 2025
131. Hollington A, Tappe O, Salverda T, Schwarz T (2015) Introduction: concepts of the Global South. Global South Studies Center Cologne. https://web.archive.org/web/20160904205139/http://gssc.uni-koeln.de/node/451. Accessed 19 Jun 2025
132. United Nations Conference on Trade and Development (2022) Handbook of statistics 2022. United Nations Publications. https://unctad.org/system/files/official-document/tdstat47_en.pdf. Accessed 18 Jun 2025
133. Hogan E, Patrick S (2024) A closer look at the Global South. Carnegie Endowment for International Peace. https://carnegieendowment.org/research/2024/05/global-south-colonialism-imperialism?lang=en. Accessed 25 Jul 2025
134. United Nations Conference on Trade and Development (2025) Classification update—March 2025. https://unctadstat.unctad.org/EN/Classifications/ClassificationsNewsletter_March2025_US_EN.pdf. Accessed 28 Jul 2025
135. Adachi T, El-Hattab AW, Jain R, Crespo KAN, Lzo CIQ, Scarpa M, Summar M, Wattanasirichaigoon D (2023) Enhancing equitable access to rare disease diagnosis and treatment around the world: a review of evidence, policies, and challenges. Int J Environ Res Public Health 20(6):4732–4768. https://doi.org/10.3390/ijerph20064732
136. Campbell A, Fane B, Wastl J (2023) A tale of two pharmas—global north and global south. Digital Science. https://www.digital-science.com/tldr/article/a-tale-of-two-pharmas-global-north-and-global-south/. Accessed 19 Jun 2025
137. Oria V (2023) More north-south collaborations needed for global health programs to succeed in sub-Saharan Africa. Alliance for Science. https://allianceforscience.org/blog/2023/04/more-north-south-collaborations-needed-for-global-health-programs-to-succeed-in-sub-saharan-africa/. Accessed 20 Jun 2025
138. Radich JP, Briercheck E, Chiu DT, Menon MP, Torra OS, Yeung CCS, Warren EH (2022) Precision medicine in low- and middle-income countries. Annu Rev Pathol Mech Dis 17(1):387–402. https://doi.org/10.1146/annurev-pathol-042320-034052
139. Gerke S, Minssen T, Cohen G (2020) Ethical and legal challenges of artificial intelligence-driven healthcare. In: Bohr A, Memarzadeh K (eds) Artificial intelligence in healthcare. Elsevier Inc., Amsterdam, pp 295–336. https://doi.org/10.1016/B978-0-12-818438-7.00012-5
140. World Economic Forum (2023) Global health and healthcare strategic outlook: A shared vision for 2035 (China edition). 2023 World Economic Forum. https://www3.weforum.org/

docs/WEF_Global_Health_and_Healthcare_Strategic_Outlook_2023_CN.pdf. Accessed 18 Jun 2025
141. United Nations Industrial Development Organization (2024) The importance of local pharmaceutical production. United Nations Industrial Development Organization. https://www.unido.org/investing-technology-and-innovation-competitiveness-business-environment-and-upgrading/pharmaceutical-production-developing-countries. Accessed 18 Jun 2025
142. Director-General, World Health Organization (2021) Global strategy and plan of action on public health, innovation and intellectual property. World Health Organization, Executive Board. https://apps.who.int/gb/ebwha/pdf_files/EB148/B148_10-en.pdf. Accessed 18 Jun 2025
143. United Nations Human Rights Council (2019) Access to medicines and vaccines in the context of the right of everyone to the enjoyment of the highest attainable standard of physical and mental health: resolution / adopted by the Human Rights Council on 11 July 2019. United Nations Digital Library. https://digitallibrary.un.org/record/3834569?v=pdf. Accessed 18 Jun 2025
144. World Health Organization (2018) Statement for rare disease day. WHO. https://www.who.int/news-room/detail/27-02-2018-statement-for-rare-disease-day. Accessed 18 Jun 2025
145. World Health Organization (2019) WHO prequalifies first biosimilar medicine to increase worldwide access to life-saving breast cancer treatment. WHO. https://www.who.int/news/item/18-12-2019-who-prequalifies-first-biosimilar-medicine-to-increase-worldwide-access-to-life-saving-breast-cancer-treatment. Accessed 18 Jun 2025
146. World Health Organization (2025) Universal health coverage (UHC). WHO. https://www.who.int/news-room/fact-sheets/detail/universal-health-coverage-(uhc). Accessed 18 Jun 2025
147. High-level Meeting on Universal Health Coverage (2019) Political declaration of the high-level meeting on universal health coverage "universal health coverage: moving together to build a healthier world". United Nations. https://www.un.org/pga/73/wp-content/uploads/sites/53/2019/07/FINAL-draft-UHC-Political-Declaration.pdf. Accessed 20 Jun 2025
148. Austin CP, Cutillo CM, Lau LPL, Jonker AH, Rath A, Julkowska D, Thomson D, Terry SF, de Montleau B, Ardigò D, Hivert V, Boycott KM, Baynam G, Kaufmann P, Taruscio D, Lochmüller H, Suematsu M, Incerti C, Draghia-Akli R, Norstedt I, Wang L, Dawkins HJS (2018) Future of rare diseases research 2017–2027: an IRDiRC perspective. Clin Transl Sci 11(1):21–27. https://doi.org/10.1111/cts.12500
149. NGO Committee for Rare Diseases (2018) The "right to health" in rare diseases: a practical contribution to implementing and achieving the sustainable development goals. https://www.rarediseasesinternational.org/wp-content/uploads/2018/05/NGO-Committee-for-Rare-Diseases-Submission-The-Right-to-Health-in-Rare-Diseases_Feb-15-2018-.pdf. Accessed 26 Jul 2025
150. Rare Diseases International (2022) Agreement with the WHO. RDI. https://www.rarediseasesinternational.org/rdi-signs-memorandum-of-understanding-with-the-world-health-organization/. Accessed 20 Jun 2025
151. Rare Diseases International (2024) Global network for rare diseases. RDI. https://www.rarediseasesinternational.org/collaborative-global-network/. Accessed 20 Jun 2025
152. Xinhua News Agency (2023) Achievements of the global development initiative and contributions to the world. https://www.mfa.gov.cn/web/ziliao_674904/zt_674979/dnzt_674981/qtzt/qqfzcy/zywj/202310/P020231018368202971513.pdf. Accessed 18 Jun 2025
153. Xu X, Zhao M, Niu H (2023) Enhancing the voice of the "Global South" and working together towards a development community. Wenhui Daily. https://brgg.fudan.edu.cn/articleinfo_2788_9.html/articleinfo_5589.html. Accessed 18 Jun 2025
154. Yin Z (2023) The "Global South" and the new globalization. People's Forum Academic Frontier 23:35–49. https://doi.org/10.16619/j.cnki.rmltxsqy.2023.23.004
155. China Pharmaceutical Innovation and Research Development Association (2021) Discussing scientific supervision for people's health | 2021 China pharmaceutical innovation policy

forum was successfully held. http://www.phirda.com/artilce_24894.html?cId=13. Accessed 18 Jun 2025
156. Gong X (2025) Biopharmaceuticals rising: China's strategic pivot to Southeast Asia amid great power tech competition. Carnegie Endowment for International Peace. https://carnegieendowment.org/research/2025/01/biopharmaceuticals-rising-chinas-strategic-pivot-to-southeast-asia-amid-great-power-tech-competition. Accessed 29 Jun 2025
157. Ministry of Foreign Affairs of China (2023) International science and technology cooperation initiative. Ministry of Foreign Affairs of the People's Republic of China. https://www.mfa.gov.cn/web/ziliao_674904/1179_674909/202311/t20231117_11182238.shtml. Accessed 20 Jun 2025
158. China Chamber of Commerce of Medicines & Health Products Importers & Exporters (2019) 2019 annual report on China-Africa pharmaceutical cooperation. http://images.mofcom.gov.cn/ml/202005/20200520041237963.pdf. Accessed 18 Jun 2025
159. Embassy of the People's Republic of China in the Kingdom of Denmark (2023) Ambassador to Denmark Feng Tie of Denmark published a signed article in Danish media titled "China is a firm member of the 'Global South'". Embassy of the People's Republic of China in the Kingdom of Denmark, Bilateral News and Embassy Activities http://dk.china-embassy.gov.cn/zdjl/202307/t20230731_11120392.htm. Accessed 19 Jun 2025
160. People's Daily (2023) Foreign aid medical services contribute to building a community of health for mankind. https://paper.people.com.cn/rmrb/html/2023-12/30/nw.D110000renmrb_20231230_2-05.htm. Accessed 26 Jul 2025
161. Chen M (2024) Warmth touches hearts: China stem cell group affiliated hospital follows up with thalassemia treatment program patients. Shanghai Securities News. https://finance.sina.com.cn/roll/2024-09-27/doc-incqqiwe1706270.shtml. Accessed 25 Jul 2025
162. Power Construction Corporation of China (2015) The African sickle cell anemia treatment research center built by PowerChina is completed. State-owned Assets Supervision and Administration Commission of the State Council. http://www.sasac.gov.cn/n2588025/n2588124/c3796465/content.html. Accessed 20 Jun 2025
163. Guangdong Institutes of Biomedicine and Health, Chinese Academy of Sciences (2025) Introduction to international cooperation. https://gibh.cas.cn/jlhz/gjjl/. Accessed 28 Jul 2025
164. State Council Information Office of China (2021) New era of China-Africa cooperation. General Office of the State Council Operation and Maintenance Unit. https://www.gov.cn/zhengce/2021-11/26/content_5653540.htm. Accessed 27 Jul 2025
165. State Council Information Office of China (2023) Jointly building the Belt and Road Initiative: a major practice in building a community with a shared future for mankind. The Third Belt and Road Forum for International Cooperation. http://www.beltandroadforum.org/n101/2023/1010/c134-915.html. Accessed 20 Jun 2025
166. Tang L, Zhao W, Li X (2022) China's international development cooperation under the international public goods perspective. World Outlook 1:95–114. https://cohd.cau.edu.cn/module/download/downfile.jsp?classid=0&filename=97b8249e9d724cda8c57fc5745e9c187.pdf. Accessed 27 Jul 2025
167. He J, Zhang X (2022) On regulatory issues of health data cross-border flow in China. Chinese Review of International Law 6:62–74. https://guojifayanjiu.ajcass.com/Admin/UploadFile/Issue/201707170003/2022/11/20221128073915WU_FILE_0.pdf. Accessed 28 Jul 2025
168. Sun C (2023) The practice and thinking of the cross-border flow security of data. Open J Leg Sci 11(3):1293–1301. https://doi.org/10.12677/OJLS.2023.113185
169. Ye C, Yan W (2024) Current situation, problems, and relief path of China's cross-border data system. J Beijing Univ Aeronaut Astronaut (Soc Sci Ed) 37(1):57–71. http://lti.ruc.edu.cn/home/xsgd1/sjygrys__fe3ee36bbfb645b19b9fc83c0ce73792/93c33fd0348e471b9790228516d9bdc4.htm. Accessed 18 Jun 2025
170. Institute for Stem Cell and Regeneration, Chinese Academy of Sciences (2024) International organizations. Institute for Stem Cell and Regeneration, CAS. http://iscr.ac.cn/kxyj/jlhz/gjzz/. Accessed 19 Jun 2025

171. Global Alliance for Genomics and Health: Organisational Members (2025) GA4GH, Toronto. https://www.ga4gh.org/our-community/organisational-members/. Accessed 19 Jun 2025
172. Okoth D (2022) 'Time to invest in genomics' in poorer countries—WHO. VaccinesWork. https://www.gavi.org/vaccineswork/time-invest-genomics-poorer-countries-who. Accessed 20 Jun 2025

Regenerative Medicine and Advanced Therapy Medicinal Products in Ukraine: Research, Regulation, and Challenges in a Time of War

11

Oksana Piven, Sergii Martynenko, and Volodymyr Shablii

Abstract

Regenerative biology stands at the forefront of biomedical research, driven by advancements in cell biology, biochemistry, materials science, 3D printing, and other cutting-edge technologies. The field's interdisciplinary nature supports its rapid translation into practical applications. However, the successful integration of scientific breakthroughs into medical practice depends on several key factors, including human capital, scientific infrastructure, innovation climate, legislative frameworks, and ethical regulations governing the application of scientific progress in medicine. In this chapter, we focus on general trends and advances in regenerative biology using cell and tissue technologies, while also outlining the main categories of regenerative biology products. We examine relevant biotechnological and legal norms, comparing existing legislation in Ukraine with European practices and the experiences of neighboring countries. Furthermore, we analyze Ukraine's human potential and scientific infrastructure, which are

O. Piven (✉)
The Department of Human Genetics of the Institute of Molecular Biology and Genetics of the National Academy of Sciences of Ukraine, Kyiv, Ukraine

Laboratory of Molecular Medical Biochemistry, Nencki Institute of Experimental Biology, Polish Academy of Sciences, Warsaw, Mazowieckie, Poland
e-mail: o.o.piven@imbg.org.ua

S. Martynenko
Institute of Cell Therapy, Kyiv, Ukraine
e-mail: martynenko@stemcellclinic.com.ua

V. Shablii
Department of Protein Synthesis Enzymology, Institute of Molecular Biology and Genetics of the National Academy of Sciences of Ukraine, Kyiv, Ukraine

Placenta Stem Cell Laboratory, Institute of Cell Therapy, Kyiv, Ukraine
e-mail: v.shabliy@stemcellclinic.com.ua

U. Ogbogu et al. (eds.), *Regenerative Medicine and the Bioethics of Globalization, Justice and Hegemony*, Collaborative Bioethics 5,
https://doi.org/10.1007/978-3-032-07289-4_11

essential for developing new technologies and methods and for implementing widely known cell- and tissue-based regenerative therapies and their eventual integration into clinical practice. Ukraine's experience reflects not only the common challenges faced by developing countries (and the Global South) but also the complex difficulties resulting from ongoing Russian aggression.

Keywords

Regenerative medicine · Advanced therapy medicinal products (ATMPs) · Biomedical research in Ukraine · Stem cell research · Wartime research ethics · Regulatory harmonization · Human cells and tissues (HCTs)

Introduction and Chapter Overview

Regenerative medicine (RM) and cell therapy (CT) represent rapidly advancing areas of modern biomedical science. Over recent decades, considerable progress has been made in stem cell biology, including stem cell isolation, identification of cell markers, and understanding the signaling pathways that govern stemness and cell fate. A major breakthrough came with the work of Shinya Yamanaka, who demonstrated that differentiated somatic cells could be reprogrammed into induced pluripotent stem cells, leading to a paradigm shift in the understanding of cell plasticity [1, 2].

Stem cells, including totipotent, pluripotent, and multipotent types, are of particular interest not only for their differentiation potential but also for their ability to act as therapeutic agents [3, 4]. Among these, mesenchymal stem cells (MSCs) have shown promise in modulating inflammation, promoting regeneration, and stimulating endogenous repair processes, which has led to their wide use in clinical and preclinical research. In addition to their direct cellular effects, stem and progenitor cells can deliver therapeutic molecules through extracellular vesicles or be modified for use in gene therapy.

Alongside progress in cell biology, new classes of therapeutic products have emerged. These include cell-based therapeutic products (CTPs) and tissue-based therapeutic products (TTPs), which differ significantly from conventional pharmaceuticals and require specific regulatory and production frameworks. These products are further classified into categories such as human cells and tissues (HCTs) for medicinal use and advanced therapy medicinal products (ATMPs), with varying degrees of manipulation and complexity. At the same time, the global regulatory landscape for these products remains fragmented, with national authorities applying different standards and procedures [5]. Nonetheless, jurisdictions such as the United States of America (USA), European Union (EU), Japan, and China play leading roles in both research activity and regulatory developments pertaining to RM and CT [1, 5, 6].

In Ukraine, RM and CT have gained particular relevance in light of the country's strategic location, well-developed medical infrastructure, and established scientific institutions. While approaches focused on tissue regeneration are more common than complex tissue engineering, there is growing interest in both autologous and allogeneic cell-based therapies. The urgency of developing effective regenerative treatments has only increased in recent years, especially due to the impact of the ongoing war and the rising number of civilian and military injuries. Despite limited funding and significant challenges caused by military aggression, Ukraine continues to build on its existing scientific capacity to meet emerging therapeutic needs.

This chapter aims to critically examine the development of RM and CT in Ukraine, with particular attention to the regulatory frameworks governing ATMPs and HCTs. It explores the historical and current conditions of RM research and development in Ukraine, identifies associated regulatory developments and challenges, and situates Ukraine's efforts within broader European and international contexts. The analysis is framed and contextualized against the backdrop of the impact of wartime conditions, the need for regulatory harmonization, and the role of international collaboration in shaping the future of RM in Ukraine.

This chapter is structured as follows:

- The section titled "RM in Ukraine: Historical Overview" provides a historical overview of RM research and development in Ukraine, including early research and clinical applications.
- The section titled "Current RM Research and Applications in Ukraine: Status, Gaps, and Challenges" examines the current state of RM research in Ukraine, outlining institutional capacity, funding sources, and the war's disruptive impact on scientific continuity.
- The section titled "Development and Oversight of ATMPs in Ukraine" analyzes the development, classification, and governance of ATMPs and HCTs in Ukraine. The discussion covers the preclinical evaluation, clinical investigation, and manufacturing and product development phases of the biomedical research lifecycle, with particular attention to regulatory and ethical oversight and the challenges posed by wartime conditions. This section also integrates a detailed assessment of existing legal instruments, identifies regulatory gaps, and reviews efforts to harmonize Ukrainian frameworks with EU and international standards.
- The section titled "Comparative Assessment of Regulatory Governance for ATMPs: Ukraine and the EU" provides a side-by-side comparison of Ukrainian and EU regulatory frameworks governing ATMPs and HCTs. It highlights shared principles such as Good Clinical Practice (GCP) and Good Manufacturing Practice (GMP) compliance, while analyzing divergences in clinical trial oversight, registration procedures, pharmacovigilance, and post-trial safety reporting. The section emphasizes Ukraine's ongoing legislative reforms aimed at greater regulatory alignment with the EU.
- The section titled "The Future of RM Innovation and Regulation in Ukraine" explores the future of RM research in Ukraine, with a focus on national and international initiatives that support scientific capacity during wartime. It discusses strategies for global integration of Ukrainian researchers, restoration of scientific infrastructure, and long-term policy development for post-conflict recovery.

- The Conclusion summarizes Ukraine's evolving RM regulatory landscape. It offers recommendations for strengthening legal frameworks, improving research infrastructure, and fostering innovation through international collaboration, particularly in light of Ukraine's experience in conducting clinical research under crisis conditions.

Together, these sections aim to present a comprehensive view of Ukraine's current position in the field of RM and to assess both the challenges and opportunities facing the country as it seeks to advance RM and CT development and regulation.

RM in Ukraine: Historical Overview

Ukraine has a long history of research in biomedical science, with early efforts in areas foundational to RM and CT. In the 1970s and 1980s, Ukrainian scientists and officials in the former Ukrainian Soviet Socialist Republic developed approaches for the clinical use of cell and tissue transplants. Institutions leading this work included the Shalimov National Institute of Surgery and Transplantation and the V. K. Gusak Institute of Emergency and Reconstructive Surgery, both operating under the auspices of the National Academy of Medical Sciences of Ukraine (NAMSU). Both played a central role in shaping biomedical research and applications during that period.

In parallel, since the 1970s, the Filatov Institute of Eye Diseases and Tissue Therapy, also part of NAMSU, has contributed significantly by developing advanced methods of allogeneic amniotic membrane transplantation, particularly for the treatment of various types of corneal ulcers.

A major development occurred in 1972 with the establishment of the Institute of Problems for Cryobiology and Cryomedicine (the Institute for Problems) within the National Academy of Sciences of Ukraine (NASU). Its formation brought together several entities: the Department of Biophysics, the Department of Cryogenic Equipment Production of FTINT of NASU, and the Problem Laboratory of Low-Temperature Preservation of Bone Marrow and Blood at the Ukrainian Institute for the Improvement of Doctors (the Problem Laboratory). Under the leadership of Professor Mykola Pushkar, the Problem Laboratory developed methods for the preservation of bone marrow, thyroid tissue, and lymphoconcentrate from human peripheral blood, along with associated therapeutic protocols [7].

Since its founding, scientists at the Problem Laboratory have focused on the structural and functional state of cells and tissues at low temperatures, with particular emphasis on minimizing cryodamage and improving cryopreservation technologies. Their research has explored how cryopreservation protocols affect stem cell viability, differentiation capacity, and post-transplant therapeutic performance.

In 1998, a Department of Cryobiology was established at the Institute for Problems with the support of the United Nations Educational, Scientific and Cultural Organization (UNESCO). The Department's research expanded to include the development of cryobanks for human and animal cells, tissues, and organs, as well as the biobanking of genetic material from rare and endangered plant and animal species. Since the 1980s, the Department has cryopreserved sperm samples from more than 30 fish species across Eurasia, including carp, sturgeon, and salmon. This Cryobank is regarded as one of the first of its kind and plays an important role in preserving fish species diversity.

Among its scientific accomplishments, the Institute for Problems developed cryopreservation techniques for hematopoietic stem cells derived from human fetal liver. A search of the Scopus database reveals 906 publications authored by institute researchers. The majority fall into biochemistry, genetics, and molecular biology (39.1%), followed by medical sciences (31.2%). The remainder covers chemistry, physics, agriculture, mathematics, and other disciplines, reflecting the Institute's interdisciplinary approach to cryobiological research.

Today, the Institute for Problems is recognized as a global leader in its field. It employs over 200 staff members, including numerous doctoral degree holders and candidates. Its research laboratories and offices bring together physicists, biophysicists, chemists, biochemists, mathematicians, engineers, biologists, and physicians who collaborate across disciplines to advance research in cryobiology.

Ukraine's efforts in clinical RM research have also evolved alongside institutional and laboratory research. The country has developed methods for cryopreserving ovarian tissue for heterotopic transplantation in the treatment of ovarian dysfunction. Numerous clinical trials have investigated the therapeutic use of cryopreserved human fetal liver-derived stem cells in various diseases.

Since the early 2010s, clinical CT research in Ukraine has increasingly focused on using cord blood cells, hematopoietic stem cells, and MSCs for treating somatic diseases. A substantial portion of this research has focused on improving outcomes in severe or refractory conditions. Several clinical trials have investigated systemic and local delivery of cryopreserved allogeneic total nucleated cord blood cells (TNCBCs) and umbilical cord tissue to promote tissue regeneration in patients with pancreatic necrosis [8–11]. Further trials have evaluated the intravenous infusion of TNCBCs in patients with refractory congestive heart failure [12], MSC-based treatments for vascular and inflammatory conditions in patients with lower limb peripheral artery disease [13–15], therapies for acute respiratory distress syndrome (ARDS) caused by COVID-19 [16], and treatment of large bone defects [17].

The clinical use of cryopreserved human amniotic membrane has also been examined in patients with diabetic foot ulcers [18] and in cases of infectious and neurotrophic keratitis [19].

All these studies were approved by the Ministry of Health of Ukraine and conducted in accordance with national legislation governing the use of HCTs. However, some of these studies were not in full alignment with international guidelines such as Good Clinical Practice (GCP) and Good Manufacturing Practice (GMP). For instance, the ATMPs used in some of the trials were not manufactured under GMP conditions, and the clinical trials were typically nonrandomized, single-center, open-label studies involving a limited number of participants and no expansion into larger-scale phase II or III trials as required by GCP.

Nonetheless, these historical developments laid a strong foundation for Ukraine's ongoing contributions to RM and CT, combining early scientific innovation with growing clinical experience across a range of therapeutic applications.

Current RM Research and Applications in Ukraine: Status, Gaps, and Challenges

Research Infrastructure and Current Institutional Activity

In Ukraine, scientific institutes and universities traditionally operate in separate domains. Scientific institutes prioritize research, while universities primarily focus on training students at the bachelor's, master's, and doctoral levels. Consequently, research in the fields of cell biology, molecular biology, and RM is mainly conducted at research institutes embedded within NASU and NAMSU.

Under NASU, the Department of Biochemistry, Physiology and Molecular Biology is most directly engaged in work related to RM. Several institutes operate within this department, including the Institute for Problems of Cryobiology and Cryomedicine of the National Academy of Sciences of Ukraine (IPCC NASU), the State Scientific Institution Center for Innovative Medical Technologies of the National Academy of Sciences of Ukraine, the State Institution Department of Biotechnical Problems of Diagnostics of the Institute of Problems of Cryobiology and Cryomedicine of the NAS of Ukraine, the Institute of Cell Biology, the O. V. Palladin Institute of Biochemistry, the R. E. Kavetsky Institute of Experimental Pathology, Oncology and Radiobiology, the D. K. Zabolotny Institute of Microbiology and Virology, the Institute of Molecular Biology and Genetics (IMBG), and the Bogomoletz Institute of Physiology. While each of these institutes contributes to the broader field of RM, the most significant contributions have come from the IPCC NASU, the IMBG, and the Institute of Cell Therapy. The latter is a high-tech center that integrates a Cryobank, a medical unit, and a scientific division. The center provides stem cell banking services, manufactures cellular and tissue-based products, and offers clinical treatments using these products [20].

The IPCC NASU houses departments and laboratories engaged in RM-related work across several areas, including stem cell research, reproductive technologies such as the cryopreservation of sperm and embryos for infertility treatment, and the development of blood products, notably lyophilized leukoconcentrate from human cord blood. The latter has shown potential for clinical use in the context of the COVID-19 pandemic [7].

The State Scientific Institution Center for Innovative Medical Technologies of the National Academy of Sciences of Ukraine has developed experimental treatments involving autologous stem stromal cells derived from adipose tissue. These therapies have targeted degenerative-dystrophic diseases such as deforming osteoarthritis, traumatic cartilage defects, and soft tissue injuries, including ligament, tendon, and muscle damage [21].

At IMBG, scientists have primarily focused on test system development, biomarker identification, and drug discovery. However, their contributions to RM include the following notable initiatives:

1. Transplantation of preconditioned umbilical cord-derived MSCs as a treatment for acute pancreatitis and its complications [10, 11].
2. Development of skin equivalents using human stem cells or their derivatives, which show therapeutic efficacy in rodent models of burn injuries [22].

IMBG researchers are also pursuing new approaches using cellular and animal models. These include the following:

- Combined bone tissue implants incorporating autologous stem stromal cells [23, 24].
- Therapeutic applications of stem cells in models of ischemic injury [25], peritonitis [26], and neuroinflammation [27, 28].

The Institute of Genetic and Regenerative Medicine (IGRM), a subdivision of the M. D. Strazhesko National Scientific Center Institute of Cardiology and Clinical and Regenerative Medicine of NASU (Institute of Cardiology), is also engaged in the development and implementation of RM. IGRM scientists focus on isolation, two- and three-dimensional cultivation, directed differentiation, cryopreservation, and transplantation of various types of stem cells. Their aim is to assess regenerative outcomes in experimental models of myocardial injury, nervous system damage, ischemic lesions, and immune dysfunction [28–31]. IGRM collaborates closely with IMBG, the Institute of Cell Therapy, and clinicians at the Institute of Cardiology to conduct clinical trials.

The scientific division of the Institute of Cell Therapy, which includes the Research and Development Centre and the Placenta Stem Cell Laboratory, continually updates its technologies to incorporate recent innovations and conducts both applied and fundamental research. Special focus is placed on developing clinical-grade placental- and umbilical cord-derived MSCs (UC-MSCs) in accordance with current GMP and ATMPs guidance, as well as on establishing quality control approaches to evaluate the functional and therapeutic properties of ATMPs. In addition, the Institute's Cryobank has established a low-temperature storage facility for clinical-grade perinatal MSC-based products for human use, developed in line with Ukrainian legislation and approved by the Ministry of Health of Ukraine. The Institute also actively investigates the properties of prenatal MSCs in various small animal models, aiming to determine their therapeutic effects in conditions such as osteoarthrosis, colon cancer, heart failure, and soft tissue and skin injuries. Currently, the Institute is conducting two clinical trials involving the application of UC-MSCs and placental-derived MSCs (p-MSCs) for the treatment of acute respiratory distress syndrome caused by COVID-19 [16] and knee osteoarthritis [32]. A clinical trial on the use of cryopreserved human amniotic membrane for patients with infectious and neurotrophic keratitis has also been recently completed [19, 33]. Following the full-scale Russian invasion of Ukraine, the Institute, in collaboration with IMBG and the O. Shalimov National Scientific Centre for Surgery and Transplantology, launched a project focused on developing clinical protocols for the treatment of patients with mine-explosive injuries using cryopreserved perinatal MSCs and amniotic membrane [34].

In addition to the research institutes, Ukraine has a large number of universities with departments that train personnel in the fields of cell biology, genetic engineering, and RM. Institutions such as Taras Shevchenko National University of Kyiv, the National Technical University of Ukraine (Kyiv Polytechnic Institute, NTUU KPI), Kyiv Academic University, Ivan Franko National University of Lviv, the National University of Kyiv-Mohyla Academy, Vasyl Stefanyk Precarpathian National University, and Zaporizhzhia National University are among those that offer bachelor's, master's, and doctoral-level education in cell biology, biochemistry, genetics, and biotechnology. Many scientific research institutes also train candidates of sciences and often maintain their own doctoral schools or faculty.

These institutes and programs offer a glimpse into Ukraine's extensive and well-established research capacity in RM. Despite the ongoing war, the country remains positioned to continue advancing the field across a range of scientific and clinical areas.

Challenges in Funding and Human Resource Development

Although Ukraine has the human capital to support the development of RM, in practice, there is a shortage of qualified specialists, particularly within the research institutes of the national academies and universities. Several factors contribute to this problem. One is the long-standing financial crisis in state support for science, which has encouraged many young researchers to leave the country for better-funded institutions with modern infrastructure in Western and Central Europe, Canada, the United States, and other countries (see Fig. 11.1) [35] Another is the war with Russia, which has further deepened the financial crisis and accelerated the migration of scientists. While the full impact of the war is addressed in the next section, its immediate consequences on the research workforce are already apparent.

Furthermore, commercial and semi-commercial companies working on RM therapies offer more attractive working conditions and often compete with public universities and institutes for talent. However, their number remains small, and their activities have also been significantly affected by the war.

A lack of strategic political planning by the Ukrainian leadership regarding the development of science and high-technology industries has contributed to the overall decline of the scientific infrastructure. This decline has reduced both the quality of research and its dissemination over the past three decades. The current war has

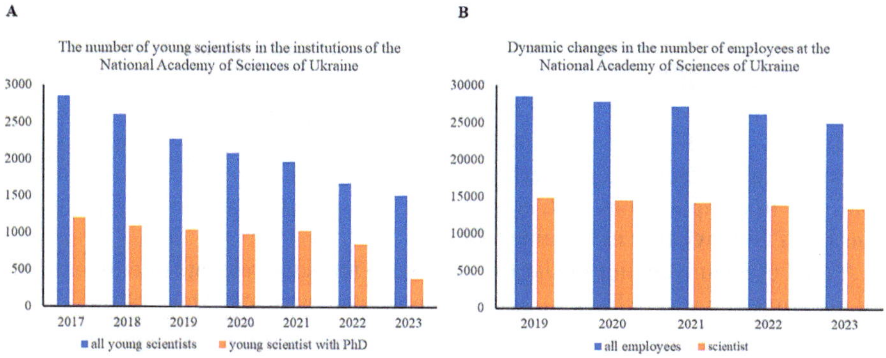

Fig. 11.1 Decline in human capital at the National Academy of Sciences of Ukraine: (**a**) number of young scientists; (**b**) total number of academy staff, including researchers. (Figure 11.1 is based on data from the National Academy of Sciences of Ukraine 2023 Annual Report [35])

made the situation worse, resulting in a deeper financial crisis that has led to the suspension of many research programs due to a lack of funds.

Despite Ukraine's strong scientific legacy and the foundation of trained professionals built over the years of independence, RM, like many other scientific disciplines, has not received adequate financial support.

Impact of the Russian–Ukrainian War on RM Research and Development

The development and translation of RM in Ukraine depend on the intensity of scientific research and preclinical studies, both of which are directly influenced by the state of the economy, political stability, and a favorable investment climate. State and private grant support for research has been sharply reduced, and international collaboration has become more complicated as a result of the Russian–Ukrainian war [36, 37]. As discussed earlier, Russian aggression has also led to a significant migration of Ukrainians, including scientists and students, to European and other countries. This migration has also resulted in a significant gender imbalance [38]. Under Ukrainian law on military service and mobilization, men of draft age are not permitted to leave the country, including male scientists, while female scientists face no such restriction. As a result, in recent years, it has largely been female scientists who have gone abroad, often receiving temporary asylum or a special war refugee-like status adopted by the Council of Europe specifically for Ukrainians. These women have also actively used various support mechanisms for Ukrainian scientists abroad. Consequently, due to this policy, there are now relatively more male scientists remaining in Ukraine. Although male scientists are not subject to mobilization because science and education are considered critical infrastructure, they, like their female colleagues, can choose to voluntarily join the armed forces, which many do.

According to a 2023 report, the number of scientists in NASU decreased by 3.7% in 2022 and a further 4.5% in 2023 [39]. As a result, NASU institutions are experiencing an acute shortage of personnel. This shortage also applies to regulatory agencies responsible for developing and implementing law and policies governing ATMPs and other RM products. Further difficulties arise from power supply issues and logistical obstacles in delivering reagents and equipment to research institutions, due to complications with cross-border transportation and the complete suspension of civil aviation in Ukraine [38].

A total of 33 universities in Ukraine have been forced to relocate after their original infrastructure was either partially or completely destroyed. Many of these institutions are now situated in territory currently occupied by the Russian Federation. Among them are Mariupol State University, Melitopol State University, and Chernihiv Polytechnic National University, the latter of which was nearly destroyed due to ongoing attacks.

The Ministry of Education and Science of Ukraine has documented widespread damage to universities in the regions most affected by the war. In Donetsk, six universities were damaged; in Zaporizhzhia, four were destroyed; and in both Chernihiv and Mykolaiv, three universities in each region were damaged. In Kharkiv, where Russian attacks have been especially persistent, the Karazinsky Sports Complex and the Faculty of Economics building were completely destroyed [39]. Additional losses include student dormitories, several institute buildings, the historic Scientific Library, and the university museum. This list is not exhaustive, as attacks on Ukrainian infrastructure continue. NASU facilities in Kyiv, Kharkiv, Dnipro, Sumy, and Mykolaiv have also been destroyed, including essential buildings and scientific equipment. To date, approximately 220 facilities belonging to leading universities under the Ministry of Education and Science have been damaged or destroyed [40]. The book value of this lost property exceeds half a billion hryvnias, or approximately 11 million euros.

When discussing destroyed infrastructure and relocated universities, it is important to not overlook the challenges faced by their teachers and students. These include maintaining the educational process, protecting the rights of both faculty and students, and ensuring access to psychological and financial support, among many other needs [41]. At the same time, the necessity of adapting to wartime conditions and relocation has also revealed a potential new model for the future of higher education, one that does not depend on a fixed physical location. This "University without Walls" model, which originated at Berdyansk State Pedagogical University in Ukraine, imagines institutions that can flexibly reallocate resources and establish new hubs for learning and research wherever needed [41]. Such dynamic organizations are better positioned to respond to the evolving needs of communities, regions, and nations, while fostering new networks of collaboration and innovation that transcend traditional boundaries. This model centers on the expertise of the academic community and the mission and values of the university, rather than its buildings and infrastructure. It may hold particular promise for training specialists in the humanities, economics, mathematics, and bioinformatics, or in other disciplines that do not require hands-on laboratory experience. However, its broader application in fields such as chemistry, biology, and genetics remains difficult to envision, where practical laboratory training remains essential.

Ukrainian journalist and scientist Khrystyna Semeryn documents Russia's war crimes in Ukraine in her multimedia project [42, 43]. Based on the data she has gathered, as of March 2024, at least 147 Ukrainian scientists have died as a result of the Russian-instigated war [43]. Some of these scientists discontinued their scientific work or postgraduate studies to join the ranks of the armed forces of Ukraine, while some were killed during the Russian bombing of civilian objects, including universities, libraries, institutes, and research centers.

Despite the complexity of the situation, researchers in Ukraine continue to carry out scientific work and foster international collaboration [39, 44]. However, in our view, Ukraine's contributions to translational research remain largely overlooked, despite the country's considerable potential in this area. The war has affected not only the physical integrity of scientific infrastructure and the availability of funding but also the direction of scientific priorities. At present, there is a marked shift toward research that addresses military needs [38]. In the field of RM, this includes increased focus on the development of implants and regenerative therapies using stem cells, offering some hope for further progress of Ukraine's scientific efforts.

Due to years of underfunding and the ongoing war, Ukraine's scientific potential and contributions, including within the field of RM, are now facing a direct threat of physical destruction of both infrastructure and human potential. The severe shortage of funds in Ukraine has made it difficult for universities and other research institutions to even pay for elementary needs such as heating and electricity, let alone to make strides in the fields of regenerative biology and translational research. As a result, the main task at the moment is to preserve and protect the existing scientific and research infrastructure and, most importantly, human capital.

Current RM Products in Development in Ukraine

Over the past decade, Ukraine has made some progress in the development of ATMPs and CTPs, particularly using autologous mesenchymal stromal cells derived from peripheral blood, bone marrow, and adipose tissue. These ATMPs include peripheral mononuclear cells enriched with CD34+ stem cells, autologous MSCs, and cryopreserved or lyophilized MSC products. In addition, allogeneic cell-based products have been developed, such as cryopreserved bone marrow–derived and placenta-derived MSCs (see Table 11.1).

The primary sources of cells for these ATMPs are adipose tissue, bone marrow, and peripheral blood, while cells of placental origin occupy a more specialized niche [32]. In clinical research studies, both autologous and allogeneic MSCs have been used [32]. Three studies involved genetically modified cellular products [45]. However, two of them (NCT04675970 and NCT04642911) lacked information about the modification process, including the type of vector used or the therapeutic gene involved. In contrast, the third study utilized SB623 cells, which are allogeneic

Table 11.1 Summary of clinical trials using cell therapy products (CTPs) in Ukraine, based on data from ClinicalTrials.gov

Identifier number	Status	Study title	Cell source	Disease/condition	Therapeutic outcome	Sponsor	Phase
NCT04461925	Unknown	Treatment of Coronavirus COVID-19 Pneumonia (Pathogen SARS-CoV-2) With Cryopreserved Allogeneic P-MMSCs and UC-MMSCs	Placenta-derived MMSCs and cryopreserved placenta-derived multipotent MSCs	COVID-19 pneumonia	Clinical effects	Institute of Cell Therapy, Kyiv	2
NCT04453111	Unknown	Efficacy of Bone-marrow-derived and Placenta-derived Multipotent Mesenchymal Stem/Stromal Cells for Osteoarthritis	Bone-marrow-derived (BM-MMSCs) and placenta-derived multipotent mesenchymal stem/stromal cells (P-MMSCs)	Knee osteoarthritis	Clinical effects	Institute of Cell Therapy, Kyiv	2
NCT01615250	Unknown	Implantation of Peripheral Stem Cells in Patient with Ischemic Cardiomyopathy (ISCIC)	Peripheral mononuclear cells enriched with CD34+ stem cells	Ischemic cardiomyopathy (ISCIC)	Efficacy and safety	Odessa National Medical University	1
NCT05080465	Active, not recruiting	Long Term Follow up Mesenchymal Stem Cell Therapy for Patients Virus-related Liver Cirrhosis	Autologous mesenchymal stem cells, including bone marrow mesenchymal stem cells (MSC)	Liver cirrhosis in patients caused by Hepatitis C and Hepatitis B or non-alcoholic Steatohepatitis	Safety and preliminary clinical activity	Ukraine Association of Biobank, Kharkiv	3
NCT04675970	Active, not recruiting	Long Term Follow up Patients with Premature Ovarian Failure ex Vivo Gene Therapy (UB-OVF)	Autologous MSC and MSC Lyophilizate	Premature ovarian failure	Long-term safety and efficacy	Ukraine Association of Biobank, Kharkiv	–

NCT04642911	Active, not recruiting	Longterm Follow-up of Subjects with Diabetes 2 Type Treatment with ex Vivo Gene Therapy	Autologous MSC	Type 2 diabetes	Long-term safety and efficacy	Ukraine Association of Biobank, Kharkiv	–
NCT03113747	Unknown	Allogeneic ADSCs and Platelet-Poor Plasma Fibrin Hydrogel to Treat the Patients with Burn Wounds (ADSCs-BWs)	allogeneic cultured adipose-derived multipotent mesenchymal stromal cells (ALLO-ADSCs)	Second- or third-degree burn wounds	Safety and efficacy	A.A. Partners, LLC, Kyiv	2
NCT03103295	Unknown	3D Tissue Engineered Bone Equivalent for Treatment of Traumatic Bone Defects (3-D-TEBE)	Autologous cultured bone marrow-derived multipotent mesenchymal stromal cells (BM-MSCs)	Critical-sized bone defects	Safety and efficacy	A.A. Partners, LLC, Kyiv	2
NCT02416492	Completed	A Study of Modified Stem Cells in Traumatic Brain Injury (TBI) (STEMTRA)	Modified stem cells (SB623)	Chronic motor deficit from traumatic brain injury	Efficacy, safety, and tolerability	SanBio, Inc. Clinical Hospital Feofaniia, Kyiv	2

bone marrow–derived MSCs modified by transient transfection with a plasmid encoding the human Notch-1 intracellular domain [45].

To date, only three studies, NCT02416492 [45], NCT04461925 [16], and NCT04453111 [32] have reported results in international peer-reviewed journals. Another study, NCT03103295, described the use of a 3D tissue-engineered bone equivalent (3D-TEBE), which incorporates autologous cultured bone marrow–derived multipotent mesenchymal stromal cells.

Generally, most studies do not provide sufficient detail about the technological processes used to develop ATMPs or CTPs. Important information is often missing, such as the cell passage number, whether cells were cryopreserved or reformulated, and whether GMP and GCP standards were followed. Despite these limitations, a consistent pattern emerges in terms of conditions treated, modes of administration, and dosing [46]. This suggests that a certain level of clinical practice standardization may already be developing around CTP use.

Development and Oversight of ATMPs in Ukraine

Overview of ATMPs Classification

In the EU, ATMPs are regulated and classified by the European Medicines Agency (EMA) under Regulation (EC) No 1394/2007 of the European Parliament and of the Council, adopted on 13 November 2007 [47]. This regulation, which amends Directive 2001/83/EC and Regulation (EC) No 726/2004, establishes the legal framework for the development, evaluation, and approval of ATMPs [47, 48]. In addition, applicants in the EU may request a formal classification procedure through the Committee for Advanced Therapies. This procedure enables the EMA, in close cooperation with the European Commission, to decide whether a given product falls under the definition of an ATMP [49].

For developers of new gene, cell, or tissue-based therapies seeking approval in the EU, there is also a specific classification procedure available [50]. This procedure is overseen by the Committee for Advanced Therapies, a body within the EMA. Through this procedure, applicants, typically pharmaceutical companies, research institutions, or other product sponsors, can request an official determination about whether their product qualifies as an ATMP under EU law. This classification is made in close cooperation with the European Commission and helps guide the appropriate regulatory and development pathway for the product.

In Ukraine, the classification of ATMPs is conducted by the State Expert Center (SEC) of the Ministry of Health of Ukraine, based on the information provided by the developer of the product. This process is governed by the SEC Guidelines CT-H МОЗУ 42–9.0:2024 on the Classification of Advanced Therapy Medicines [51]. The outcome of the classification is specific to each product. As such, the examples included in the CT-H МОЗУ 42–9.0:2024 are illustrative only and should not be applied directly to other products. Each case must be assessed individually, following the guidance and expert recommendations of the SEC.

The classification of ATMPs in Ukraine includes four categories, based on product characteristics and components, as follows: gene therapy medicinal products (GTMPs), somatic cell therapy medicinal products (sCTMPs), tissue-engineered products (TEPs), and combined ATMPs (cATMPs):

1. *GTMPs* contain or consist of recombinant nucleic acids used in or administered to humans to modify genetic sequences, with therapeutic or diagnostic effects directly related to the sequence or its expression. Vaccines and extracellular vesicles loaded with proteins, noncoding RNA, or mRNA are excluded from this category [52].
2. *sCTMPs* consist of substantially manipulated cells or tissues, or those used for different essential functions in donor and recipient, intended for treating or diagnosing disease through pharmacological, immunological, or metabolic action. Bone marrow transplants and blood transfusions are not included [47, 52].
3. *TEPs* consist of engineered cells or tissues used to regenerate, repair, or replace human tissue. Non-engineered procedures like direct skin grafting are excluded [47].
4. *cATMPs* combine any of the above with a medical device as an integral part. The cellular component must contain viable cells or, if non-viable, must exert a primary effect on the body. Devices and non-cell-based scaffolds alone do not qualify [47].

Preclinical Evaluation of ATMPs in Ukraine

Due to the structural and biological complexity of ATMPs, traditional pharmacological and toxicological testing methods are often inadequate. Their preclinical development differs considerably from that of conventional drugs, especially in study duration and methodology. In Ukraine, the Ministry of Health Order No. 460 of 2015 outlines requirements for preclinical evaluation of ATMPs submitted for registration [53]. Preclinical studies aim to assess the safety and efficacy of investigational ATMPs using animal and ex vivo models. Regulatory requirements governing preclinical studies of all medicinal products [54] and mandating adherence to Good Laboratory Practice (GLP) [55] also apply to ATMPs. However, due to the specific characteristics of ATMPs, it is not always feasible to perform all studies strictly under GLP conditions. To address this, regulatory guidance from the EMA on the application of GLP to ATMPs is also considered.

Selecting an appropriate animal model for preclinical evaluation of ATMPs in Ukraine requires careful consideration of several factors, including donor-recipient histocompatibility, immune system involvement in the product's mechanism of action, the duration of product persistence, the potential for genetic material transfer to somatic or germ cells, and the risk of carcinogenesis. Because the genetic distance between humans and animal models complicates assessment, researchers may use immunocompromised, humanized, or transgenic models to better replicate the target disease and human physiology and to address species-specific differences in pathophysiology and receptor interactions. Conventional toxicity studies in healthy animals are often insufficient for ATMPs; instead, disease-specific models offer more clinically relevant safety data. Small animal models are commonly used due to their accessibility and ease of genetic manipulation, but when anatomical or immunological differences hinder translational relevance, larger or homologous models are recommended [53, 56, 57]. Where suitable animal models are unavailable, alternative approaches such as in vitro, ex vivo, or in silico methods may be employed to support safety evaluations.

Preclinical data should support key clinical decisions, such as identifying a safe and effective dose, confirming administration routes and methods, flagging potential toxicities and target organs, and defining safety monitoring parameters. The required scope of data varies depending on product-specific risks, available scientific knowledge, and experience with similar therapies. Factors such as cell type, level of manipulation, vector and transgene design, and intended clinical application all influence preclinical study design. The degree and duration of exposure also shape the anticipated clinical risk.

Ukrainian regulatory authorities have developed a framework informed by European standards, including the EMA's *Guideline on Quality, Non-Clinical and Clinical Aspects of Medicinal Products Containing Genetically Modified Cells* [56] and the *Guideline on quality, non-clinical and clinical requirements for investigational advanced therapy medicinal products in clinical trials* [57]. These guidelines help ensure that Ukrainian oversight of such products is consistent with international regulatory practices.

Pharmacological and pharmacokinetic evaluations for ATMPs are tailored to the product type. For sCTMPs, biodistribution, cell viability, and differentiation are critical, while for GTMPs, persistence and protein expression from the transgene must be studied. The scope of required studies depends on the risks and duration of action of the product. Toxicity assessments must consider potential genotoxicity, carcinogenicity, and germline integration, especially for GTMPs. In vitro and in vivo studies are used to detect adverse effects, particularly those stemming from transgene products. For GTMPs with potential reproductive effects, additional studies are required to assess fetal and perinatal risks unless otherwise justified. However, more explicit regulatory guidance is needed to standardize methodologies for these preclinical assessments in Ukraine.

In Ukraine, preclinical trials are a mandatory prerequisite for conducting clinical trials involving human cell and tissue transplants [58]. However, Ukrainian regulations do not provide detailed procedures or specific requirements for preclinical studies, which presents a significant regulatory gap. The regulations lack a defined minimum list of preclinical tests, such as pharmacodynamic, pharmacokinetic, and toxicity studies, and do not offer guidance on how these should be carried out. The regulations also fail to clearly outline the expectations for selecting and using animal models in preclinical testing. That said, they do permit both in vitro and in vivo experimental studies, including carcinogenicity testing and studies using immunodeficient mice.

In contrast, EU regulations and guidelines offer more comprehensive guidance on preclinical study requirements for HCTs. For example, EMA guidance emphasizes the importance of quality control across both preclinical and clinical phases and a risk-based approach for assessing the behaviour and potential safety concerns surrounding the administration of cell-based products in novel biological environments [50]. EMA guidance also provides specific recommendations for selecting animal models, conducting dose-dependent studies, and evaluating biodistribution, migration, persistence, and toxicity environments [50].

Regulation of RM Clinical Trials in Ukraine

To initiate a clinical trial involving ATMPs in Ukraine, researchers must comply with a structured approval process outlined in Order No. 690 of 2009 [59]. The trial must receive approval from the SEC and from the local ethics committee at the sponsoring institution. Additionally, all participants must be insured for life and health, and the trial must be registered with the Ministry of Health. Compliance with GCP is mandatory for all clinical trials, including bioequivalence studies. Personnel involved in any stage of clinical trials must be trained and periodically re-certified in the principles of GMP and GCP. Order No. 690 also contains detailed provisions regarding trial design, participant protection, facility and researcher requirements, product labeling, safety monitoring, and reporting. The main components required for approval are submitted as part of the investigational product dossier.

Clinical testing of HCTs is governed by Order No. 630 of 2007 [58], which mandates both ethics committee approval and a positive opinion from the SEC. Before beginning a trial, investigators must submit preclinical study results, manufacturing protocols referencing standard operating procedures, information on tissue procurement and transportation logistics, and data on the production and quality control of the HCTs. The SEC evaluates this information and designates the clinical base, stem cell bank, and responsible personnel for the trial. Order No. 630 includes additional provisions for protocol amendments, reporting of adverse events, and inspection or suspension of clinical trials when necessary.

Generally, Ukrainian regulations governing clinical trials of medicinal products incorporate widely recognized international ethical standards, including the International Council for Harmonisation of Technical Requirements for Pharmaceuticals for Human Use *Guideline for Good Clinical Practice* [60], the World Medical Association *Declaration of Helsinki* [61], the Council for International Organizations of Medical Sciences (CIOMS), *International Ethical Guidelines for Health-Related Research Involving Humans* [62], the Belmont Report [63], and the *Nuremberg Code* [64]. SEC recommends that ethics review bodies follow the core ethical principles established in these international guidelines.

Ukrainian regulatory oversight of clinical trials also encompasses a robust legal framework aimed at safeguarding research participants. These protections are embedded in national legislation that addresses key ethical obligations, such as participant consent, risk disclosure, and the rights of vulnerable populations. For example, participant data privacy is protected by the *Law of Ukraine About Personal Data Protection* [65], which sets out clear rules for how personal information collected during clinical trials must be handled. Informed consent is another key requirement. Under Article 8 of the Ukrainian law on medicinal products [66], a clinical trial cannot begin until a participant, or their legal representative, has given written consent after being fully informed about the trial, including its purpose, potential outcomes, the properties of the investigational product, expected benefits, and possible health risks.

Ukrainian law imposes safeguards for vulnerable populations in clinical trials. According to Article 141 of the Criminal Code of Ukraine, conducting clinical testing without the written consent of a patient or their legal representative, or conducting such testing on a minor or a legally incapable person, is punishable by law if it results in death or other grave consequences [67]. Article 142 establishes criminal liability for the illegal conduct of biomedical, psychological, or other experiments that expose a person's life or health to danger, with increased penalties, including bans from professional practice, if the acts involve minors, coercion, deception, multiple victims, or result in lasting health damage [67]. While not explicitly set out in Article 141, a number of practical requirements related to clinical trials are understood to be supported by its provisions and have been implemented accordingly. All clinical trials must be supervised by qualified medical personnel, who bear full responsibility for participant safety. Research involving vulnerable groups is only permitted if it directly addresses their specific health needs and offers a prospect of benefit. Adults must provide written informed consent, and trials involving minors require parental consent along with age-appropriate information communicated to the child. Consent and trial documentation must also be submitted to guardianship authorities in accordance with national health regulations. Clinical trials involving minors without parental care, such as orphans or adopted children, are strictly prohibited.

Ukraine has long been considered an attractive location for clinical research due to its relatively well-developed medical infrastructure, a highly centralized healthcare system, qualified medical personnel, and a patient population generally open to participating in clinical trials. This appeal was reflected in ClinicalTrials.gov data, which listed around 400 active clinical studies in Ukraine as of April 2022 [68]. However, the Russian-instigated war has drastically altered these conditions. Ongoing shelling has severely disrupted patient care, data collection, and storage, while also posing serious risks to those working or receiving treatment in hospitals.

Many companies, including Cromos, an international contract research organization, have begun relocating clinical trials to countries such as Poland and Georgia, while continuing to support a limited number of patients still enrolled in trials within Ukraine [68]. Major biotech firms such as Kintor, Merck, and Pfizer have suspended the recruitment of new participants. Hospitals in the eastern and southern regions, where some facilities have been destroyed, are no longer fully functional. At the same time, clinics in central and western Ukraine are overwhelmed with both civilian and military patients. Russia's attacks on civilian infrastructure and residential buildings, in violation of the Geneva Convention, have further strained the healthcare system.

The war has also resulted in the mobilization and enlistment of many doctors and clinical staff who previously worked on trials, creating a shortage of qualified personnel to carry out research. Delivering medications to trial sites has become difficult due to Ukraine's logistical isolation. Nevertheless, tools and practices developed during the COVID-19 pandemic, such as remote monitoring and data collection, have helped maintain a degree of trial continuity. A 2023 study by Gujinović et al. [69] analyzed 888 trials either conducted exclusively in Ukraine (5.2%) or involving Ukraine as part of multinational studies (94.8%). The study found that the vast majority of trials (99.6%) were sponsored by non-Ukrainian entities. As of February 24, 2023, 267 trials (30.1%) had no registry updates following the start of the war, and Ukraine was removed as a study location in 15 multinational trials (1.7%), an average of 9.4 months after the invasion began. While overall clinical trial activity declined, improvements in security, logistics, and healthcare operations have allowed most ongoing studies to continue, though no new trials have been initiated. A similar situation exists for RM clinical trials. The combination of ongoing conflict, widespread infrastructure damage, an overloaded healthcare system, and limited funding has made active clinical research in RM extremely difficult.

Regulation of ATMP Manufacturing Protocols and Quality Controls in Ukraine

The manufacturing and quality control of ATMPs in Ukraine is governed by national legislation that mandates compliance with established procedures for the manufacture of investigational medicinal products, including comparators and placebos [59]. These procedures require the submission of data from preclinical and clinical studies, as well as relevant clinical experience with the product, as part of the investigational dossier. This framework is intended to ensure the safe production and evaluation of ATMPs before clinical use.

Regulations concerning HCTs fall under national laws on transplantation and economic licensing [70, 71], along with government resolutions that define procedures for stem cell procurement and the operation of tissue and cell banks [72, 73]. These resolutions establish basic organizational, technological, and personnel requirements for facilities handling such products. However, the standards currently in place for cord blood and tissue banks do not meet internationally accepted GMP or GLP conditions.

The collection, transportation, processing, and storage of HCTs, such as fetal tissue, umbilical cord blood, placenta, and amniotic membrane, are regulated by national orders [74]. Still, there are notable gaps in quality control requirements. With the exception of umbilical cord blood and anatomical material, Ukrainian legislation lacks detailed provisions for the procurement of human tissues and cells, including donor eligibility criteria, required laboratory testing, and standards for donation, procurement, and transplantation procedures.

Cord blood donor screening protocols require testing for HIV 1/2, hepatitis B and C (including nucleic acid amplification testing), and syphilis [74]. EU directives also recommend screening for antibodies against hepatitis B core antigen, HTLV-1 (in high-prevalence areas), and other infections as indicated by donor history [75]. Further testing may be necessary depending on donor risk factors and the characteristics of the cells or tissues involved [75]. While these provisions partially align with international expectations, the Ukrainian framework remains incomplete in several areas, particularly in terms of harmonization with EU and global standards. International standards such as those set out by NetCord and the Foundation for the Accreditation of Cellular Therapy (FACT) offer more comprehensive guidance on the collection, testing, and release of cord blood for clinical use [76]. Ukraine could adopt these standards to strengthen national protocols and better align its practices with internationally recognized benchmarks.

Ukrainian laws and regulations require that manufacturing processes comply with GMP and GCP [77, 78]. The production, handling, and use of ATMPs are subject to rules established under Order No. 690 [59] and the national GMP guidelines, which are harmonized with international standards. These rules include specific procedures for receiving, quarantining, testing, storing, labeling, and packaging investigational medicinal products. Any materials used in a trial must be documented in line with written protocols, and all deviations or damage must be reported to the responsible quality control officer.

Current Regulatory Challenges and Gaps

The ongoing war in Ukraine presents unique challenges that require rapid and flexible approaches to patient care, diagnosis, treatment, and rehabilitation, especially due to the increase in combat-related injuries. Traditional global research ethics rules are designed for stable, peacetime conditions. However, Ukraine's current situation demands a crisis-responsive model of clinical research that addresses urgent societal and patient needs under exceptional circumstances [79–81].

Although such a model has not yet been fully established in Ukraine, examples from other global emergencies may offer relevant insights. For instance, the ongoing COVID-19 pandemic continues to demonstrate the importance of adaptable clinical trial protocols. Large-scale cardiovascular studies conducted during the pandemic have employed innovative methods to maintain participant safety and scientific rigor despite difficult conditions [79]. These adaptations, such as decentralized trial management, flexible visit schedules, modified statistical endpoints, and alternative drug delivery methods, required close coordination among researchers, sponsors, clinical sites, and regulators, and may provide a useful framework for Ukraine and other countries facing crisis conditions [79].

In crisis settings, protocols must also plan for disruptions such as damaged infrastructure, restricted movement, or limited resources. Timelines and endpoints should remain flexible, using adaptable or floating metrics to ensure that research can continue and remain relevant despite unstable conditions.

Regulatory authorities elsewhere currently recommend that hospital ethics committees refer to guidelines developed during the COVID-19 pandemic for conducting expedited ethical evaluations of new drug and vaccine research [79]. These procedures may serve as a model for organizing the work of ethics committees under martial law in Ukraine. In the context of ongoing military aggression, it is likely that the Ministry of Health and SEC will propose further guidance for ethics committees based on emerging wartime experiences.

When evaluating the effectiveness of current legislation and ethical oversight mechanisms, it is clear that Ukraine faces challenges similar to those experienced globally. Ethics committees responsible for reviewing clinical trial protocols rely on the same core international frameworks used in other countries. Like many other nations, Ukraine encountered regulatory gaps during the COVID-19 pandemic, especially in handling urgent and complex situations. In response, the healthcare system implemented broader online monitoring of clinical trials and addressed logistical difficulties [79].

A key strategy for improving future trial resilience could involve incorporating flexible endpoints and reporting requirements, particularly in situations where regular monitoring is disrupted. It is important to emphasize that the current decline in clinical trial activity is not due to failures in the regulatory system itself. Rather, it is a direct result of the ongoing war, which has disrupted healthcare delivery, infrastructure, and public services across the country.

Due to the rapidly evolving nature of wartime conditions, it is difficult to conduct a comprehensive assessment of the effectiveness of Ukraine's research and clinical trials governance framework. However, regulatory system experts have observed that the system demonstrates resilience and is adapting in ways that sustain a commitment to ethical research practices [82].

At the same time, the war has introduced new ethical and logistical challenges that require flexible solutions. These include constraints such as military blockades and limited access to certain regions. In critical situations, such as when urgent medical decisions must be made but the patient is unable to give consent and no legal representative is available, current ethical standards may need to be reinterpreted. In particular, bioethical principles and associated practices may need to be adapted to reflect the realities of war. Such revisions could help support ethical and practical decision-making during extreme and time-sensitive conditions.

Lastly, product traceability remains a key gap. A system for tracking cells and tissues used in ATMPs is needed to ensure traceability from donor material selection through manufacturing and final delivery to the recipient. This need arises from the complex and sensitive nature of ATMP components. In line with international guidance, traceability should cover all raw materials and substances that come into contact with cells, tissues, or other ATMP components [83].

Comparative Assessment of Regulatory Governance for ATMPs: Ukraine and the EU

In both Ukraine and the EU, clinical trials involving HCTs and ATMPs must comply with GCP standards. The manufacturing of ATMPs for clinical use is also subject to GMP requirements. The role and responsibility of regulatory authorities overseeing clinical trials of HCTs and ATMPs in Ukraine is broadly similar to that of EU regulators. In both contexts, these authorities are tasked with the same level of oversight and competence. Ukraine and the EU share several foundational principles in regulating ATMPs, including compliance with GCP and GMP standards during clinical trials and product manufacturing. Both systems require risk-based approaches and adherence to product-specific instructions. Ukrainian GMP guidelines for ATMPs are generally harmonized with EU GMP rules. However, unlike the EU, Ukraine does not yet have ATMP-specific GMP guidelines, though development is underway as part of the ongoing synchronization with EU frameworks.

In the EU, ATMP clinical trials are governed by Regulation (EU) No 536/2014 [84] and supported by national legislation in EU member countries that may impose additional requirements. Oversight is carried out by national authorities or specialized bodies such as Germany's ATMP-specific agency [85]. Additional EU efforts are underway to develop legal instruments that cover quality control and testing requirements tailored to various categories of ATMPs [52, 56, 84, 86, 87]. By contrast, Ukraine does not have specific legal instruments that regulate the manufacturing, preclinical studies, and clinical trials of ATMPs. Clinical trials of ATMPs are currently governed by general laws applicable to all medicinal products, which do not fully address the unique risks associated with ATMPs. Ukraine does not have a dedicated institution for ATMPs. Instead, oversight of clinical trials is handled by the SEC and the Ministry of Health, rather than by a specialized agency focused solely on ATMPs.

Ukraine has not yet adopted important EU-level documents such as the EMA's 2014 *Guideline on the quality, non-clinical and clinical aspects of gene therapy medicinal products* [87]. Current Ukrainian legislation lacks detailed provisions for pharmacokinetic, pharmacodynamic, and toxicological evaluations that reflect the specific properties of different subtypes of ATMPS, including GTMPs, sCTMPs, TEPs, and cATMPs. There are also no clear standards for the selection and use of animal models in preclinical research on these products.

To begin addressing these gaps, it would be beneficial for Ukraine to establish protocols for in vivo pharmacokinetic profiling of expression products in genetically modified somatic cell therapies. Guidance is also needed on whether to require evaluation of the distribution, persistence, and toxicity of degradation products derived from ATMPs. These elements are not addressed in existing Ukrainian orders governing preclinical research procedures.

Ukraine is in the process of developing legislation on ATMPs aimed at aligning its regulatory framework with EU and broader international standards, including guidance from the EMA, the US Food and Drug Administration (FDA), and the World Health Organization. As part of this process, a national guideline on the classification of ATMPs has been introduced, drawing on EMA, FDA, and WHO frameworks [88]. Ukraine has also adopted GMP for ATMPs through domestic regulations modeled on the EU's specialized GMP rules. Current regulatory reforms are focused on building comprehensive frameworks for quality control and on harmonizing preclinical and clinical standards, particularly for products containing genetically modified cells, in accordance with relevant EMA and European Commission guidelines.

Regarding HCTs, Ukraine has a separate procedure for clinical trials and transplants but lacks specialized guidance on quality and manufacturing. EU regulations and guidelines provide detailed standards for donor selection, tissue procurement, and non-clinical study requirements [50, 75]. In contrast, Ukraine's legal framework remains underdeveloped in this area, with the exception of umbilical cord blood products.

Furthermore, Ukraine registers ATMPs using the same process as other medicinal products, whereas the EU follows a centralized procedure involving the EMA's Committee for Advanced Therapies. The Committee issues opinions on ATMP quality, safety, and efficacy that inform final decisions binding across all EU member states, making the EU process more coordinated and specialized.

Regarding safety monitoring, Ukraine lacks a legal requirement to report adverse events that occur after a clinical trial concludes, in contrast to EU legislation, which mandates post-trial reporting [47]. Under Ukrainian regulations, sponsors are only required to report serious adverse events that occur during a clinical trial [51]. A similar disparity exists in the broader pharmacovigilance systems of the two jurisdictions, particularly in the level of specialization and infrastructure. In the EU, while ATMPs are subject to all standard pharmacovigilance legislation and guidelines, they are additionally governed by ATMP-specific pharmacovigilance requirements issued by the EMA [89]. In Ukraine, post-registration safety monitoring is governed by a general pharmacovigilance order (Order No. 898) rather than a specific framework for ATMPs [90]. Reporting responsibilities fall to the marketing authorization holder in both jurisdictions. While the EU uses the EudraVigilance electronic system for submitting adverse event reports, Ukraine still relies on paper-based systems and its national pharmacovigilance database.

Ukrainian timelines for reporting adverse events are aligned with EU standards: serious adverse reactions must be reported within 15 days, and non-serious ones within 90 days. For adverse events occurring abroad, only serious unexpected reactions require notification within 15 days, while others are included in periodic safety updates [52].

In summary, Ukraine follows core EU principles in ATMP oversight but lacks the same level of specialization and institutional infrastructure. Pharmacovigilance systems in the EU are significantly more robust and supported by multiple legal instruments, while Ukraine relies on a single order that applies generally to all medicinal products.

The Future of RM Innovation and Regulation in Ukraine

Current Initiatives for Supporting Biomedical Research

Despite the extreme difficulties caused by the Russian invasion, Ukrainian research institutions have continued to carry out scientific work, largely with the support of the National Research Fund of Ukraine (NRFU), which resumed its operations in 2022. One notable initiative was the 2022 competition titled *Science for the Reconstruction of Ukraine in the War and Post-War Periods*, which funded various projects, including 17 in the biomedical field [91, 92]. A support program for scientists from relocated institutions was also launched with the assistance of the University of Cambridge through the *Cambridge—NRFU 2022* grant scheme, which included one biomedical project [93].

In 2024, the NRFU funded 81 projects through the *Science for Strengthening the Defense Capability of Ukraine* competition, 30 of which are in the fields of biology, medicine, and agricultural sciences [94]. Several other programs were also launched in 2023, including *Research Infrastructures for Conducting Advanced Scientific Research*, *Advanced Science in Ukraine*, and *Joint Ukrainian–Swiss Research Projects*, the last of which was co-funded by the Swiss National Science Foundation [95, 96]. Four of the selected projects focused on biomedical research.

During 2022 and 2023, Ukrainian scientific institutions developed close cooperation with a number of foreign academies, including the Polish Academy of Sciences, the National Academy of Sciences of the United States, the Royal Danish Academy, the Royal Society of Great Britain, and the national academies of Sweden, Germany, and Australia. These collaborations were aimed at supporting both displaced scientists and those still working in Ukraine. The goal of many of these initiatives is long-term integration of Ukrainian scientists into international research networks, not just short-term support.

Several important programs have emerged from these efforts, including joint Ukrainian–Polish scientific laboratories and multilateral research collaborations through the IMPRESS-U initiative with the United States, Poland, Lithuania, Estonia, and Latvia [97]. In 2024, the NRFU and the German Research Foundation began work on the terms for a joint competition to fund bilateral research projects [98]. Additionally, long-standing cooperative agreements between the Ukrainian Ministry of Education and Science and its counterparts in Germany, Lithuania, and Poland continue to provide infrastructure and training support for Ukrainian scientists.

International bilateral grants are another key mechanism supporting Ukrainian scientific cooperation. These programs are especially important because they expand and strengthen scientific networks, facilitate new collaborations, and help sustain existing partnerships, all of which are critical for rebuilding Ukrainian science in the post-war period.

The growing body of international support during the war has had a direct impact on the ability of Ukrainian scientists to continue their work. These programs not only sustain scientific activity within Ukraine but also offer resources and institutional backing to refugee scientists. Such international cooperation has played a central role in stabilizing and preserving Ukraine's scientific landscape during a period of profound disruption.

In addition, Ukraine's ongoing integration with the EU includes its participation in key EU research initiatives. In 2023, Ukraine joined Horizon Europe (2021–2027) and the Euratom research and training programs (2021–2025), granting Ukrainian research institutions equal access to EU funding [99]. The Horizon Europe Office in Ukraine began operations in December 2023, supported by NASU and the Government of Ukraine in cooperation with the European Commission. The office was established through a grant agreement and provides advisory services to researchers, national contact points, and institutions involved in EU scientific programs [100].

Despite these substantial international efforts, domestic funding remains insufficient to support competitive biomedical research, including work in RM. While current programs preserve core research capacity, they do not fully meet the needs of modern, internationally competitive scientific work.

While various national and international initiatives have helped preserve Ukraine's scientific capacity, it is important to note that there are currently no programs specifically dedicated to advancing discoveries in RM. As a result, existing strategies for scientific cooperation and international integration reflect broader developments across Ukrainian science rather than targeted efforts in this field. Over the past decade, Ukrainian scientists have increasingly engaged with the European research landscape, a trend that predates the war and was formalized through Ukraine's accession to Horizon Europe and Euratom programs.

This process of integration is continuing and expanding, supported by additional agreements involving NASU, the Ministry of Education and Science, and international partners. Looking ahead, it is essential that post-war recovery efforts include the full incorporation of Ukrainian science into global research infrastructure. This includes not only participation in large-scale EU and international programs but also the creation of joint laboratories, coordinated research initiatives, and long-term partnerships between Ukrainian institutions and their counterparts abroad. Such efforts will be critical for rebuilding Ukraine's scientific landscape and ensuring the sustainability of fields like RM in the years to come.

Prospects for Legislation and Policy Development Post-Conflict

The future development of RM in Ukraine will depend largely on the outcome of the war, including the preservation of Ukrainian statehood and institutional independence. As previously noted, Ukraine already possesses a foundational research infrastructure, with an established network of universities, hospitals, and scientific institutions capable of training specialists and conducting clinical research in RM. This system offers a strong platform for advancing research and innovation.

In the post-war period, demand for RM applications, such as cell-based therapies, tissue patches, and combined implants, is expected to rise significantly, particularly to support the recovery of civilians and military personnel. This growing demand may help stimulate innovation, but demand alone will not be sufficient. The development of targeted programs focused on rebuilding scientific capacity will be essential. These programs must include sustained investment in research centers, restoration of laboratory infrastructure, and the implementation of state-supported and internationally funded grant mechanisms.

Strategic initiatives should include grants aimed at attracting or repatriating scientists, expanding dedicated funding for RM research, and introducing competitive funding structures to drive high-quality outcomes. Equally important is the continued harmonization of Ukrainian legislation with international standards governing the clinical research, manufacturing, and quality control of stem cell products and ATMPs. Such alignment will help facilitate Ukraine's full integration into the global scientific and regulatory landscape.

In the long term, Ukraine has the potential to contribute significantly to the field of RM, not only through scientific advances in stem cell technologies and ATMPs, but also by offering regulatory and practical insights from conducting clinical research under crisis conditions. Ukraine's experience navigating challenges such as system overload, infrastructure damage, and logistical disruption may provide valuable lessons for the global scientific community. These insights, combined with scientific progress, could position Ukraine as a meaningful contributor to international collaboration and innovation in regenerative medicine.

Conclusion

This chapter has examined the development of RM and the regulation of ATMPs and HCTs in Ukraine, with reference to relevant European Union and international standards. Despite the ongoing war and related disruptions, Ukraine has demonstrated resilience in maintaining its scientific infrastructure. Through the support of national grant programs and international partnerships, Ukrainian researchers have been able to continue their work under difficult conditions, preserving core capacities across institutions.

At the regulatory level, Ukraine has taken meaningful steps toward harmonizing its legal framework with that of the EU, particularly through its participation in Horizon Europe and the Euratom research and training programs. Areas such as GCP, GMP, and pharmacovigilance have seen full or partial alignment with EU models. However, as this chapter has shown, the overall legislative and regulatory landscape remains fragmented. Regulation is distributed across multiple governance instruments, many of which are outdated or lack the specificity required to address the complexities of ATMPs and stem cell-based therapies.

Clinical trials involving ATMPs in Ukraine are currently conducted under general drug regulations. These do not adequately account for the distinct properties and associated risks of cell- and gene-based products. Additionally, post-trial safety monitoring remains underregulated. Unlike in the EU, there is no comprehensive legal requirement in Ukraine to report adverse events that occur after the conclusion of clinical trials. This gap limits the effectiveness of pharmacovigilance and poses challenges for the safe long-term use of RM products.

The need for RM will likely increase in the post-war period, particularly to support the rehabilitation of civilians and military personnel with complex injuries. This demand presents an opportunity to prioritize investment in RM through dedicated funding, infrastructure rebuilding, and legal reform. It also highlights the importance of developing a clear and unified legal framework that distinguishes between conventional drugs, ATMPs, and HCTs, and that supports both innovation and patient safety.

Finally, international collaboration remains essential. The integration of Ukrainian research institutions into broader scientific networks provides access to equipment, training, and infrastructure that are often unavailable domestically. Continued cooperation through bilateral grants, multilateral programs, and institutional partnerships will be crucial for rebuilding the country's research capacity. As Ukraine moves forward, its experiences conducting research under wartime conditions may also contribute to international discussions on clinical trial design, regulatory flexibility, and ethical oversight in crisis settings.

In sum, Ukraine has the scientific foundation and institutional potential to play a meaningful role in the development of RM modalities and products. Realizing this potential will require a coordinated effort to support researchers, update regulatory frameworks, and deepen international engagement in the years ahead.

References[1]

1. Charitos IA, Ballini A, Cantore S et al (2021) Stem cells: a historical review about biological, religious, and ethical issues. Stem Cells Int 2021:9978837. https://doi.org/10.1155/2021/9978837
2. Lendahl U (2022) 100 plus years of stem cell research—20 years of ISSCR. Stem Cell Reports 17(6):1248–1267. https://doi.org/10.1016/j.stemcr.2022.04.004
3. Poliwoda S, Noor N, Downs E et al (2022) Stem cells: a comprehensive review of origins and emerging clinical roles in medical practice. Orthop Rev 14(3). https://doi.org/10.52965/001c.37498
4. Vazin T, Freed WJ (2010) Human embryonic stem cells: derivation, culture, and differentiation: a review. Restor Neurol Neurosci 28(4):589–603. https://doi.org/10.3233/RNN-2010-0543
5. Hirai T, Yasuda S, Umezawa A, Sato Y (2023) Country-specific regulation and international standardization of cell-based therapeutic products derived from pluripotent stem cells. Stem Cell Rep 18(8):1573–1591. https://doi.org/10.1016/j.stemcr.2023.05.003
6. Zhai X, Qiu R (2013) Regulations/ethical guidelines on human adult/mesenchymal stem cell clinical trial and clinical translation. In: Zhao RC (ed) Essentials of mesenchymal stem cell biology and its clinical translation. Springer, New York, pp 279–313. https://doi.org/10.1007/978-94-007-6716-4_19
7. Institute of Problems of Cryobiology and Cryomedicine of the NAS of Ukraine (n.d.) Historical milestones. https://cryonas.org.ua/en/historical-milestones. Accessed 24 Jul 2025

[1] Some references in this list were originally published in Ukrainian and have been translated into English using Google Translate. The accuracy of the translations may be limited.

8. Kebkalo AB (2012) Surgery and cellular technologies in the complex treatment of patients with pancreatic necrosis. Ukr J Clin Surg 10:53–56
9. Kebkalo AB (2013) Evaluation of clinical efficacy of complex treatment for necrotic pancreatitis using cord blood stem cell transplantation and cryopreserved cord tissue. Ukr J Surg 2(21):47–53
10. Kebkalo AB, Mamchych VI, Lobyntseva GS, Shabliy VA (2011) The effect of cord blood stem cells on the system of haemostasis in patients with different stages of necrotic pancreatitis. Ukr J Clin Lab Med 6(1):67–73
11. Kebkalo AB, Lobintseva GS, Seminog VI, Shabliy VA (2012) The cord blood and umbilical cord use in the complex treatment of patients with pancreatic necrosis. Ukr J Emerg Med named after G O Mozhaev 13(1):77–85
12. Driuk NF, Kirimov VI, Barna IE et al (2012) Application of autologous aspirates as well as multipotent stromal cells of the bone marrow and adipose tissue in vascular surgery. Ukr J Clin Surg 12:24–29
13. Dombrovs'kyĭ DB (2010) Autotransplantation of multipotent stromal cells from adipose tissue as a method of stimulation of angiogenesis in the limb ischemia in experiment. Ukr J Clin Surg 1:46–50
14. Poliachenko IV, Driuk MF, Dombrovs'kyĭ DB (2010a) Ultrastructural changes of vascular endothelium in patients with chronic ischemia of the extremities after conduction of multipotent stromal cells from adipose tissue transplantation. Ukr J Clin Surg 6:50–53. https://pubmed.ncbi.nlm.nih.gov/20734820/
15. Poliachenko IV, Driuk MF, Dombrovs'kyĭ DB (2010b) Histological and immunohistochemical characteristics of stimulated angiogenesis in transplantation of multipotent stem cells derived from the adipose tissue in patients with chronic ischemia of the extremities. Ukr J Clin Surg 5:40–43. https://pubmed.ncbi.nlm.nih.gov/20623978/
16. Bukreieva T, Svitina H, Nikulina V et al (2023) Treatment of Acute Respiratory Distress Syndrome caused by COVID-19 with human umbilical cord mesenchymal stem cells. Int J Mol Sci 24(5):4435. https://doi.org/10.3390/ijms24054435
17. Vasyliev R, Oksymets V, Rodnichenko A et al (2017) Tissue engineering-based approach for restoration of combat related critical sized bone defects. Cytotherapy 19(5):S223–S224. https://doi.org/10.1016/j.jcyt.2017.02.311
18. Ustymenko A, Nemtinov P, Bolgarska S et al (2021) The clinical effectiveness of cryopreserved human amniotic membrane in diabetic foot syndrome. Cell Organ Transplantol 9(2):80–87. https://doi.org/10.22494/cot.v9i2.129
19. Sereda KV, Drozhzhyna GI, Gaidamaka TB (2016) The effect of different techniques of cryopreserved amniotic membrane transplantation on the repair processes of the cornea on the bacterial keratitis model. Ophthalmol Eastern Europe 6(2):229–239
20. Institute of Cell Therapy (n.d.) Institute of cell therapy. https://instituteofcelltherapy.com/en/. Accessed 24 Jul 2025
21. National Academy of Sciences of Ukraine (2021) Orthopedists of the Center for Innovative Medical Technologies of the National Academy of Sciences of Ukraine use the latest treatment method using stem cells. https://old.nas.gov.ua/UA/Messages/Pages/View.aspx?MessageID=7696. Accessed 24 Jul 2025
22. Papuga OY, Macewicz LL, Lukash LL (2023) From mesenchymal stem cells to their secretoms as basic components of dermal coverings for the treatment of massive burns. Biopolym Cell 39(2):90–109. https://doi.org/10.7124/bc.000A91
23. Shevchenko OM, Kulak LD, Kuzmenko MM et al (2021) Investigation of influence of heat treatment on structure and properties of biocompatible Ti–18Nb–xSi alloys. Metallofizika i Noveishie Tekhnologii 43(7):887–907. https://doi.org/10.15407/mfint.43.07.0887
24. Skliarenko Y, Kolomiiets VV, Balatskyi VV et al (2024) The impact of the physicochemical properties of calcium phosphate ceramics on biocompatibility and osteogenic differentiation of mesenchymal stem cells. BMC Res Notes 17:295. https://doi.org/10.1186/s13104-024-06937-y

25. Konovalov S, Moroz V, Konovalova N et al (2023) Multipotent mesenchymal stromal cells of various origins reduce reactive gliosis in the hippocampal CA1 area during acute ischemia-reperfusion of the rat brain. Cell Organ Transplantol 11(2):122–129. https://doi.org/10.22494/cot.v11i2.159
26. Pikus PO, Rymar SY, Shuvalova NS, Kordium VA (2022) Study of the interaction between macrophages and human umbilical cord MSCs in vivo on the model of perotonitis in mice. Biopolym Cell 38(4):231–241. https://doi.org/10.7124/bc.000A80
27. Lykhmus O, Kalashnyk O, Skok M et al (2024) Extracellular vesicles derived from mesenchymal stem cells ameliorate cognitive impairment caused by neuroinflammation in young but not aged mice. Explor Neurosci 3:207–218. https://doi.org/10.37349/en.2024.00045
28. Skok M, Deryabina O, Lykhmus O et al (2022) Mesenchymal stem cell application for treatment of neuroinflammation-induced cognitive impairment in mice. Regen Med 17(8):533–546. https://doi.org/10.2217/rme-2021-0168
29. Chepurnyi YV, Kustrjo TV, Korsak AV et al (2018) Influence of adult neural crest-derived multipotent stem cells on regeneration of orbital soft tissue content after experimental injury. Probl Cryobiol Cryomed 28(1):59–63. https://doi.org/10.15407/cryo28.01.059
30. Kyryk V, Ustymenko A, Lutsenko T et al (2022) Regenerative effects of mouse aortic endothelial cells in a murine model of critical limb ischemia. Cell Organ Transplantol 10(2):90–96. https://doi.org/10.22494/cot.v10i2.143
31. Labunets I, Ustymenko A, Panteleymonova T, Rodnichenko A (2024) Colony-forming potential of murine bone marrow-derived progenitor cells in experimental neurodegenerative pathology and under the impact of melatonin in vivo or in vitro. Cell Organ Transplantol 12(1):52–58. https://doi.org/10.22494/cot.v12i1.165
32. Holiuk Y, Birsa R, Bukreieva T et al (2025) Effectiveness and safety of multiple injections of human placenta-derived MSCs for knee osteoarthritis: a nonrandomized phase I trial. BMC Musculoskelet Disord 26:418. https://doi.org/10.1186/s12891-025-08664-2
33. Drozhzhyna G, Sereda K, Vit V, Molchanuk N (2016) Trasplante de membrane amniotica para el tratamiento de queratitis ulcerative recurrente por virus herpes simple. Horiz Med 16(3):79–82
34. Science in Ukraine (2025) Competition for joint Ukrainian–Lithuanian scientific research projects for implementation in 2026–2027. https://nauka.gov.ua/information/ul2025/. Accessed 24 Jul 2025
35. National Academy of Science of Ukraine (2023) Annual Report. https://files.nas.gov.ua/PublicMessages/Documents/0/2024/04/240416150722348-4306.pdf. Accessed 24 Jul 2025
36. Gaind N, Liverpool L (2023) War shattered Ukrainian science—its rebirth is now taking shape. Nature 618:900–901. https://doi.org/10.1038/d41586-023-02031-8
37. UNESCO (2024) Analysis of war damage to the Ukrainian science sector and its consequences. UNESCO. https://www.unesco.org/en/articles/analysis-war-damage-ukrainian-science-sector-and-its-consequences/. Accessed 24 Jul 2025
38. Hondermarck H, Finiuk N, Jiang CC, Stoika R (2024) Standing strong: war-related challenges in Ukrainian biomedical research and opportunities for support. FASEB Bioadv 6(10):401–405. https://doi.org/10.1096/fba.2024-00072
39. Editorial (2023) Rebuilding Ukrainian science can't wait—here's how to start. Nature 614:593–594. https://doi.org/10.1038/d41586-023-00505-3
40. Ministry of Science and Education of Ukraine (n.d.) Save Schools Ukraine. https://saveschools.in.ua/en/. Accessed 24 Jul 2025
41. Tsybuliak N, Suchikova Y, Hurenko O et al (2023) Ukrainian universities at the time of war: from occupation to temporary relocation. Torture 33(3):39–64. https://doi.org/10.7146/torture.v33i3.136256
42. Semeryn K (n.d.) Personal website. https://khrystynasemeryn.wordpress.com/. Accessed 24 Jul 2025
43. Semeryn K (2024) Ukrainian scientists at war: a project that preserves the memory of scientists killed by Russia. Media Sapiens. https://ms.detector.media/trendi/

post/36292/2024-09-30-ukrainski-vcheni-na-viyni-proiekt-shcho-zberigaie-pamyat-pro-vchenykh-ubytykh-rosiieyu/. Accessed 24 Jul 2025
44. Dolgin E (2022) War in Ukraine disrupts trials, cancer care. Cancer Discov 12(5):1178. https://doi.org/10.1158/2159-8290.CD-ND2022-0004
45. Kawabori M, Weintraub AH, Imai H et al (2021) Cell therapy for chronic TBI: interim analysis of the randomized controlled STEMTRA trial. Neurology 96(8):E1202–E1214. https://doi.org/10.1212/WNL.0000000000011450
46. Gindraux F, Hofmann N, Agudo-Barriuso M et al (2022) Perinatal derivatives application: identifying possibilities for clinical use. Front Bioeng Biotechnol 10. https://doi.org/10.3389/fbioe.2022.977590
47. European Parliament and Council of the European Union (2007) Regulation (EC) No. 1394/2007 of the European Parliament and of the Council of 13 November 2007 on advanced therapy medicinal products and amending Directive 2001/83/EC and Regulation (EC) No. 726/2004. https://eur-lex.europa.eu/legal-content/EN/TXT/PDF/?uri=CELEX:32007R1394. Accessed 24 Jul 2025
48. European Medicines Agency (2007) Marketing-authorisation procedures for advanced-therapy medicinal products. https://www.ema.europa.eu/en/human-regulatory-overview/marketing-authorisation/advanced-therapies-marketing-authorisation/marketing-authorisation-procedures-advanced-therapy-medicinal-products. Accessed 24 Jul 2025
49. European Medicines Agency (n.d.-a) Advanced therapy medicinal products: overview. https://www.ema.europa.eu/en/human-regulatory-overview/advanced-therapy-medicinal-products-overview. Accessed 24 Jul 2025
50. European Medicines Agency (2017) Development of non-substantially manipulated cell-based ATMPs: flexibility introduced via the application of the risk-based approach. https://www.ema.europa.eu/en/documents/regulatory-procedural-guideline/development-non-substantially-manipulated-cell-based-advanced-therapy-medicinal-products-flexibility-introduced-application-risk-based-approach_en.pdf. Accessed 24 July 2025
51. State Expert Center of the Ministry of Health of Ukraine (2024) Guidelines CT-H MO3Y 42–9.0:2024 on the Classification of Advanced Therapy Medicines. https://www.dec.gov.ua/?ZG93bmxvYWQ=d3AtY29udGVudC91cGxvYWRzLzIwMjQvMDMvNTQ0NzYtZG5fNDgwXzIwMDMyMDI0X2RvZC5wZGY=. Accessed 24 Jul 2025
52. European Commission (2021) Directive 2001/83/EC of the European Parliament and of the Council of 6 November 2001 on the community code relating to medicinal products for human use. EUR-Lex Access to European Union Law. https://eur-lex.europa.eu/LexUriServ/LexUriServ.do?uri=OJ:L:2001:311:0067:0128:en:PDF. Accessed 24 Jul 2025
53. Verkhovna Rada of Ukraine (2015a) Ministry of Health of Ukraine Order 07/23/2015 No. 460 on amendments to the procedure for conducting an examination of registration materials for medicinal products submitted for state registration (re-registration), as well as examination of materials on making changes to registration materials during the validity of the registration certificate and approval of the procedure for checking materials attached to the application for state registration of individual medicinal products, regarding their volume. https://zakon.rada.gov.ua/laws/show/z1210-15#Text. Accessed 24 Jul 2025
54. Verkhovna Rada of Ukraine (2009a) Ministry of Health of Ukraine Order 14.12.2009 No. 944 on approval of the Procedure for conducting preclinical studies of medicinal products and examination of materials for preclinical studies of medicinal products. https://zakon.rada.gov.ua/laws/show/z0053-10#Text. Accessed 24 Jul 2025
55. Prodanchuk GM (2024) Good Laboratory Practice (GLP) in modern Ukrainian legislation. Literature review. PhD dissertation, Bukovinian State Medical University. https://dspace.bsmu.edu.ua/bitstream/123456789/24615/1/Prodanchuk%20G.M.%20Good%20laboratory%20practice_221-222.pdf. Accessed 24 Jul 2025
56. European Medicines Agency (2020) Guideline on quality, non-clinical and clinical aspects of medicinal products containing genetically modified cells. https://www.ema.europa.eu/en/documents/scientific-guideline/

guideline-quality-non-clinical-and-clinical-aspects-medicinal-products-containing-genetically-modified-cells-revision-1_en.pdf. Accessed 24 Jul 2025
57. European Medicines Agency (2025) Guideline on quality, non-clinical and clinical requirements for investigational advanced therapy medicinal products in clinical trials EMA/CAT/22473/2025. https://www.ema.europa.eu/en/documents/scientific-guideline/guideline-quality-non-clinical-clinical-requirements-investigational-advanced-therapy-medicinal-products-clinical-trials_en.pdf. Accessed 24 Jul 2025
58. Verkhovna Rada of Ukraine (2007) Ministry of Health of Ukraine Order 10.10.2007 No. 630 on approval of the Procedure for conducting clinical trials of tissue and cell transplants and examination of clinical trial materials and amendments to the Procedure for conducting clinical trials of medicinal products and examination of clinical trial materials. https://zakon.rada.gov.ua/laws/show/z1206-07#Text. Accessed 24 Jul 2025
59. Verkhovna Rada of Ukraine (2009b) Order 23.09.2009 No. 690 on approval of the Procedure for conducting clinical trials of medicinal products and examination of clinical trial materials and the model regulation on ethics committees. https://zakon.rada.gov.ua/laws/show/z1010-09#Text. Accessed 24 Jul 2025
60. International Council for Harmonisation of Technical Requirements for Pharmaceuticals for Human Use (2016) Guideline for Good Clinical Practice E6(R3). https://database.ich.org/sites/default/files/ICH_E6%28R3%29_Step4_FinalGuideline_2025_0106.pdf. Accessed 24 Jul 2025
61. World Medical Association (2013) WMA declaration of Helsinki—ethical principles for medical research involving human subjects. https://www.wma.net/policies-post/wma-declaration-of-helsinki-ethical-principles-for-medical-research-involving-human-subjects/. Accessed 24 Jul 2025
62. Council for International Organizations of Medical Sciences (CIOMS) (2016) International ethical guidelines for health-related research involving humans. https://www.who.int/docs/default-source/ethics/web-cioms-ethicalguidelines.pdf?sfvrsn=f62ee074_0. Accessed 24 Jul 2025
63. National Commission for the Protection of Human Subjects of Biomedical and Behavioral Research (1979) The Belmont Report. Ethical principles and guidelines for the protection of human subjects of research. U.S. Department of Health, Education, and Welfare. https://www.hhs.gov/ohrp/regulations-and-policy/belmont-report/read-the-belmont-report/index.html. Accessed 24 Jul 2025
64. Office of Research Integrity, U.S. Department of Health and Human Services (n.d.) Nuremberg Code: Directives for human experimentation. https://ori.hhs.gov/content/chapter-3-The-Protection-of-Human-Subjects-nuremberg-code-directives-human-experimentation. Accessed 24 Jul 2025
65. Verkhovna Rada of Ukraine (2010) Law of Ukraine About Personal Data Protection. https://zakon.rada.gov.ua/laws/show/2297-17#Text. Accessed 24 Jul 2025
66. Verkhovna Rada of Ukraine (1996a) On the procedure for the entry into force of the Law of Ukraine "On Medicinal Products". https://zakon.rada.gov.ua/laws/show/124/96-вр#Text. Accessed 24 Jul 2025
67. Verkhovna Rada of Ukraine (2001) The Criminal Code of Ukraine. https://zakon.rada.gov.ua/laws/show/2341-14?lang=en#Text. Accessed 24 Jul 2025
68. Rubin R (2022) Clinical trials disrupted during war in Ukraine. JAMA 327(16):1535–1536. https://doi.org/10.1001/jama.2022.5571
69. Gujinović D, Viđak T, Melnikova M et al (2023) Changes in registration parameters for ongoing clinical trials in Ukraine after 2022 Russian invasion. JAMA Netw Open 6(6):e2320202. https://doi.org/10.1001/jamanetworkopen.2023.20202
70. Verkhovna Rada of Ukraine (2015b) Law of Ukraine on licensing of types of economic activities. https://zakon.rada.gov.ua/laws/show/222-19#Text. Accessed 24 Jul 2025
71. Verkhovna Rada of Ukraine (2018) Law of Ukraine on the application of transplantation of anatomical materials to humans. https://zakon.rada.gov.ua/laws/show/2427-19#Text. Accessed 24 Jul 2025

72. Verkhovna Rada of Ukraine (2016) Cabinet of Ministers of Ukraine Resolution No. 286 on approval of the licensing conditions for conducting business activities of banks of umbilical cord blood, other human tissues and cells according to the list approved by the Ministry of Health. https://zakon.rada.gov.ua/laws/show/286-2016-п#Text. Accessed 24 Jul 2025
73. Verkhovna Rada of Ukraine (2020) Cabinet of Ministers of Ukraine Resolution No. 257 on approval of the procedure for obtaining and providing hematopoietic stem cells and exchanging information on available human anatomical materials intended for transplantation. https://zakon.rada.gov.ua/laws/show/257-2020-%D0%BF#Text. Accessed 24 Jul 2025
74. Verkhovna Rada of Ukraine (2014) Ministry of Health of Ukraine Order 10.07.2014 No. 481 on approval of the procedure for collection and temporary storage of umbilical cord (placental) blood and/or placenta. https://zakon.rada.gov.ua/laws/show/z0869-14#Text. Accessed 24 Jul 2025
75. Commission of the European Communities (2006) Commission Directive 2006/17/EC of 8 February 2006 implementing Directive 2004/23/EC of the European Parliament and of the council as regards certain technical requirements for the donation, procurement and testing of human tissues and cells. EUR-Lex. https://eur-lex.europa.eu/eli/dir/2006/17/oj. Accessed 24 Jul 2025
76. NetCord, Foundation for the Accreditation of Cellular Therapy (FACT) (2024) NetCord-FACT International Standards for Cord Blood Collection, Banking, and Release for Administration, Eighth Edition, Version 8.2
77. Kundiiev YI, Vitte PN, Lyzogub VG et al (2008) Creating a national infrastructure for good clinical practice in clinical trials in Ukraine. Pharm Med 22:355–358. https://doi.org/10.1007/BF03256732
78. Verkhovna Rada of Ukraine (1996b) Law of Ukraine about medicines. https://zakon.rada.gov.ua/laws/show/123/96-%D0%B2%D1%80#Text. Accessed 24 Jul 2025
79. Bhatt AS, Lindholm D, Nilsson A et al (2023) Operational challenges and mitigation measures during the COVID-19 pandemic–lessons from DELIVER. Am Heart J 263:133–140. https://doi.org/10.1016/j.ahj.2023.05.013
80. Kerpel-Fronius S, Kurihara C, Crawley FP et al (2022) The ethical responsibility to continue investigational treatments of research participants in situation of armed conflicts, economic sanctions or natural catastrophes. Front Med 9. https://doi.org/10.3389/fmed.2022.950409
81. Kurihara C, Crawley FP, Baroutsou V et al (2022) The continuation of clinical trials in times of war: a need to develop ethics and situationally adaptive clinical research guidelines. Front Med 15(9). https://doi.org/10.3389/fmed.2022.966220
82. Herasymchuk T, Rasputniak S, Dobrova V (2023) Clinical trials continuity: Ukraine state expert center perspective and recommendations. In: DIA Global Forum Special Section: Clinical Research in Ukraine, March 2023. https://globalforum.diaglobal.org/issue/april-2023/clinical-trial-continuity-ukraine-state-expert-center-perspective-and-recommendations/. Accessed 24 Jul 2025
83. World Health Organization (2023) Considerations in developing a regulatory framework for human cells and tissues and for advanced therapy medicinal products. https://www.who.int/publications/m/item/considerations-in-developing-a-regulatory-framework-for-human-cells-and-tissues-and-for-advanced-therapy-medicinal-products. Accessed 24 Jul 2025
84. European Commission (2014) Clinical trials—Regulation EU No 536/2014. https://health.ec.europa.eu/medicinal-products/clinical-trials/clinical-trials-regulation-eu-no-5362014_en. Accessed 24 Jul 2025
85. Pizevska M, Kaeda J, Fritsche E et al (2022) Advanced therapy medicinal products' translation in Europe: a developers' perspective. Front Med (Lausanne) 3(9):757647. https://doi.org/10.3389/fmed.2022.757647
86. European Commission (2019) Guideline on good clinical practice specific to advanced therapy medicinal products. https://health.ec.europa.eu/system/files/2019-10/atmp_guidelines_en_0.pdf. Accessed 24 Jul 2025
87. European Medicines Agency (2014) Guideline on the quality, non-clinical and clinical aspects of gene therapy medicinal products. https://www.ema.europa.eu/en/

documents/scientific-guideline/guideline-quality-non-clinical-and-clinical-aspects-gene-therapy-medicinal-products_en.pdf. Accessed 24 Jul 2025
88. Andriichuk B, Samchuk-Kolodiazhna Z (2022) Legal regulation of the manufacturing practice for advanced therapy medicinal products in the context of changes in the current legislation of Ukraine. Law Innov 4(40):23–29. https://doi.org/10.37772/2518-1718-2022-4(40)-4
89. European Medicines Agency (n.d.-b) Pharmacovigilance for advanced therapies. https://www.ema.europa.eu/en/human-regulatory-overview/post-authorisation/advanced-therapies-post-authorisation/pharmacovigilance-advanced-therapies. Accessed 24 Jul 2025
90. Verkhovna Rada of Ukraine (2006) Ministry of Health of Ukraine Order 27.12.2006 No. 898 on approval of the procedure for pharmacovigilance. https://zakon.rada.gov.ua/laws/show/z0073-07#TexT. Accessed 24 Jul 2025
91. National Research Foundation of Ukraine (2022) Call 2022.01 'Science for the Recovery of Ukraine in the War and Post-War Periods'. https://nrfu.org.ua/en/news-en/annocement-of-the-call-2022-02-science-for-the-recovery-of-ukraine-in-the-war-and-post-war-periods/. Accessed 24 Jul 2025
92. National Research Fund of Ukraine (2023) The National Research Foundation of Ukraine: general overview and challenges at wartime. https://scienceeurope.org/news/nrfu-general-overview-and-challenges-at-wartime/. Accessed 24 Jul 2025
93. Drohobych State Pedagogical University (2023) The National Research Fund of Ukraine announces a new competition "Cambridge—NFDU 2022". https://international.dspu.edu.ua/en/grants-scholoarships-programs/the-national-research-fund-of-ukraine-announces-a-new-competition-cambridge-nfdu-2022/. Accessed 24 Jul 2025
94. National Research Foundation of Ukraine (2024) Перелік проєктів, що рекомендуються до реалізації за рахунок грантової підтримки НФДУ із зазначенням рекомендованого обсягу фінансування [List of projects recommended for implementation with grant support from the NRFU indicating the recommended funding amount]. Approved by the NRFU Scientific Council (Protocol No. 21 dated 28.06.2024) and the Competition Commission (Protocol No. 19 dated 28.06.2024). https://nrfu.org.ua/wp-content/uploads/2024/07/perelik-proyektiv-shho-rekomenduyutsya-do-realizacziyi_nr_28.06.2024.pdf. Accessed 24 Jul 2025
95. National Research Foundation of Ukraine Scientific Council (2024) Call 'Ukrainian-Swiss Joint Research Projects: Call for Proposals 2023'. https://nrfu.org.ua/wp-content/uploads/2024/08/peremozhczi_shnnf_nfdu.pdf. Accessed 24 Jul 2025
96. Swiss National Science Foundation (2023) Joint Call for Proposals for Research and Development Projects: "Ukrainian-Swiss Joint Research Projects: Call for Proposals 2023". https://www.snf.ch/media/fr/DxrMqCg3fOFyG9hp/Call_document_NRFU-SNSF.pdf. Accessed 24 Jul 2025
97. Narodowe Centrum Nauki (2023) IMPRESS U: International Multilateral Partnerships for Resilient Education and Science System in Ukraine (Call launched August 2, 2023). Narodowe Centrum Nauki, Poland. https://www.ncn.gov.pl/ogloszenia/konkursy/impress-u. Accessed 24 Jul 2025
98. Deutsche Forschungsgemeinschaft (German Research Foundation) (2024) National Research Foundation of Ukraine Delegation Visits the DFG. https://www.dfg.de/en/about-us/international-cooperation/news/reports/2024/241122-delegation-ukraine-en. Accessed 24 Jul 2025
99. European Commission (2022) Ukraine's association agreement to Horizon Europe and Euratom Research and Training Programmes enters into force. https://research-and-innovation.ec.europa.eu/news/all-research-and-innovation-news/ukraines-association-agreement-horizon-europe-and-euratom-research-and-training-programmes-enters-2022-06-09_en. Accessed 24 Jul 2025
100. Horizon Europe Office in Ukraine (n.d.) Horizon Europe Office in Ukraine. https://horizon-europe.org.ua/en/heo-in-ua/. Accessed 24 Jul 2025